# Novel Material and Technological Solutions in Foundry Engineering

# Novel Material and Technological Solutions in Foundry Engineering

Editor

**Tomasz Wróbel**

MDPI • Basel • Beijing • Wuhan • Barcelona • Belgrade • Manchester • Tokyo • Cluj • Tianjin

*Editor*
Tomasz Wróbel
Department of Foundry
Engineering
Silesian University
of Technology
Gliwice
Poland

*Editorial Office*
MDPI
St. Alban-Anlage 66
4052 Basel, Switzerland

This is a reprint of articles from the Special Issue published online in the open access journal *Materials* (ISSN 1996-1944) (available at: www.mdpi.com/journal/materials/special_issues/novel_material).

For citation purposes, cite each article independently as indicated on the article page online and as indicated below:

LastName, A.A.; LastName, B.B.; LastName, C.C. Article Title. *Journal Name* **Year**, *Volume Number*, Page Range.

**ISBN 978-3-0365-4394-9 (Hbk)**
**ISBN 978-3-0365-4393-2 (PDF)**

# Contents

# About the Editor

**Tomasz Wróbel**

Professor Tomasz Wróbel is an Associate Professor at the Department of Foundry Engineering, Faculty of Mechanical Engineering, Silesian University of Technology in Gliwice, Poland. He is a researcher dealing with the field of material science and a specialization in foundry engineering, in particular on the topics: layered castings; continuous casting; heat treatment; the inoculation of metals and alloys; electromagnetic stirring; stainless steel casting; cast iron; and pure Al and its alloys. Moreover, his scientific interests also concern the methodology of research in the field of chemical composition studies of metallic materials and studies of microstructure with the use of electron microscopes. Professor Tomasz Wróbel is the author or co-author of approximately 160 scientific publications, and the co-owner of 2 patents. Among others, he is a reviewer for dozens of scientific journals from around the world, e.g., *Materials & Design, Journal of Alloys and Compounds, Materials, Metals, Journal of Materials Engineering and Performance,* and *International Journal of Cast Metals.* Moreover, is a member of the Reviewer Board for the journal *Metals,* the Organizing Committee of an International Scientific Conference entitled "Solidification and Crystallization of Metals", the Foundry Commission of Polish Academy of Sciences, the Polish Foudrymens Association, and the Polish Federation of Engineering Associations.

# Preface to "Novel Material and Technological Solutions in Foundry Engineering"

Foundry engineering is still a strongly developing field of material science, and represents an important branch in the global production of metallic materials. Publications concerning foundry engineering are very popular and applicable for many academic scientists and engineering work in many industries. Therefore, given the great potential for development in this field of science and industry, this Special Issue was created for the journal *Materials*, entitled "Novel Material and Technological Solutions in Foundry Engineering". This Special Issue contains very interesting papers concerning, among others, cast composites, layered castings, selected aspects of the crystallization of alloys and the technology of cast and heat treatments of Al alloys and cast iron, the properties of moulding sands, properties of Ni-base superalloys, and the technology of repairing castings using welding.

All of the papers published in this Special Issue have been peer-reviewed according to the journal standard. My special thanks and sincere gratitude go to all the authors and reviewers for their hard work, timely responses, and careful revisions. All authors have made tremendous contributions and offered generous support, thus ensuring the success of this Special Issue.

**Tomasz Wróbel**
*Editor*

MDPI

*Article*

# Surface Phenomena at the Interface between Silicon Carbide and Iron Alloy

**Mirosław Cholewa, Tomasz Wróbel *, Czesław Baron and Marcin Morys**

Department of Foundry Engineering, Silesian University of Technology, 7 Towarowa Street, 44-100 Gliwice, Poland; miroslaw.cholewa@polsl.pl (M.C.); czeslaw.baron@polsl.pl (C.B.); marcin.morys@gmail.com (M.M.)

* Correspondence: tomasz.wrobel@polsl.pl

**Abstract:** The paper discusses a potential composite produced using the casting method, where the matrix is gray cast iron with flake graphite. The reinforcement is provided by granular carborundum (β-SiC). The article presents model studies aimed at identifying the phenomena at the contact boundary resulting from the interaction of the liquid matrix with solid reinforcement particles. The scope of the research included, primarily, the metallographic analysis of the microstructure of the resulting composite, carried out by using light (LOM) and scanning electron (SEM) microscopy with energy dispersive X-ray spectroscopy (EDS) analysis. The occurrence of metallic phases in the boundary zone was indicated, the contents and morphology of which can be optimized in order to achieve favorable functional properties, mainly the tribological properties of the composite. In addition, the results obtained confirm the possibility of producing similar composites based on selected iron alloys.

**Keywords:** casting composite; silicon carbide; gray cast iron; graphite; pearlite; reinforcement particles; metallic matrix

**Citation:** Cholewa, M.; Wróbel, T.; Baron, C.; Morys, M. Surface Phenomena at the Interface between Silicon Carbide and Iron Alloy. *Materials* **2021**, *14*, 6762. https://doi.org/10.3390/ma14226762

Academic Editor: Aniello Riccio

Received: 4 October 2021
Accepted: 8 November 2021
Published: 10 November 2021





## 1. Introduction

The concept of the paper and direction of the conducted research is based on the use of gray cast iron as a matrix and silicon carbide (SiC) particles as reinforcement. Ultimately, the aim is to obtain a material resistant to frictional wear with high coefficient of friction. A hypothetically hard reinforcement phase promotes an increase in wear resistance. Graphite present in the matrix microstructure can act as a solid lubricant, and thus, it can stabilize the frictional conditions, providing the ability to damp mechanical vibrations, typical for cast iron. In addition, graphite in cast iron increases thermal conductivity compared to iron alloys without the graphite phase. It is predicted that the potential applications of the composite produced as part of the paper include brake discs, as well as working parts of mining and construction machines. The condition for further operational research is to obtain favorable micro- and macrostructural features of the presented composite. In addition, it should be initially noted that the analysis of surface phenomena presented in the paper will become the basis for further structural and mechanical studies of the innovative composite gray cast iron—SiC.

Proper understanding of surface phenomena at the interface between ceramics and metal matrix is the basis for designing new materials—e.g., showing the properties of composites. The widely tested and used composites have a matrix of light Al and Mg alloys, which can be obviously justified—these are materials with favorable performance properties—in this case showing minimum mass density and maximum mechanical properties.

The subject of the study is a slightly different association of potential components. They include silicon carbide in the allotropic β form, i.e., not completely pure, relatively cheap and available, and an iron alloy similar to cast iron with flake graphite. The metallic matrix is a cheap, well-known material with versatile functional properties and the lowest processing temperature of iron alloys.

The subject of the research is therefore the combination of materials with possibly minimally differentiated carbon concentration in the ceramic–metallic matrix system. Here, attention should be paid to the driving force of the diffusion process. One of the several important factors is the level of the concentration difference at the component boundary. Carbon, having a small atomic radius, is easily diffusible. In addition, silicon carbide belongs to ceramics featuring maximum thermal conductivity and very high hardness, which bodes well, for example, when creating potential new wear-resistant composite materials. According to the adopted concept, silicon carbide remains in contact with the liquid iron alloy. Such a system is highly chemically reactive. In many definitions of composites, chemical reactions between components are disqualifying in terms of a classical composite. Nevertheless, the indicated solution seems to be attractive in terms of technology and applications provided. The intention is to control the physicochemical reactions and create a favorable combination of ceramics with metal in order to manufacture a good material, perhaps even only showing the characteristics of a composite.

The above-mentioned concept components constitute the content of the paper. The phenomena at the interface between the components are considered: cast iron matrix–technical SiC as reinforcement. It is expected to form a material showing potentially high wear resistance, based on cheap and well-known engineering materials and using foundry technology.

Therefore, the aim of the study is to check the possibility of creating a material with composite properties based on components showing physicochemical reactions.

During chemical reactions SiC is decomposed, while physical reactions consist in forming solutions and phases with the components of the matrix alloy. The control and steering of the reactions should ensure their stopping in a timely manner—so as to keep the ceramic particles in the solidified matrix. Foundry processes provide similar conditions, additionally allowing to control crystallization processes to optimize the microstructure particularly in the interface of the components.

The use of silicon carbide in the metallurgy and foundry of ferrous alloys is quite common and fundamentally different from the concept presented in the paper. In particular, SiC is used effectively during the melting of gray cast irons of any form of graphite—from flake to nodular form. Its acts in two ways. It increases the content of carbon and silicon in the alloy and modifies the microstructure, thus improving the functional properties of cast irons. In terms of technology, it is an excellent alternative to commonly used carburizers in the form of e.g., coal coke, petroleum coke, anthracite or natural and synthetic graphites. Increasing the silicon content in the alloys is generally carried out by adding ferrosilicon (FeSi). In many cases, it is necessary to apply both metallurgical treatments to the liquid alloy. Moreover, both treatments have a significant impact on the microstructural features of cast irons. They have a positive effect on the shape, number and dispersion of the precipitated particles of graphite. They also affect the distance between the ferrite and cementite plates in the pearlite. The treatments of metallurgical modification optimize the microstructure of cast irons. This is generally achieved by increasing the number of nuclei [1–4]. It is the silicon derived from SiC that has a modifying effect, protecting the alloy from excessive crystallization of cementite ($Fe_3C$), which limits the crystallization of graphite, increasing the hardness and brittleness of cast irons—usually locally in areas with high cooling rate. Considering the above, SiC is an excellent component in metallurgical processes for the production of cast iron, providing the desired microstructure and reducing the sensitivity of cast iron to the cooling rate that varies across the section of castings with different wall thicknesses. In the case under consideration, this makes an important advantage, because fragmentation/modification of microstructure directly at the interface between the components of composite is to be expected.

It seems that the modification treatments with SiC are gaining importance for the so-called synthetic cast iron obtained e.g., by remelting steel scrap with low carbon content.

Graphite morphology may be disadvantageous when using some carburizers. In such cases, SiC has a modifying effect. An important advantage of SiC is that modification effect is extended in time—greater stability of the crystallization nuclei, which is probably due to the longer dissolution process compared to the alternative use of ferrosilicon (FeSi). The mechanism of action is hypothetical. The formation of graphite clusters around SiC particles was observed. This is the result of the local supersaturation of the bath with carbon and silicon. Eutectic graphite clusters can be also found in the presence of ferrosilicon (FeSi). However, their stability over time lowers [2]. It can also be hypothetically assumed that the different diffusion of Si and C is not only local in nature, but also temporary and periodic. Similar phenomena can be minimized by introducing SiC particles with the minimum technologically achievable dimensions. The trend is obvious: maximization of the contact surface of the reactants, maximization of dispersion (also in purely mathematical terms), high reaction rate in the volume of the metal. The limitations include: a high degree of homogenization, i.e., intensive mixing—e.g., in an arc furnace or a dedicated induction furnace, and probably a relative shortening of the modification effect. An example of an attempt to solve the problem is the procedure described in the patent [5], where nanoparticles bound to microparticles can effectively act as a nucleator. The object is a nucleating agent for casting ductile or vermicular cast iron using silicon nanocarbon. According to the patent, the mixture of nano- and microscale SiC powder comprises nano-SiC particles attached to the surface of micrometer-sized SiC particles.

Critically speaking, the information provided about the reactions between the key components—the reactants, are obvious, and can be controlled by many factors. They include temperature, component shares, granularity and time. Such a state opens up many technological possibilities of producing a technologically advantageous and useful material containing SiC in a matrix containing Fe.

Considering the composite production, the solutions presented above can be used for melting the matrix in form of cast iron with the use of specialized treatments, also for melting "synthetic" cast iron from steel scrap.

Despite the fact that SiC is thermodynamically stable, it is subject to oxidation resulting in covering with $SiO_2$ oxide on the surface. Among the reinforcement materials in composites, this surface phase ($SiO_2$) in some cases, in general, the oxide phase may facilitate wetting and affect the microstructure of the transition zone depending on the metallic matrix of the composite used. Composites based on alloys, e.g., Al with SiC particles, are commonly known and used. Composites with dispersion reinforcement in the form of SiC microparticles dominate here. It is common to try to minimize their size, inter alia, in order to limit the gravitational segregation of components. This can be carried out by increasing the share of internal friction forces and viscosity, which depend on the particle development area. The phases of the transition zone determine the site of proper composite formation. It is theoretically and practically the most sensitive microarea of each composite.

With large particles, the oxide surface represents a relatively smaller share and its possible removal should be easier. Slag-forming materials or dedicated surfactants can thus facilitate the production of the composite in the liquid state. The controlled process of crystallization is aimed at providing proper modification—generally the fragmentation of the microstructure in the boundary zone of the components.

The presented system of components (FeC–SiC) basically meets all theoretical assumptions for the production of a correct composite in the liquid and solid state, with the controlled primary crystallization of the matrix and phases in the transition zone. It is advisable to provide heat treatment, despite a significant difference in thermal expansion of the components.

The available literature shows few similar concepts of $MMC_p$ (Fe–SiC) composite production and significantly diversified in terms of technology and practical applications.

For example, the composite material [6] produced from silicon carbide and iron has high strength, hardness, wear resistance, heat resistance and oxidation resistance. The method comprises the following steps: silicon carbide and iron powder are mixed and evenly ground to obtain a homogeneous sample, then the material is formed using a hydraulic press and the sample is coated with iron powder and then sintered. Thus, the process consists of multiple stages and does not apply to the casting techniques.

Another similar solution in the field of sintering is the composite with the ferrosilicon phase in situ [7] obtained by using the carbothermal reduction method, thanks to which a porous composite material SiC–Fe–Si can be produced. The ingredients include: 35–73 %wt. silicon carbide, 5–12 %wt. carbon powder and 22–53 %wt. iron oxide; the added binder weight ranges from 1 to 5 %wt. of the total weight of the raw materials. The composite is produced by mixing, pressing and molding, as well as sintering in argon. As in the previous case, the process consists of multiple stages and the obtained material is porous, which, in principle, leads to its mechanical weakening compared to monolithic materials.

Another example of creating a similar composite in terms of materials used refers to a composite material based on iron and aluminum, enriched with silicon carbide [8]. The product has good high temperature strength and nonoxidation properties, as well as a low thermal expansion coefficient. The silicon carbide reinforcement is 0.1–1.0 %wt. of the total weight and the particle size ranges from 1 to 5 nm. The manufacturing method comprises the following steps: pretreating a silicon carbide reinforcement, melting, homogenizing annealing, forging, hot rolling, wire drawing and heat treatment. Thus, the intended use and manufacturing technology, despite the fact that they differ from the presented concept, indicate alternative applications and manufacturing possibilities. Furthermore, they indicate potential directions for development by alloying the composite with aluminum.

Potentially, the chemical components of the matrix can perfectly regulate surface phenomena. The material quoted below was also made by using sintering techniques with reinforcement in the form of SiC particles. This is an Fe-based composite material reinforced with SiC particles [9]. The matrix material contains the following components in percent by weight: 3.4–3.6 %wt. C, 1.9–2.1 %wt. Si, 0.3–0.5 %wt. Mn, 0.5–0.8 %wt. Cr, 0.1–0.2 %wt. Ti, 0.05–0.10 %wt. Te, 0.03–0.05 %wt. Mg, 0.01–0.03 %wt. Re, at most 0.05 %wt. P, at most 0.02 %wt. S and balance Fe. According to the authors, silicon carbide particles can be well combined with the substrate to effectively improve the mechanical properties of the material subjected to sintering and quenching. After quenching and tempering heat treatment from 1250 °C, the maximum hardness and tensile strength can reach 45.4 HRC and 1859 MPa. Because of its good mechanical properties, its practical applications relate to specialized technical applications that are not necessarily common, including tribological applications.

The composite presented in [10], made by using an iron alloy containing the following components in percent by weight: 1.5–5.5 %wt. Al, 3.6–7.0 %wt. Mo, 2.0–3.3 %wt. Cu, 0.2–1.5 %wt. Cr, 12.0–22.0 %wt. Zr, 2.5–4.5 %wt. Mg, 1.0–2.0 %wt. Mn, 1.2–1.4 %wt. Si, 0.08–0.14 %wt. B and SiC whiskers, is conceptually most similar to the one presented in this paper. Whiskers as the main hard phase can be well combined with matrix, thus effectively improving the mechanical properties of the sintered and quenched material. After the tempering and heat treatment is carried out, the hardness and strength of the alloy increase significantly. There is no doubt that similar materials can be used in kinematic pairs exposed to abrasive wear. However, the process still bypasses casting techniques, ensuring freedom of shape and minimizing production procedures.

On the one hand, the presented examples show the high potential of Fe–SiC material association, while on the other hand, they show specialized and precisely dedicated technologies. The examples should be a rationale for the effective use of components that are available and technologically friendly. By design, the presented references are not substantially theoretically underpinned, but with the knowledge available, they can focus many research projects on similar areas.

The practical combination of cast iron with flake graphite, together with silicon carbide, places this material in the category of wear-resistant materials. A typical combination of hard and soft phases should provide the effect of minimal wear even in conditions of little or no technical lubrication. The soft graphite phase has a lubricating effect and makes a storehouse for wear products. The hard SiC phase increases the wear resistance, and the ferritic-pearlitic matrix transfers loads mainly compressive in nature. Compared to typical composites, e.g., on the Al matrix, the suggested material has a higher mass density, but also a hypothetically greater potential for frictional wear under many different conditions of cooperation of kinematic pairs, e.g., under conditions of erosive or abrasive wear.

Overall, the technology of casting a composite on a cast iron matrix with SiC particles may be a favorable solution in terms of simplicity of the technology and the attractiveness of tribological practical applications. The prerequisite is to obtain the correct surface phenomena at the interface of the components, which was adopted as the main aim of the study.

## 2. Materials and Methods

This research project presents a model that shows the matrix of the composite in the form of the simplest gray cast irons with flake graphite, approximately eutectic with a typical saturation factor and eutectic equivalent. The research ignores the influence of typical alloy components increasing mechanical properties, in particular tribological ones, e.g., carbide forming or improving castability and lowering the matrix melt viscosity. The purpose of their use may be to increase the hardness of the metallic matrix and to facilitate the wetting processes of SiC particles. Apart from Fe and C, this type of component includes Si, Cr, Mo and V.

The second component, i.e., the composite reinforcement, is SiC, which is not as obvious as the matrix material. In classic production processes, there are three classes of commercial SiCs differing in the level of thermodynamic stability. Conditionally thermally stable α-SiC, thermally unstable β-SiC and reactive metallurgical SiC. β-SiC is black in color and contains more α-SiC than the green variant that has such impurities as $Fe_2O_3$, $Al_2O_3$, CaO, $SiO_2$, $MnO_2$ (the terms "black", "green" and "metallurgical" are trade names). The chemical composition differs depending on the producer, and an example of the differentiation of the composition of individual varieties is given in Table 1.

**Table 1.** Properties of silicon carbide [11].

| SiC, %wt. | $Fe_2O_3$, %wt. | C, %wt. | Magnetic Fraction, %wt. | Density, g/cm³ |
|---|---|---|---|---|
| | | Green SiC | | |
| ≥97.0 | ≤1.0 | ≤0.2 | ≤0.4 | 3.1 ÷ 3.2 |
| | | Black SiC | | |
| ≥96.0 | ≤1.0 | ≤0.3 | ≤0.4 | 3.1 ÷ 3.2 |
| | | Metallurgical SiC | | |
| ≥88.0 | ≤1.0 | ≤0.3 | ≤0.5 | 3.1 ÷ 3.2 |

Silicon carbide together with boron carbide ($B_4C$) are covalent carbides. The tetrahedral system of covalent bonds enforces the tetrahedral ordering of Si and C atoms. Thus, their typical structures are the analogs of sphalerite β-SiC and wurtzite α-SiC (according to the Somerfeld rule). The β form is considered to be unstable in the entire temperature range, but it most often crystallizes under conditions where the primary synthesis product does not recrystallize [12]. In the literature [13], over 100 SiC polytypes have been described. The differences of polytypes are manifested by the different sizes of unit cells. The causes of polytypism have not yet been clearly explained.

From among the available SiC varieties, the thermally unstable β-SiC was selected for testing, most often occurring in one basic structure of 3C sphalerite, counting on the possibility of controlling physicochemical reactions at the contact boundary: Fe–(β-SiC) alloy. Confirmation of the possibility of obtaining limited reactions is the not fully satisfied stoichiometry requirement. It is expressed by the molar ratio Si:C = 1.03:1.05. Thanks to this, SiC undergoes slow thermal decomposition, and the resilience of the decomposition products reach measurable values, starting from the temperature of 1500 °C [14].

Summing up, the tests assumed gray cast iron with flake graphite as a metallic matrix, with the chemical composition presented in Table 2 and determined using the LECO GDS500A (LECO Corporation, St. Joseph, MI, USA) emission spectrometer. On the other hand, the reinforcement is β-SiC with a significant model particle size, facilitating the wetting of the components, with a significant reserve for physicochemical reactions reducing the volume of the reinforcement particles. The β-SiC particles with a grain size of 0.8–1.6 mm were used, while the $Fe_{met}$ content presented in Table 3 in fact represents the presence of iron oxide range from FeO to $Fe_2O_3$. In general, $Fe_2O_3$ oxide is the dominant one.

**Table 2.** Chemical composition of the metallic matrix: gray cast iron with flake graphite.

| C, %wt. | Si, %wt. | Mn, %wt. | Cr, %wt. | Cu, %wt. | Ni, %wt. | P, %wt. | S, %wt. | Fe, %wt. |
|---|---|---|---|---|---|---|---|---|
| 2.81 | 1.24 | 0.61 | 0.11 | 0.09 | 0.05 | 0.62 | 0.10 | rest |

**Table 3.** Chemical composition of β-SiC silicon carbide used in the tests according to the manufacturer's data [15].

| SiC, %wt. | C, %wt. | Si, %wt. | $SiO_2$, %wt. | $Fe_{met}$, %wt. |
|---|---|---|---|---|
| ≥97.0 | ≤0.3 | ≤1.0 | ≤1.0 | ≤0.3 |

The composite subjected to testing was produced by casting technology. Liquid gray cast iron at a temperature of 1530 °C was poured by gravity into a sand mold with a granular preform β-SiC placed in its cavity. Then, microscopic metallographic tests were carried out on samples etched with Nital with the following composition: 5 mL $HNO_3$ + 95 mL $C_2H_5OH$ using the NIKON ECLIPSE LV150N (NIKON Metrology Europe NV, Leuven, Belgium) as OM (optical microscope) and on nonetched samples using the Phenom ProX (Phenome-World, Eindhoven, The Netherlands) as SEM (scanning electron microscope) with BSE (backscattered electron) imaging, 10 kV electron beam accelerating voltage and EDS (energy dispersive spectroscopy) analysis.

## 3. Research Results

An example of a cross-section of a gray cast iron–silicon carbide composite produced by the casting method is shown in Figure 1.

On the other hand, Figures 2–7 and Table 4 show the results of metallographic tests. It was found that the obtained chemical composition of the cast iron matrix and its crystallization conditions in the sand form determined the final microstructure consisting of flake graphite in a pearlitic matrix (Figure 2). In addition to the above-mentioned phases, the microstructure of the tested gray iron contained a small amount of steadite (phosphorus eutectic) and nonmetallic inclusions in the form of MnS sulfides.

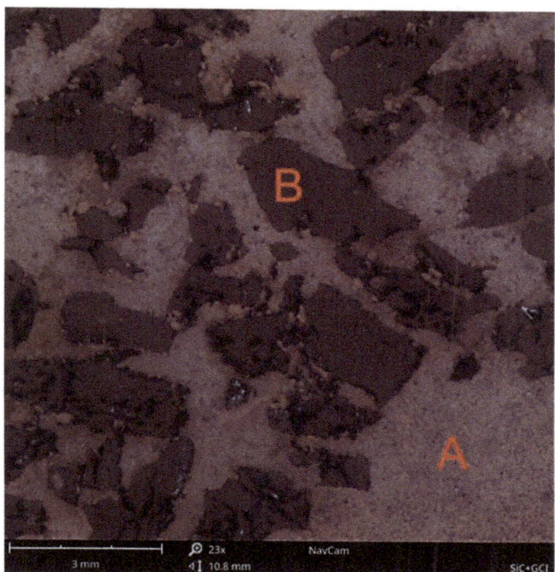


**Figure 1.** An example of a composite macrostructure of gray cast iron (**A**) and silicon carbide β-SiC (**B**), mag. 23×.

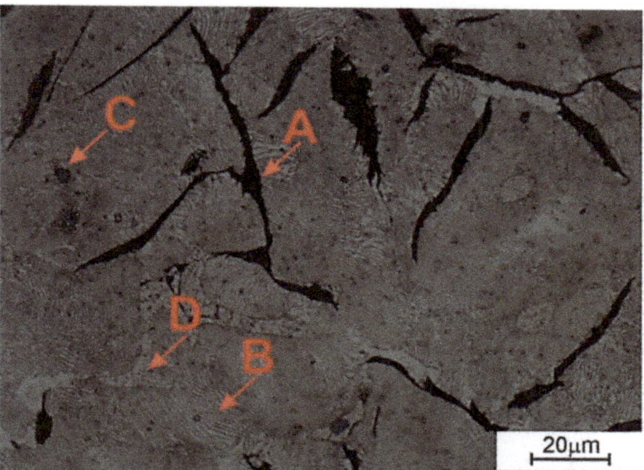


**Figure 2.** The microstructure of gray cast iron constituting the matrix of the composite, OM, mag. 500×, etching: Nital. (**A**) flake graphite; (**B**) pearlite; (**C**) MnS sulfides; (**D**) steadite.

In the area of contact between the cast iron matrix with the particle reinforcement β-SiC, the dispersion of flake graphite grew. The area of the fragmentation of graphite extended into the matrix by a distance calculated in fractions of a millimeter (Figure 3a). In addition, a significant ability of cast iron to infiltrate into the spaces between β-SiC particles was observed. Figure 3b shows the gap filled with a metallic matrix with a thickness of tens of μm. This effect confirms that the flowability of the cast iron used was particularly high. That was probably caused by a local increase in the concentration of C and Si at the front of the metal stream as a result of the component dissolution reaction in the contact of the matrix with the reinforcement. Certain confirmation of the above statement comes from the results of the microanalysis of the chemical composition in point 2 in Figure 4, presented in the form of a spectrogram in Figure 5b and in Table 4. By contrast, Figure 3c shows the transition zone locally occurring in the area of the matrix contact with the reinforcement, which should be classified as a group of nonwall crystals. Its location was typical for this type of phases occurring in in situ composites. They show a significant diversity of chemical composition and a high degree of fragmentation (Figure 6). An example of the chemical composition of the transition zone, i.e., point 3 in Figure 4, is presented in the form of a spectrogram in Figure 5c and Table 4. In fact, this is not a chemical compound, but rather a phase formed at the interface between physical and chemical reactions. Precipitations of this phase appeared in the places where the reactions occurred the earliest, and diffusion into the matrix dis not degrade it. The chemical elements appearing in this phase included Fe, Si, and C in descending order, while its structural components were made up of very fine graphite in a metallic matrix (Figures 6 and 7).

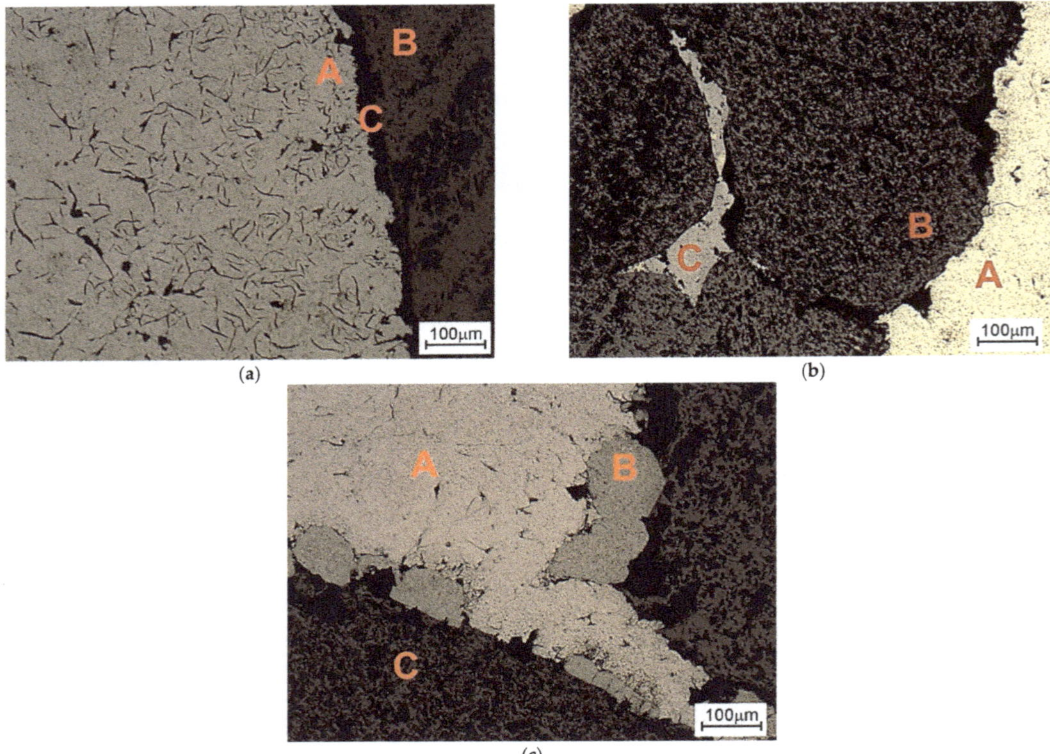


**Figure 3.** Composite microstructure of gray cast iron–β-SiC in the area of the matrix contact with the reinforcement, OM, mag. 100×, etching: Nital. (**a**) A—Fragmented graphite zone; B—β-SiC reinforcement; C—transition zone 2; (**b**) A—metallic matrix (gray cast iron); B—β-SiC reinforcement; C—gap filled with a metallic matrix; (**c**) A—metallic matrix (gray cast iron); B—transition zone 1; C—β-SiC reinforcement.

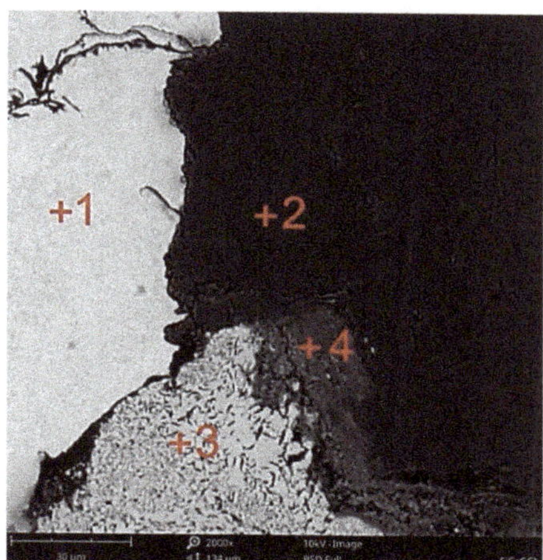


**Figure 4.** Composite microstructure of the gray cast iron–β-SiC in the area of the matrix contact with the reinforcement containing marked EDS analysis points, SEM, mag. 2000×, nonetched sample; 1—β-SiC reinforcement; 2—metallic matrix (gray cast iron); 3—transition zone 1; 4—transition zone 2.

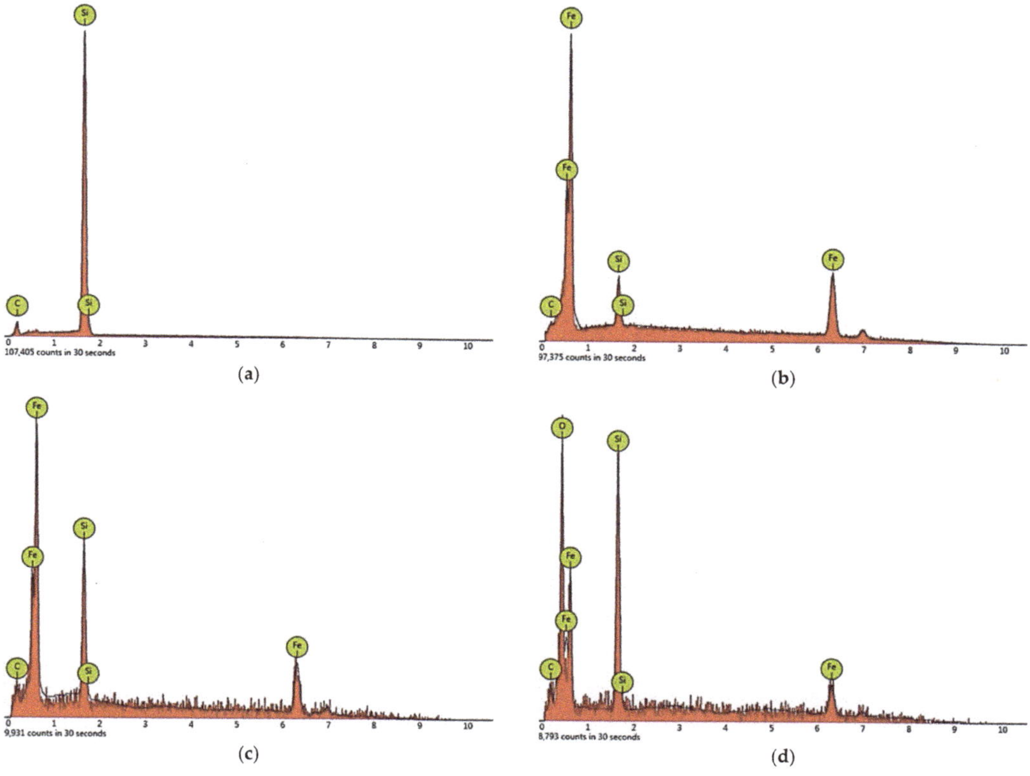

107,405 counts in 30 seconds

(a)

97,375 counts in 30 seconds

(b)

9,931 counts in 30 seconds

(c)

8,793 counts in 30 seconds

(d)

**Figure 5.** EDS point analysis spectrograms: (**a**) point 1 in Figure 4; (**b**) point 2 in Figure 4; (**c**) point 3 in Figure 4; (**d**) point 4 in Figure 4.

**Table 4.** The result of the EDS point analysis (marking points according to Figure 4).

| C | | O | | Si | | Fe | |
|---|---|---|---|---|---|---|---|
| **%at.** | **%wt.** | **%at.** | **%wt.** | **%at.** | **%wt.** | **%at.** | **%wt.** |
| | | | Point 1 in Figure 4 | | | | |
| 52.03 | 31.69 | - | - | 47.97 | 68.31 | - | - |
| | | | Point 2 in Figure 4 | | | | |
| 12.75 | 3.19 | - | - | 8.25 | 4.83 | 79.00 | 91.97 |
| | | | Point 3 in Figure 4 | | | | |
| 18.49 | 5.19 | - | - | 17.94 | 11.78 | 63.57 | 83.02 |
| | | | Point 4 in Figure 4 | | | | |
| 14.87 | 6.19 | 37.23 | 20.63 | 20.23 | 19.68 | 27.66 | 53.50 |

The transition zone under analysis, due to the phase composition, had mechanical properties located between the ceramic reinforcement and the cast iron matrix. This phase increased at the expense of the β-SiC particles, manifested by its rounded contact surface with the reinforcement particles. This surface was different from the raw β-SiC surface shaped by mechanical crushing. With a favorable and optimal proportion of morphology and dispersion, this type of transition phase may minimize the effects of the differential expansion of the two components of the composite and thus may affect its thermal properties. The differentiation of the degree of fragmentation of the flake graphite (Figure 6) indicates a locally variable intensity of the reaction. In addition to the effect of different wetting processes, this may also suggest a different reactivity of the β-SiC particle surface, which may also be justified by different specific surface, edge and tip energies of silicon carbide crystals forming polycrystalline β-SiC particles.

The second phase of the transition zone, the chemical composition of which, i.e., point 4 in Figure 4, is shown as a spectrogram in Figure 5d and Table 4, probably composed of iron and silicon oxide with an admixture of carbon. The presence of carbon in a significant share (even with the error of the EDS measurement method) excludes the existence of $Fe_2SiO_4$ fayalite at the moment of capture. This was the moment when the reaction was stopped due to cooling and crystallization of the system.

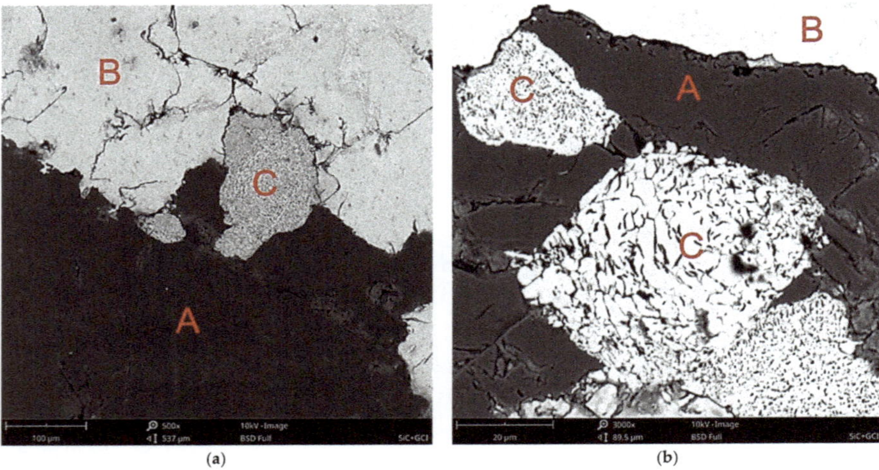


(a) (b)

**Figure 6.** *Cont.*

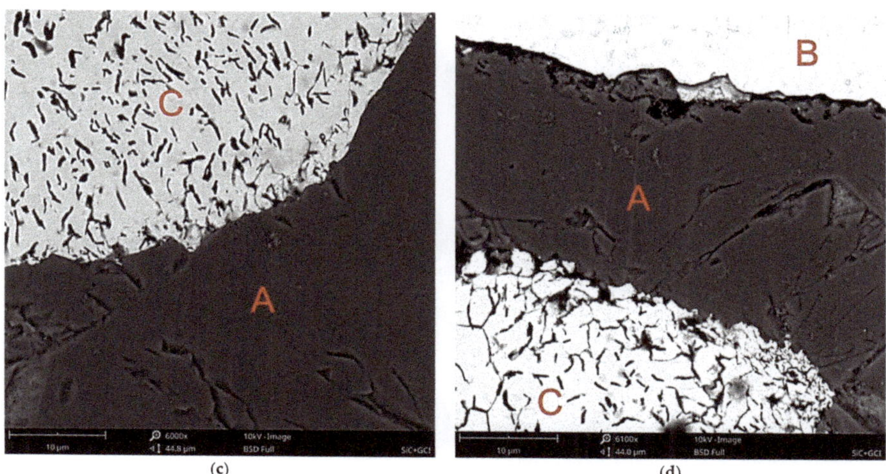


Figure 6. Composite microstructure of gray cast iron–β-SiC in the area of the matrix contact with the reinforcement, SEM; (a) mag. 500×; (b) mag. 3000×; (c) mag. 6000×; (d) mag. 6100×, nonetched sample; A—β-SiC reinforcement; B—metallic matrix (gray cast iron); C—transition zone 1.

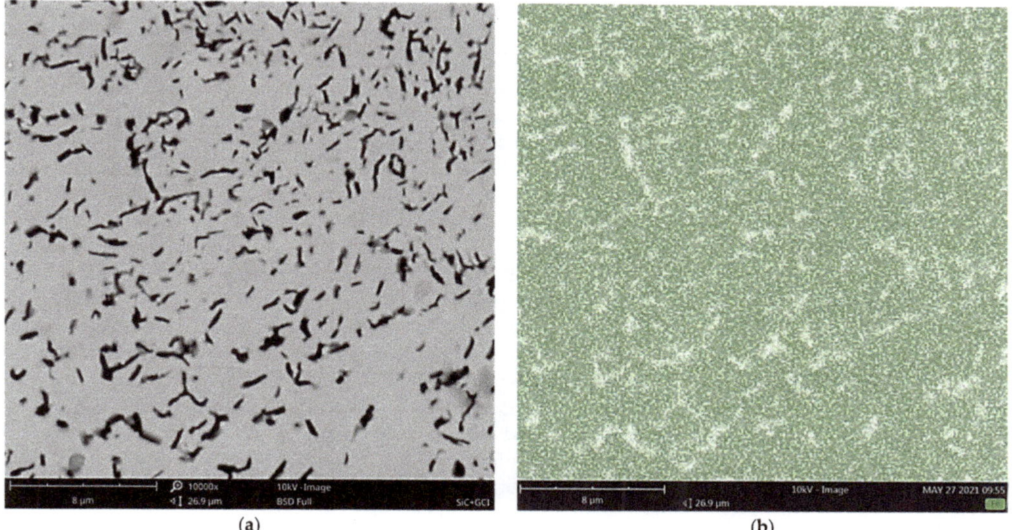

Figure 7. Cont.

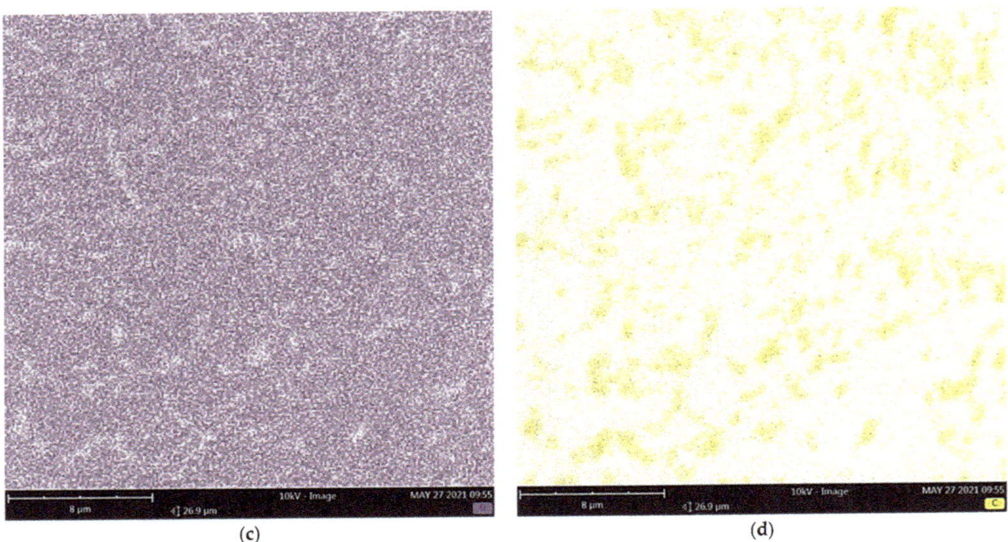

**Figure 7.** The microstructure of the transition zone 1 and the distribution map EDS of elements in this area: (**a**) tested area, SEM, mag. 10,000×, nonetched sample; (**b**) distribution of Fe at concentrations in the tested area 50.81 %at./77.01 %wt.; (**c**) distribution of Si at concentrations in the tested area 15.93 %at./12.15 %wt.; (**d**) distribution of C at concentrations in the tested area 33.25 %at./10.84 %wt.

## 4. Discussion of Research Results

Based on the obtained research results, the following hypothesis can be formulated on the creation of a connection at the boundary of the contact between the cast iron matrix and the silicon carbide particles reinforcing the composite in the casting technology used. The oxidized surface of the β-SiC particles consists mainly of $SiO_2$. There are two compounds in the composition of silicon carbide in the amount of up to approx. 1%wt. $SiO_2$ and $Fe_2O_3$. However, in the metal bath there are iron oxides with a significant share of $Fe_2O_3$. This is due to classic metallurgical treatments. Research by Naro [16] showed that $Fe_2O_3$ can react with $SiO_2$ silica to form iron silicate fayalite ($Fe_2SiO_4$) that makes a physical barrier preventing the dissolution of gases in the metal. In the liquid alloy β-SiC is slowly dissolved, which is a condition for the deoxidation effect [17]. In principle, these phenomena do not apply to the case under consideration, but they should be taken into account, as their occurrence is likely at conditions above 1500 °C, i.e., at the temperature of pouring liquid cast iron into a sand mold with a β-SiC preform placed in its cavity. Chemically bound silicon on the surface of β-SiC particles stimulates subsurface enrichment of the particles with carbon, which, based on the difference in concentration, diffuses into the metallic matrix in the immediate vicinity of the silicon carbide particles. Hypothetically, the sequence of chemical reactions is as follows:

$$2Fe_2O_3 + 2SiO_2 \rightarrow 2Fe_2SiO_4 + O_2 \quad (1)$$

$$2FeO + SiO_2 \rightarrow Fe_2SiO_4 \quad (2)$$

$$3Fe_2O_3 + 2SiC \rightarrow 4Fe + SiO_2 + Fe_2SiO_4 + CO + CO_2 \quad (3)$$

$$3Fe_2SiO_4 + 2SiC \rightarrow 6Fe + 5SiO_2 + 2CO \quad (4)$$

$$FeO + SiC \rightarrow Fe + Si + CO \quad (5)$$

$$3FeO + SiC \rightarrow 3Fe + SiO_2 + CO \quad (6)$$

It should be emphasized that the effect of $Fe_3O_4$ oxide is generally ignored [16,17]. Probably, fayalite protects locally against physical migration of gases, so the released oxygen $O_2$ promotes the oxidation of another surface of exposed silicon carbide. The periodicity of the reactions and their local character seem to be justified, as they significantly depend on the temperature field and thermal conductivity of the composite components in the microarea "reinforcement particle–matrix–environment". The remaining products are Si and CO. Silicon has a local modifying effect, while CO along with the temperature decrease, in accordance with the Boudoudard reaction ($C + CO_2 \leftrightarrow 2CO$), is released with graphite "in situ", which may have a nucleating effect on its crystallization in a metallic matrix, i.e., ferritic, pearlitic or ferritic-pearlitic. Silicon dioxide together with fayalite forming physical solutions should constitute the phase components in the transition zone between the components. With a sufficiently high volume, they can segregate and constitute slag-forming components.

In addition, carburization of the matrix may locally lower its liquidus temperature. This in turn helps to lower the melt viscosity of the matrix, facilitating the wetting of the β-SiC particles. This means that the physical dissolution process is not instantaneous but progresses with the progressive thermal decomposition of β-SiC. By regulating the temperature and pouring time in the conditions of gravity casting, it is possible to stop the reaction by keeping the "frozen" β-SiC particles in the metallic matrix. Permanent active reactivity perfectly improves the conditions for creating a composite connection at the boundary of the contact of both components. In addition, graphite and pearlite are subject to modification in the contact zone of both components, i.e., in the most sensitive zone of each composite. The modification consists in refining and increasing the dispersion as well as improving the morphology of graphite and refining pearlite, i.e., reducing the distance between ferrite and cementite, while maintaining the thickness ratio in the proportion of 1:7. Such a system makes it possible to maximize the share of reinforcement in the metallic matrix well above the classic few up to dozen %.

## 5. Conclusions

By analyzing the results of the conducted research, the following substantive and application conclusions were formulated. The authors of the paper also presented the assumptions and directions of future research in the field of technological usability of the produced composite in configuration the gray cast iron matrix reinforced with SiC particles from the point of view of its mechanical properties.

### *5.1. Substantive Conclusions*

- The transitional phases at the contact point of the cast iron matrix with the silicon carbide particles constituting the reinforcement confirm the favorable wetting phenomena, conditioning the obtaining of a technologically useful composite.
- The type and morphology as well as phase dispersion in the transition zone require optimization depending on the intended use of the products made of gray cast iron–β-SiC composite. Typical foundry technological factors (casting parameters) make it possible to regulate the thermal capacity of the components, the initial conditions as well as the cooling and crystallization rate of the entire composite. Selection of technological factors, i.e., the matrix of the molding sand, as well as the initial temperature and mass of mold, depends on the massiveness of the casting expressed, e.g., by the casting solidification modulus or individual thermal centers in the casting, according to commonly used standards.
- Important for the quality of the resulting composite is the anaerobic metallic phase identified as very fine graphite in the metallic matrix. This phase, containing Fe, Si and C in its chemical composition, may be responsible for the fragmentation of the microstructure of the cast iron matrix in the vicinity of the β-SiC particles, and thus may act as a graphite modification precursor. The differentiation of the degree of fragmentation of graphite in this phase indicates the local variability of the rate

of physicochemical reactions occurring between the components. In addition, the research results confirm the possibility of controlling the speed of these reactions. The reaction rate is affected by the temperature drop in the mold in the range from the pouring temperature to the solidus temperature. The diffusion phenomena in the liquid state of the matrix are the strongest.

### 5.2. Application Conclusions

- Taking into account the mechanical properties of the gray iron matrix and the reinforcement in the form of β-SiC particles, it is possible to obtain the final material with the properties of an abrasion-resistant composite. The practical verification of this statement should be carried out on a test stand that faithfully reproduces the type of wear depending on the working conditions of the composite machine part.
- It is anticipated that the potential applications of the composite developed are working components of mining machines for the mining industry or for brake discs. The accepted hypothesis assumes that the use of the gray cast iron–silicon carbide composite for brake discs will allow us to obtain functional properties located between the commonly used materials, traditionally cast entirely of gray cast iron, and the special materials based on SiC ceramics.

**Author Contributions:** Conceptualization, M.C.; investigation, T.W., C.B. and M.M.; methodology, M.C.; resources, M.C., T.W. and C.B.; supervision, M.C.; writing—original draft, writing—review and editing, M.C., T.W. and C.B. All authors have read and agreed to the published version of the manuscript.

**Funding:** This publication was financed from the statutory subsidy of the Faculty of Mechanical Engineering of the Silesian University of Technology in 2021.

**Institutional Review Board Statement:** Not applicable.

**Informed Consent Statement:** Not applicable.

**Data Availability Statement:** The data presented in this study are available in article.

**Conflicts of Interest:** The authors declare no conflict of interest.

## References

1. Elektroschmelzwerk GmbH. Dissolution of Silicon Carbide in Iron Melts. UK Patent GB1427381A, 10 March 1976.
2. Vaśko, A. Microstructure and mechanical properties of synthetic nodular cast iron. *Arch. Foundry Eng.* **2010**, *10*, 93–98.
3. Yunes Rubio, R.; Hong, L.; Saha-Chaudhury, N.; Bush, R. Dynamic wetting of graphite and SiC by ferrosilicon alloys and silicon at 1550 °C. *ISIJ Int.* **2006**, *46*, 1570–1576. [CrossRef]
4. Perzyk, M.; Waszkiewicz, S.; Kaczorowski, M.; Jopkiewicz, A. *Foundry Engineering*; WNT: Warsaw, Poland, 2004.
5. Harbin Coredwire Metallurgy Co Ltd. Spheroidal Graphite Cast-Iron and Vermicular Cast Iron Inoculant and Preparation Method Containing Nanometer Silicon Carbide. China Patent CN107267704A, 20 October 2017.
6. Chongqing Xuxing Chemical Co Ltd. Method for Manufacturing Silicon Carbide and Iron Composite Material. China Patent CN104532100A, 22 October 2014.
7. Northeastern University. In-Situ Silicon Carbide-Iron Silicon Composite Material and Preparation Method Therefor. China Patent WO/2020/057096A1, 26 March 2020.
8. Yancheng Xinyang Electric Heat Material Co Ltd. Silicon Carbide-Enhanced Iron/Aluminum Composite Material and Preparation Method Thereof. China Patent CN104233065A, 24 December 2014.
9. Chongqing Huafu Powder Metallurgy Co Ltd. Silicon-Carbide-Particle-Reinforced Iron-Base Composite Material and Preparation Method Thereof. China Patent CN104372254A, 29 October 2014.
10. Qingdao Matt Rieu New Material Technology Co Ltd. Silicon-Carbide-Whisker-Reinforced Iron-Base Composite Material and Preparation Method Thereof. China Patent CN104561807A, 19 December 2014.
11. Stojczew, A.; Janerka, K.; Jezierski, J.; Szajnar, J.; Pawlyta, M. Melting of grey cast iron based on steel scrap using silicon carbide. *Arch. Foundry Eng.* **2014**, *14*, 77–82. [CrossRef]
12. Stobierski, L. *Carbide Ceramics*; AGH: Cracow, Poland, 2005.
13. Verma, A.; Krishna, P. *Polymorphism and Polytypism in Crystals*; John Wiley and Sons: New York, NY, USA, 1966.
14. Moser, M. *Microstructures of Ceramics*; Akademiai Kiado: Budapest, Hungary, 2002.
15. P.P.H. REWA. Available online: https://pph-rewa.pl/produkt/weglik-krzemu-czarny/ (accessed on 1 September 2021).

16. Naro, R. Porosity defects in iron castings from mold–metal interface reactions. *AFS Trans.* **1999**, *107*, 839–851.
17. Benecke, T. Metallurgical silicon carbide in electric furnace and copula furnace. *Giesserei* **1981**, *68*, 344–349.

 *materials*

MDPI

*Article*

# Comprehensive Research and Analysis of a Coated Machining Tool with a New TiAlN Composite Microlayer Using Magnetron Sputtering

**Štefan Michna [1], Iryna Hren [1,*], Jan Novotný [1], Lenka Michnová [1] and Václav Švorčík [2]**

1 Faculty of Mechanical Engineering, Jan Evangelista Purkyne University, Pasteurova 3334/7, 400 01 Usti nad Labem, Czech Republic; stefan.michna@ujep.cz (Š.M.); jan.novotny@ujep.cz (J.N.); lenka.michnova@ujep.cz (L.M.)

2 Deparment of Solid State Engineering, University of Chemistry and Technology Prague, Technická 5, 166 28 Praha 6, Czech Republic; vaclav.svorcik@vscht.cz

* Correspondence: iryna.hren@ujep.cz; Tel.: +420-475-28-5547

**Abstract:** The application of thin monolayers helps to increase the endurance of a cutting tool during the drilling process. One such trendy coating is TiAlN, which guarantees high wear resistance and helps to "smooth out" surface defects. For this reason, a new type of weak TiAlN microlayer with a new composition has been developed and applied using the HIPIMs magnetron sputtering method. The aim of this study was to analyze surface-applied micro coatings, including chemical composition (EDX) and microstructure in the area of the coatings. Microstructural characterization and visualization of the surface structures of the TiAlN layer were performed using atomic force microscopy. To study the surface layer of the coatings, metallographic cross-sectional samples were prepared and monitored using light and electron microscopy methods. The microhardness of the test layer was also determined. Analyses have shown that a 2-to-4-micron thick monolayer has a microhardness of about 2500 HV, which can help increase the life of cutting tools.

**Keywords:** magnetron sputtering; HIPIMs method; TiAlN layer; XRD analysis; EDS analysis; surface morphology; coating thickness; AFM microscopy

**Citation:** Michna, Š.; Hren, I.; Novotný, J.; Michnová, L.; Švorčík, V. Comprehensive Research and Analysis of a Coated Machining Tool with a New TiAlN Composite Microlayer Using Magnetron Sputtering. *Materials* **2021**, *14*, 3633. https://doi.org/10.3390/ma14133633

Academic Editors: Tomasz Wróbel and Sandra Maria Fernandes Carvalho

Received: 19 May 2021
Accepted: 24 June 2021
Published: 29 June 2021



## 1. Introduction

The HIPIMS method (high power impulse magnetron sputtering) is a newly developing special PVD coating technology. It features a very high power density (1000 W/cm$^2$) per sputtering electrode, with very short selectable pulse times (50–200 μs), which produce a higher plasma density than standard magnetron sputtering (>10$^{19}$ m$^3$). HIPIMS is a progressive method of applying layers with a controlled flow of material on the substrate due to ionization of the sprayed material. The spraying of the coatings is carried out using a source of particles (targets) with kinetic energy, and suitable examples include, Ar + ions, and, for layer formation, also nitrogen, carbon, oxygen, etc. This technology is used mainly for dedusting various conductive and non-conductive elements (Ti, Al, Zr, C, Cu, B, Si, etc.) in various proportions and together with supplied gases (nitrogen, oxygen, methane, hydrogen, etc.) a combination of different coatings can be produced, including non-conductive or composite layers. The elements that need to be released for coatings are dusted off from the solid state to the gaseous state through targeted bombardment of the source by argon and krypton atoms. The released atoms are subsequently ionized and, together with the gas atoms that can be brought into the coating chamber, are directed from the source target to the surface of the instruments. The result of the composition of these atoms is a homogeneous, extremely smooth coating, without the presence of microcapsules. Despite its increased energy intensity, it is widely used in many areas of industry, such as aerospace, military, automotive, etc., to provide a high degree of wear resistance and to extend tool life.

The HIPIMS method was first described in Ehiasarian et al. [1] for the pre-treatment of substrates before the application of nitride coatings. In their experiments, the high content of metal ions in the bombardment flow promoted the etching of the substrate surface. During the experiments, it was found that this bombardment contributed to the formation of a metal-implanted region that maintained a strong bond between the coating and the substrate.

Another large-scale work was carried out by Tiron [2], which focused on improving the deposition rate parameter in HIPIMs technology by changing the magnetic field and the pulse duration configurations. During the experiments, ten different materials (C, Al, Ti, Mn, Ni, Cu, Zn, Mo, TA, and W) were used, which were subjected to magnetron sputtering and HIPIMs discharge, without and using an external magnetic field. The results showed that the use of an external magnetic field increased deposition rates by approximately 40% to 140%, for only the selected materials, namely Mn, Ni, Cu, and Zn. An advanced study was carried out by Santiago [3], which was aimed at the effect of positive pulses on HIPIMS deposition of hard carbon coatings. In the experiments, different amplitudes of positive voltage (100 to 500 V) were used when applying carbon DLC coatings. The results of the experiments showed that the best voltage used was 500 V, at which point an increase in the hardness of the carbon coating was observed from 9.6 GPA to 22.5 GPA. In their study, Sarakinos et al. [4] found that the rate of carbon deposition in HIPIMs is only about 1.3–1.5 times higher than in conventional DC deposition using the same cathode and the same average current. However, the films deposited with HIPIMs had a density of 2.27–2.67 $g/cm^3$, measured by X-ray reflectivity [5]. It was found that the sp3 fraction of these films is from 20 to 40% when using a substrate bias UB = $-125$ to $-175$ V. On the other hand, there are no reports yet on the ratio of the carbon ion to the neutral carbon for HIPIMS carbon plasma. Only a few experiments have been carried out using HIPIMS for the deposition of carbon films [6–8], Cr [9,10], Nb [11], Al, Ti, W [12], and WC [13] to increase the adhesion of hard coatings, such as CrN or TiAlN [14]. On the other hand, the choice of metal for sputtering is usually made by taking into account the chemical affinity of the coating–substrate system [12]. As the main factors determining the level of adhesion, the chemical strength of the bond was identified together with the mechanical influence of the metal acting as a compliant interlayer reducing shear stress [14].

The HIPIMs method is a progressive method of layer deposition with control over the material flow to the substrate due to the ionization of the sprayed material. Therefore, extensive research and analyses of the TiAlN layer has been carried out. This layer was invented at the Faculty of Production Technologies of UJEP and has never been studied in a similar study. The aim of this work was to explore and analyze the TiAlN composite layer of a new composition. Improving the quality of the material and eliminating defects can be done by examining all factors affecting the HIPIMs process, such as, first of all, mapping the connectivity of the layer with the basic material. Increasing surface abrasion and wear resistance can be achieved by examining factors affecting the HIPIMS process, such as, in particular, mapping the thickness of the layer and its compactness on the surface of the product. The possibilities of using this alloy are extensive, from machine forging or cast aluminum alloys, milling steels, cast steels, and cast irons to machining refractory and Ti alloys.

## 2. Experimental Setup and Material

### *2.1. Experimental Material*

For the experiment, a cutting tool was used, which was based on tungsten carbide (WC) particles prepared using powder metallurgy and combined with Co (chemical composition: 88.6% W, 10.3% Co, 0.3% Cr, 0.05% Fe and 0.15% S). These materials exhibit a high hardness and their maximum application temperature is 800–900 °C. The sample was made in the form of a milling drill with the following parameters: length—100 mm and width—13.5 mm; one such sample is shown in Figure 1.

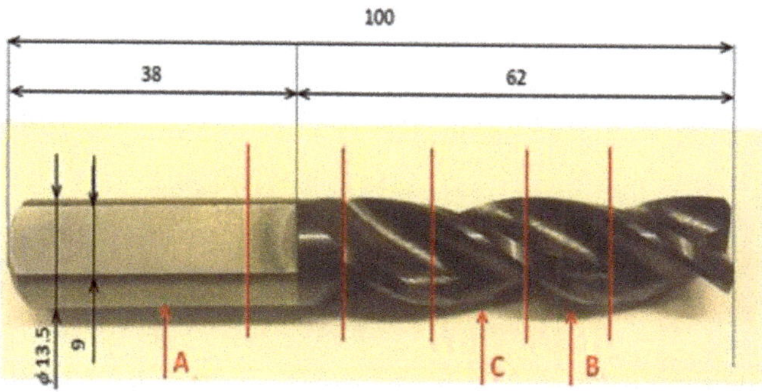


**Figure 1.** The whole body of the milling drill with a TiAlN coating applied to its working part with the determination of the places of selection for analyses (author source).

## 2.2. Experimental Setup

The principle of HIPIMS PVD sputtering is shown in Figure 2; the applied substance is the cathode (target) of the magnetron discharge. The discharge burns in a very dilute inert gas (usually Ar) in a vacuum chamber. Above a negatively charged target (500 to 1000 V), argon plasma is maintained, the positive ions of which are accelerated by the electric field to the target and, upon impact, sputter it. In front of the target, a magnetic field with a defined shape is created using an electromagnet or permanent magnets [15]. In this case, electrons that escape from the space in front of the target during classical sputtering must move along the helix, along the lines of force due to the Lorentz force [16,17]. Thus, their path in the vicinity of the target is significantly lengthened, the duration of their stay in the discharge region is also extended, and the probability of ionization of other atoms of the working gas increases. This makes it possible to maintain a discharge at a lower pressure (on the order of a tenth of a Pa) and at a lower voltage (on the order of a hundred volts). In particular, the lower pressure is positively reflected in the greater purity of the formed layers [18].

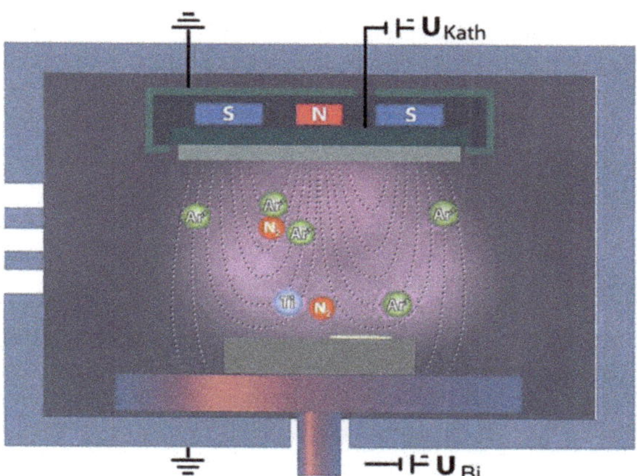


**Figure 2.** The principle of HIPIMs sputtering (source CemeCon) [18].

The process for the preparation of our studied composite coating on the milling drill source targets that were used in the creation of the studied TiAlN coatings. These are plates in which bright colored aluminum rollers were pressed. In the coating process, the source target (Figure 3) was bombarded with argon and krypton atoms. The released Ti + Al particles, together with the supplied nitrogen gas, form a TiAlN coating on the surface of the milling drill. By changing the amount of Al rollers in the titanium plate, we were able to influence the resulting chemical composition of the coating, namely the ratio between titanium and aluminum.

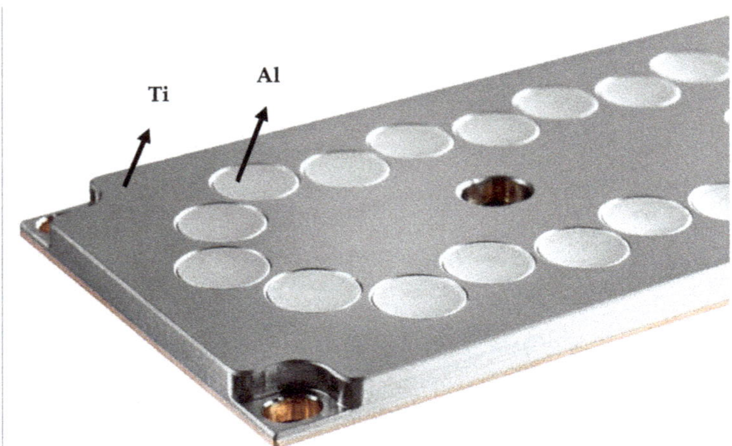


**Figure 3.** TiAl target with a nitrogen atmosphere for preparation of TiAlN coating (source CemeCon) [18].

Before the deposition processes, all samples were cleaned in an ultrasonic bath with ethanol for 10–15 min. After cleaning, the samples were placed in the chamber on a rotating table at an angle of 0°. The substrates were maintained at a constant temperature, determined before application to be 500 °C for 140 min. The TiAlN layer was applied using industrial coating device (CemeCon CC800/9XL HPPMS (CemeCon AG, Aachen, Germany)), in the HIPIMs mode. Pre-treatment with plasma etching was chosen to purify and activate the substrate surface prior to the deposition processes [19]. The spray device used had four cathodes capable of operating in the DCMS mode and two cathodes in the HIPIMS mode. The targets were installed on the HIPIMs cathode. Coatings were applied for 200 min using TiAlN targets at 5 kW, a frequency of 1000 Hz, and a pulse duration of 100 µs. To ensure uniformity of the composition, a substrate rotation speed of 10 RPM was used. After applying the TiAlN layers, the samples were cooled to 20 ± 5 °C for 30 min. A single layer of TiAlN was applied at a size of 3 ± 1 µm.

### 2.3. Experimental Methods

After plasma coating with a TiAlN layer, the structure and phase composition of the samples were analyzed using X-ray diffraction (XRD). The measurements were performed on a PANalytical X′Pert PRO X-ray diffractometer (PANalytical, Middle Watch Swavesey, Cambridge, UK) in a symmetrical Bragg–Brentano configuration under $Cu_{K\alpha}$ radiation ($\lambda$ = 1.5418 Å). The diffraction data were collected using an X′Celerator 1D detector in 2 theta range 20–80°. The concentrations of individual elements were analyzed using tube-above sequential wavelength dispersive X-ray fluorescence (WDXRF). The measurements were performed on a Rigaku ZSX Primus 4 device with a XMET 8000 Geo Expert handheld XRF analyzer (Oxford Instruments, Abingdon, UK) using the "soil" calibration. The data were evaluated using a non-standard fundamental parameters method, which allowed the determination of elements in the F-U range in concentrations from a hundredth of a ppm to 100%.

Five samples were prepared using conventional techniques—wet grinding and diamond emulsion polishing. All samples were manually prepared. The final mechanical chemical finishing was performed using a Struers OPS suspension (Struers, Ballerup, Denmark). After phosphoric acid etching, the structures of the material were observed and documented using light microscopy on a LEXT OLS 3100 confocal laser microscope from Olympus.

Scanning electron microscopy (SEM) was used for surface imaging of the coating with a SEM microscope (Tescan, Lyra3 GMU, Brno, Czech Republic) and for chemical analyses of the TiAlN micro-coating, EDS analyses were performed for the elemental distribution in the coating in terms of homogeneity and the structure of the coating. The accelerating voltage of the electrons was set to 5 kV. All the samples studied had to be dusted with metal to make the surface electrically continuous. The chemical composition and atomic abundances were studied using energy dispersive X-ray spectroscopy (EDS, X-Maxn analyzer, SDD detector 20 $mm^2$, Oxford Instruments, Abingdon, UK). The acceleration voltage of the SEM-EDS method was set to 5 kV.

The study of surface morphology and the roughness of the tool samples was also assessed using atomic force microscopy (AFM) (Dimension ICON, Bruker Corp., Billerica, MA, USA). The QNM overhead mode in air was utilized for investigations. A silicon tip was used for the holder $Si_3N_4$, SCANASYST-AIR with a spring constant of 0.4 $N{\cdot}m^{-1}$. NanoScope analysis software was employed for data processing. The mean roughness value (Ra) represented the arithmetic mean of deviation from the median plane of the sample. The samples were analyzed in the Scan-Assyst mode (tapping mode) using a nitride lever (SCANASYST-AIR) with a Si tip. The typical tip radius of the curvature was smaller than 10 nm. The effective area represents the real area with the determination difference from the "basic" area in % (growth).

The microhardness test of milling the drill bit with the applied TiAlN coating was performed according to the CSN EN ISO 6507-1 standard on a microhardness tester (Mitutoyo HM-220) [20]. To measure the hardness, a 136° pyramidal diamond indenter creating a square indent was used. The nominal value of load was HV 0.2 (F = 1961 N, 200 g), which acted on the test specimen for 10 s. Due to the thickness of the layer, emphasis was placed on the chosen diagonal of the indenter in order to respect the ISO 6507-1 standard.

## 3. Results and Discussion

### *3.1. Analysis of the Chemical Composition of the Milling Drill*

Powerful milling drills, made of high-quality high-speed steel (HSS), which is ennobled with legures-tungsten, molybdenum, and chromium, are widely used in the automotive, aerospace, and military industries. Spring steel used in the production of industrial cobalt milling drills resists bending of the milling drill when drilling into a hard metal. The combination of tungsten and cobalt, together with the optimized geometry of the tip of the milling drill, keeps the working edges of the milling drill sharp four times longer. The prepared samples were examined using XRF analyses, where the area of the examined sample, including the result of the chemical composition, is shown in Figure 4. This analysis led to the exact composition of the body of the core of the milling drill before its subsequent coating. From the results of the XRF analyses, it is evident that tungsten with an 88.6% content is the constituent material of the milling drill and there is also about 10.3% cobalt serving as the bonding material.

XRD phase analysis was performed to determine the phase composition of the tool. Figure 5 shows the obtained X-ray diffractograms with the diffraction peaks that were determined using a conventional θ-2θ X-ray diffraction scan of the studied tool. The diffractogram shows that the core of the milling drill is composed of WC. The spades belonging to the WC are wider, with a clearly larger FWHM (full width at half maximum) being the difference between the two values of the independent variable at which the dependent variable is equal to half of its maximum value. That is, the particles of the WC are fine-grained. Widespread at all peaks, WC (001), (100), (101), (110), (002), (111), (200), and (102) were approximately the same, so they were probably regular crystallites. Another peak at Co (111), in the diffractogram, opposite to the content of the WC, indicates the presence of another phase. It is noticeable that this is a cubic alpha phase, which confirms the content of Co as a binder. Thus, the quantification of the elemental composition in the conclusion is supported only by the chemical composition and not by the phase analysis.

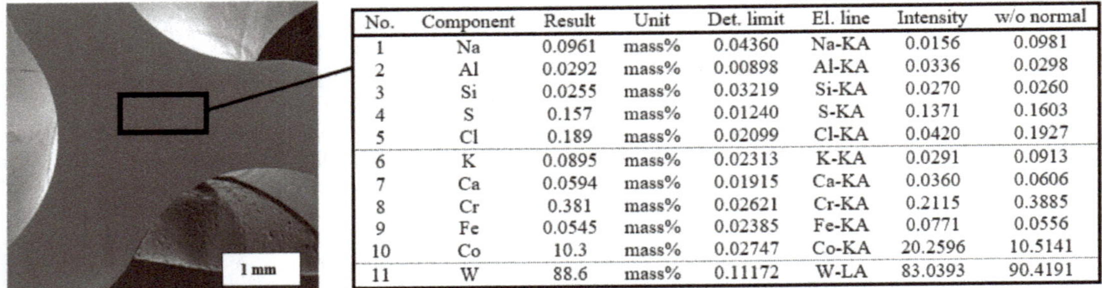

| No. | Component | Result | Unit | Det. limit | El. line | Intensity | w/o normal |
|---|---|---|---|---|---|---|---|
| 1 | Na | 0.0961 | mass% | 0.04360 | Na-KA | 0.0156 | 0.0981 |
| 2 | Al | 0.0292 | mass% | 0.00898 | Al-KA | 0.0336 | 0.0298 |
| 3 | Si | 0.0255 | mass% | 0.03219 | Si-KA | 0.0270 | 0.0260 |
| 4 | S | 0.157 | mass% | 0.01240 | S-KA | 0.1371 | 0.1603 |
| 5 | Cl | 0.189 | mass% | 0.02099 | Cl-KA | 0.0420 | 0.1927 |
| 6 | K | 0.0895 | mass% | 0.02313 | K-KA | 0.0291 | 0.0913 |
| 7 | Ca | 0.0594 | mass% | 0.01915 | Ca-KA | 0.0360 | 0.0606 |
| 8 | Cr | 0.381 | mass% | 0.02621 | Cr-KA | 0.2115 | 0.3885 |
| 9 | Fe | 0.0545 | mass% | 0.02385 | Fe-KA | 0.0771 | 0.0556 |
| 10 | Co | 10.3 | mass% | 0.02747 | Co-KA | 20.2596 | 10.5141 |
| 11 | W | 88.6 | mass% | 0.11172 | W-LA | 83.0393 | 90.4191 |



**Figure 4.** Area of the tool core examined by XRF and its elemental representation of the tool material.

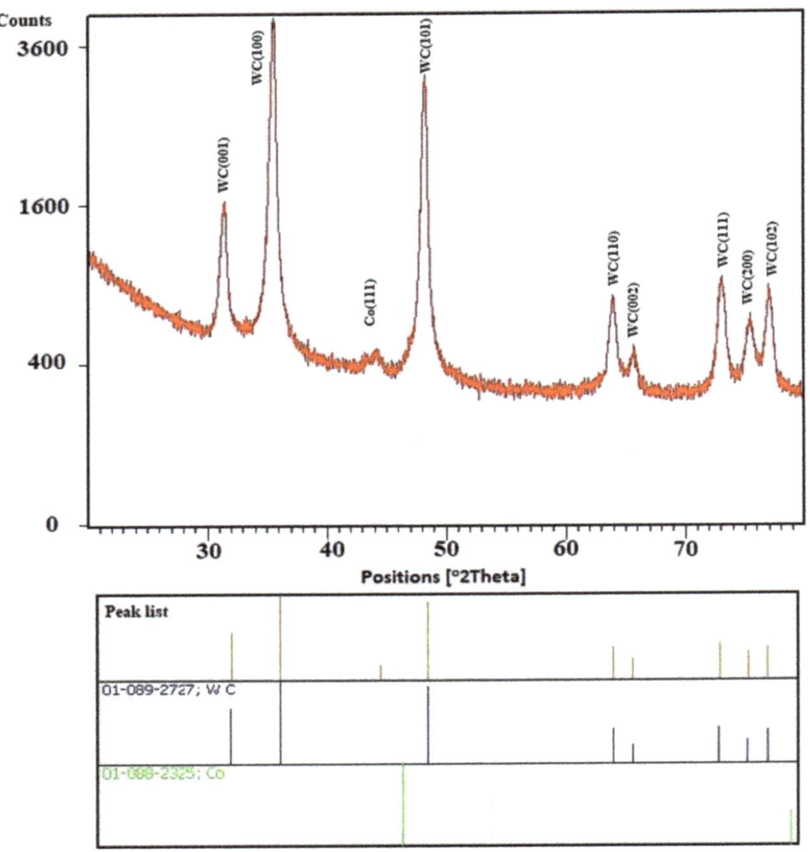


**Figure 5.** X-ray diffractograms of the tool core.

### 3.2. EDS Coating Analysis

Among the current most-often created abrasion-resistant coatings can be found layers based on TiAlN. This layer was developed with the need to increase the oxidation resistance of tin layers. TiAlN layers compared to TiN, not only have higher oxidation resistance, but also a higher hardness of about 2000 HV, which allows the use of these coatings in intensive cutting conditions, such as high-speed machining and dry machining [21]. Point EDS analysis of the selected location (Figure 6) shows a composite TiAlN coating with the following percentage of elements: Ti-38.5%, Al-30.3%, N-27.3%, C-2.5%, O-1.5%. The analysis showed that the composite TiAlN coating contained titanium, aluminum and nitrogen. We performed a standard point analysis, followed by acquiring data from the elemental maps. Thus, the presented spectrum was acquired from multiple points on the sample, and is presented as a "combined" EDS spectrum, representing the average surface chemistry of the sample's surface. The abundance of carbon in the elemental composition of the analyzed samples can be explained by the diffusion of this element from the substrate. Oxygen occurs in an air atmosphere and it is probably that during sputtering, oxygen atoms were entrained and their integration into the coating layer occurred. The ratio of titanium to aluminum can be adjusted by the aluminum content in the target [22]. In addition, carbon (2.5%) and oxygen (1.5%) were present in the coating in a small amount as polluting elements originating from the used technology, which Wagner also indicated in his study [23].

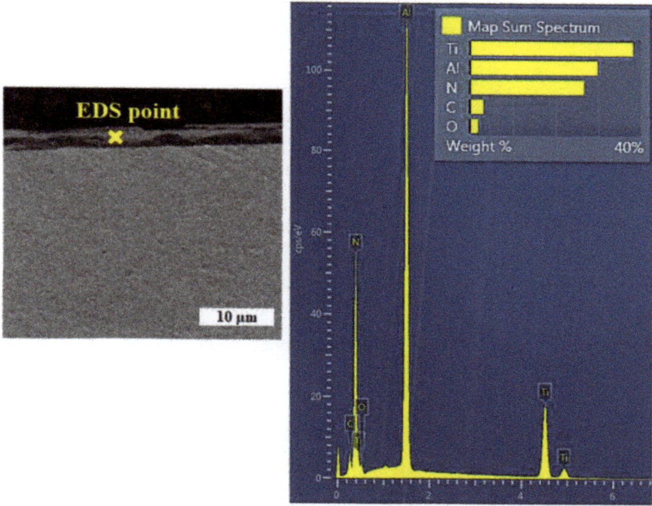


**Figure 6.** Selected site for spot analysis of the coating with the EDS recording of individual elements.

The following images (Figure 7) show a coated layer of TiAlN, marked with the place of examination (letter A). There is a gradual approximation of the area shown in Figure 7a, so that the morphology of poles can be identified from the point of view of the applied technology of magnetron sputtering in the HIPIMs mode. In the TiAlN composite coating, isolated individual pores of max. 8–10 µm in size (Figure 7b) or separate foreign parts with a size max. of 10–15 µm (Figure 7c) are present. The images show that the morphology of the TiAlN coating was formed by lamination of the lamellas next to each other, where the individual lamellas are uniform over a narrow size range and reach a thickness in the range of 0.7–1.0 µm (Figure 7d). A similar morphology was studied in the publication by Wagner [23] or Jiang [24]. The same phenomenon was studied in all five samples examined in this study.

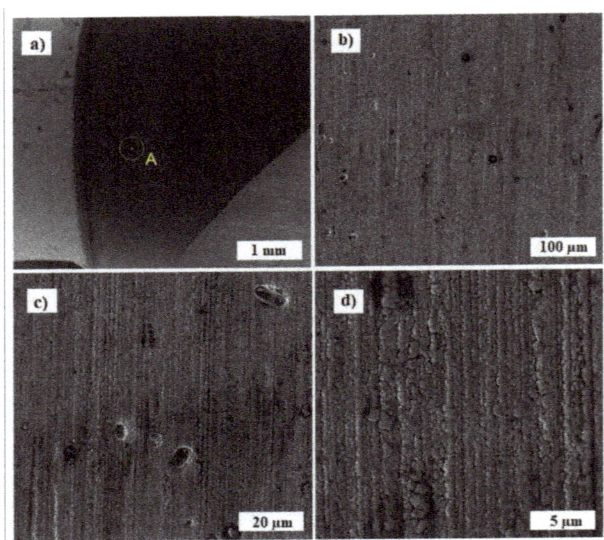


**Figure 7.** SEM secondary electron mode (SE) images of the place of examination, marked with the letter A (**a**), details (**b–d**) provide a detailed view of the morphology of the TiAlN coating.

EDS analysis was also performed using a scanning electron microscope, which showed a very uniform distribution of individual elements in a given layer. The distribution of the elements used for coating (titanium, aluminum, and nitrogen), as well as carbon as an alloying element, was monitored and shown in Figure 8. From the point of view of the documented results of the distribution of the individual elements, it can be concluded that the distribution of all elements is very homogeneous, and two areas of greater inhomogeneity are noticeable for carbon compared to the rest of the studied area.

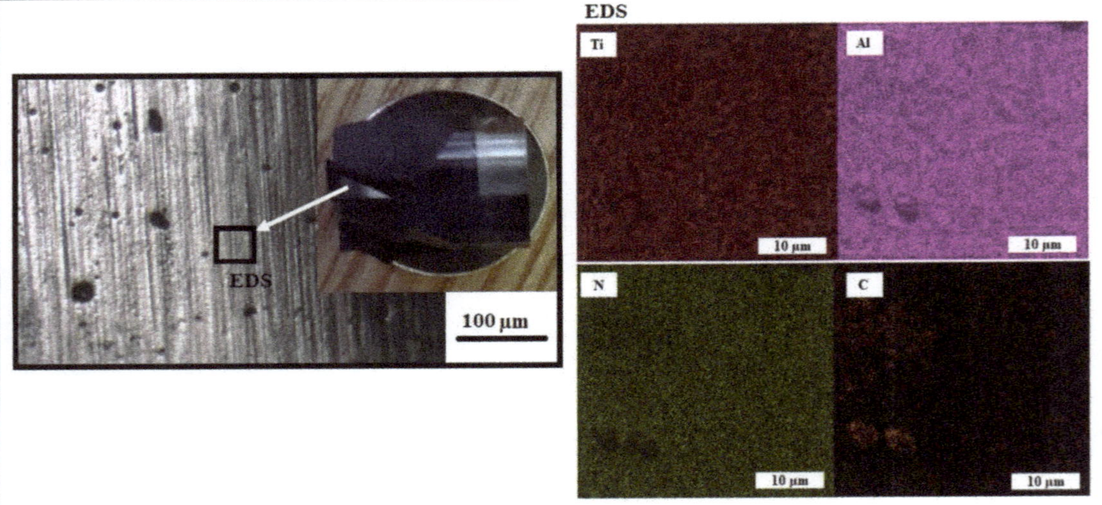


**Figure 8.** EDS analysis of the layout of individual elements in the TiAlN layer.

SEM of the TiAlN coating are provided in Figure 9. A linear EDS analysis of the individual elements located in this area was performed.

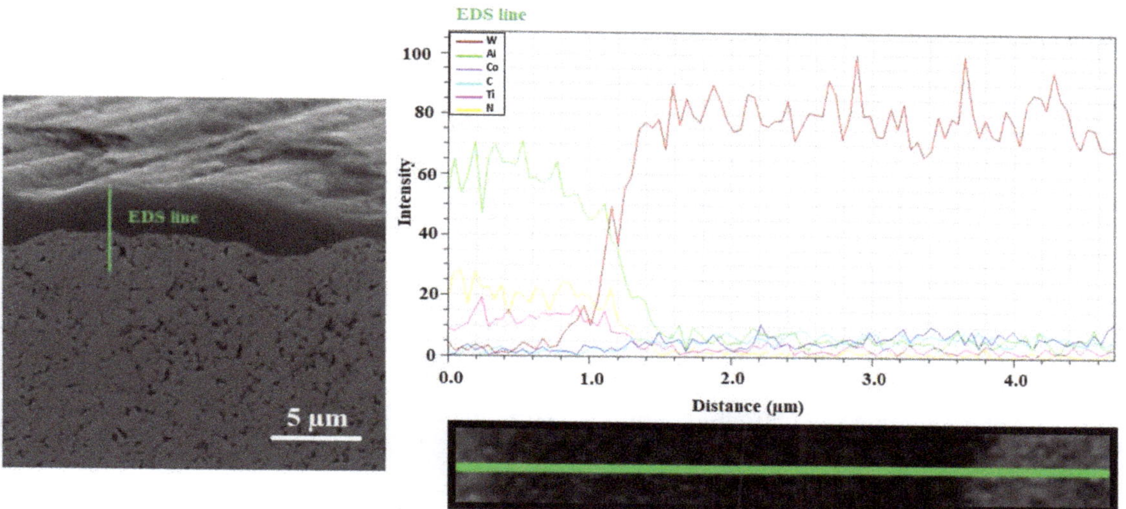


**Figure 9.** The morphology of the resulting micro-coat in a cross section (SEM) including the EDS measurement in a straight line.

The average thickness of the PVD coatings, as determined by SEM, was 1.5–1.7 μm (Figure 9). TiAlN coatings were characterized by a fine-grained microstructure with a low porosity, small internal stresses, excellent adhesion and a good thermal stability [25]. However, looking at the morphology of the micro-coating in detail, the compactness of this layer is observed to have a small number of pores. Moreover, in this area, with the help of linear EDS analysis, it can be seen that no expansion joint was found between the layer and the base substrate. Figure 9 also shows good adhesion between the layer and the base material. In the examined area of the layer (1–3 μm), a gradual decrease of Al is evident. This phenomenon was caused by the process of magnetron sputtering, in which Al was diffused into the base material [26]. A similar phenomenon can also be seen with nitrogen and very slightly with Ti (due to the fact that the differential potential of Ti compared to Al is much smaller [27]). During the magnetron sputtering process, W particles were also bombarded into the coating (at the bottom of the coating). This indicates a very good interconnection of the coating and the base material.

### 3.3. *Analysis of the Surface of the Milling Drill by AFM Microscopy*

Before carrying out the analyses, the samples were degreased with technical alcohol for 10 s. Subsequently, the surface morphology and roughness of the samples were studied using atomic force microscopy (AFM). As can be seen from the atomic force microscopy (and optical microscope) images, the surface of the milling drill also showed a sharper morphology (Figure 10) in accordance with the results from the scanning electron microscope. As introduced in [28], Ra and RMS values are dependent on the scan length; therefore, we introduced the saturated values in our paper.

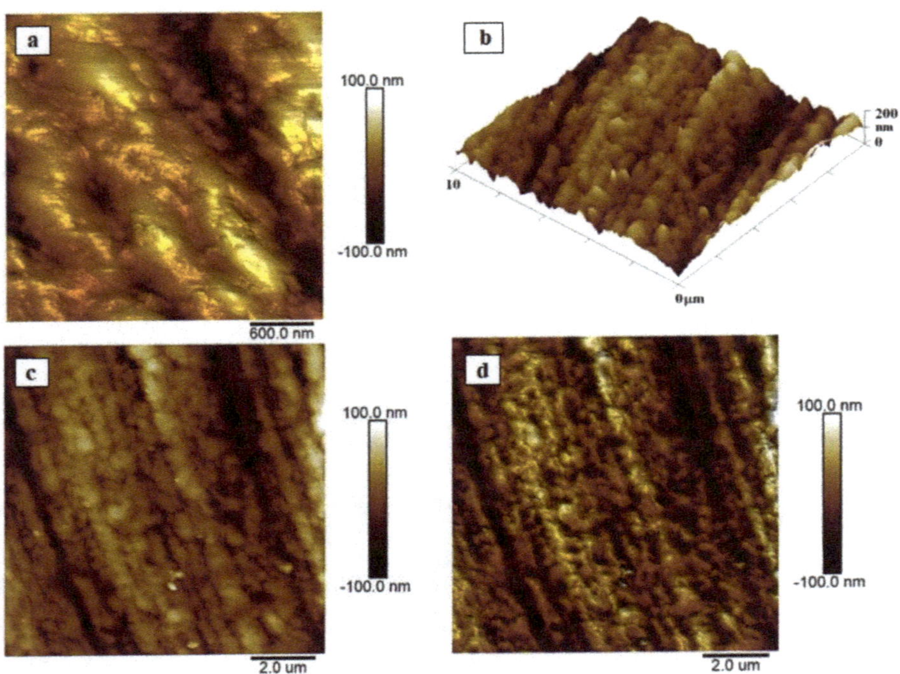


**Figure 10.** AFM analysis of the surface morphology and roughness: (**a**) analysis of tool and (**b**) 3D map of the tool-Ra = 20.5 nm, RMS = 25.8 nm, (**c**,**d**) detailed analysis of the tool-Ra = 17.5 nm, RMS = 22.4 nm.

Average roughness, *Ra*, is defined as the mean surface roughness representing the arithmetic mean of the deviations from the mean (zero) area. Compared to the interval of max/min heights (from −100 nm to 100 nm) it is commonly known that the value, as the result of *Ra* analysis, is much lower compared to Zmin and Z max.

$$R_a = \frac{\sum_{i=1}^{N} |Z_i - Z_{cp}|}{N},$$

$Z_{cp}$ is the Z value of the center (zero) area, $Z_i$ is the current Z value, and $N$ is the number of points in the scan (512 × 512 for our case).

The mean roughness ($R_a$) is the mean of deviations from the mean plane of the sample. As can be seen from the images from the atomic force microscopy (and optical microscopy), the surface of the milling drill also has a sharper morphology (Figure 10a) in accordance with the results from the scanning electron microscope, with a roughness of 20.5 nm for a square of 10 × 10 μm$^2$. In the analysis of the surface details (square 1 × 1 μm$^2$), sharper nanostructures, homogenously distributed on the surface, with sizes in the range of 15–30 nm, are visible (Figure 10c,d). Shaded images (Figure 10a–d) were also used as alternative images for all 2D samples. The shaded versions and the corresponding "normal" 2D (Figure 10c,d) and 3D (Figure 10b) versions are one and the same, only "artificial lighting" from the software was used to make the details more visible. Unfortunately, no similar study of the roughness of these surfaces has been performed, so the results cannot be compared.

### 3.4. Analysis of the Connection of the TiAlN Coating on the Working Part of the Milling Drill

Analysis and measurement of the thickness of the TiAlN coating was carried out in the working part of the milling drill, in a perpendicular section, using an OLS 3000 laser focal microscope (Figure 11a). The measurement was along the entire working length of the milling drill, in a perpendicular section from the tip, the side blade, and the side nail to the facet, in a total measured length of 6.4 mm.

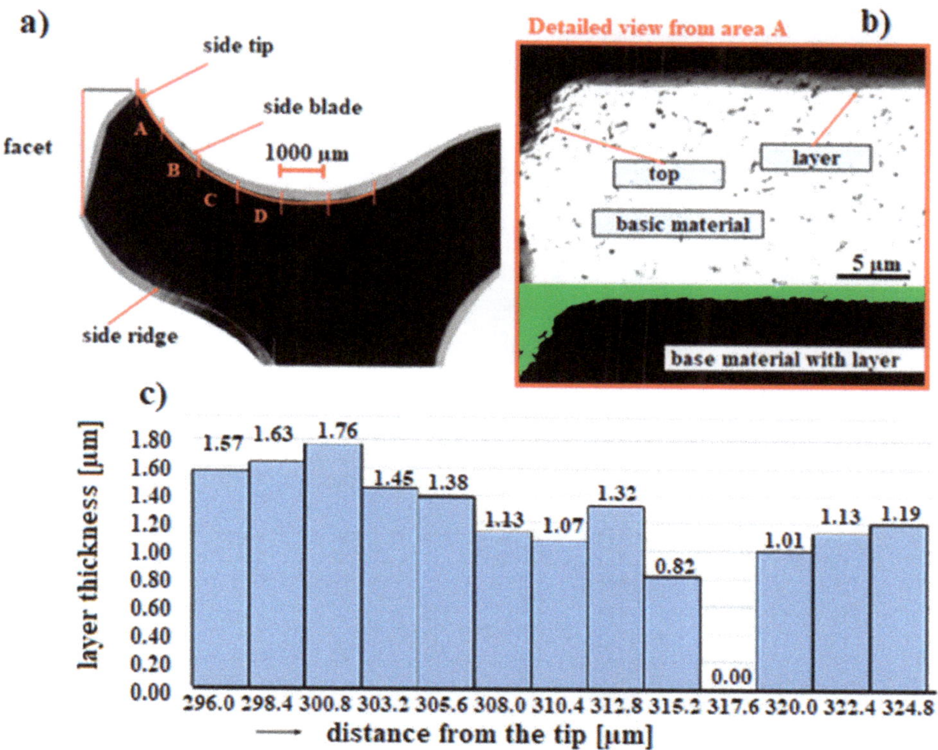


**Figure 11.** Display of the measured length of the coated TiAlN layer of the tool (**a**) showing the place of the coated layer (**b**) of the tool where the coating is almost completely absent (**c**).

The measured thickness of the TiAlN layers range from 0.57 µm to max. 1.76 µm (Figure 11b), but, for example, about 0.3 and 0.5 µm from the tip, there are places that are practically without a layer (Figure 11c). From the overall visualization of the measured section of 6.4 mm, four places (marked with the letters A, B, C, D) are visible where the layer thickness is practically zero (marked with a circle in Figure 12). Therefore, it will be necessary to change the position of the milling drill a bit more when coating, so that even shielded areas are covered and a more even coating is achieved. Another method that could allow a uniform distribution of the coating is the combination of the TiAlN/TiAlCN coating mentioned by Tillmann [29] and Xiang [30].

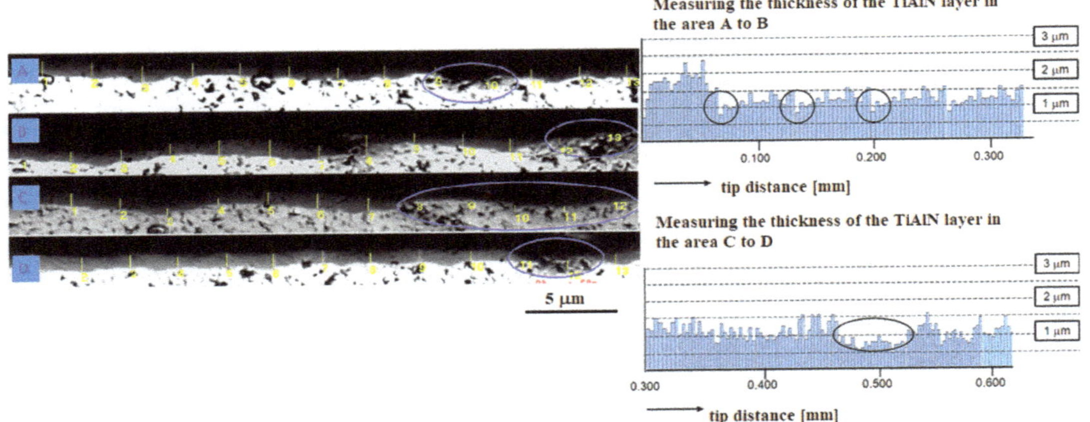


**Figure 12.** Representation of the thickness of the TiAlN layer of the tool in the measured section—indicating weak points A, B, C, D in the coated layer.

### 3.5. Hardness Measurement

Table 1 shows the average value of the Vickers hardness measurements of the substrate and the measurements of coated layer hardness. Twenty measurements were done on each specimen and the average values were taken (10 for the basic substrate and 10 for the TiAlN coating). The determined hardness of the hard WC–Co metal was 1800 HV and corresponded to the value reported in the literature [31].

**Table 1.** Representation of the hardness measurement result, HV.

| Coating | Number of Layers | Hardness, HV |
|---|---|---|
| None | - | 1800 |
| TiAlN | Single layer | 2500 |

Due to the measured values of microhardness, one can clearly observe a steep increase in the values of microhardness of the applied TiAlN coating. The applied monolayer also showed high values of microhardness, max. 2500 HV. Similar results were achieved in the work of Wei [32]. In comparison with that publication, where the micro-layer of TiN/TiAlN had a microhardness of 3000 HV, the applied monolayer described in this article reached almost the same indicators, namely 2500 HV.

## 4. Conclusions

The analyses presented in this publication focused on the study of a new multi-layer micro-permeability for cutting tools from WC alloy, intended for drilling into hard metal. The following conclusions were drawn:

- From the results of XRF analyses, it is evident that tungsten with 88.6% content together with WC carbon was the constituent material of the milling drill;
- EDS analyses of the composite TiAlN coating showed that the morphology of the TiAlN coating was formed by lamination of the lamellas next to each other, where the individual lamellas are uniform over a narrow size range and reach a thickness in the range of 0.7–1.0 μm;
- EDS analyses proved very good homogenization of individual elements and their uniform representation in a given layer (titanium, aluminum, and nitrogen);

- Based on AFM analyses, we can conclude that the TiAlN layer showed sharper nanostructures, homogeneously distributed on the surface, with sizes in the range of 15–30 nm;
- The Ra parameter for the TiAlN layer was 17.5 to 20.5 nm;
- In some places, about 0.3 and 0.5 mm from the tip, there were places with practically no layer and, overall, in the 6.4 mm section, there are four places where the layer thickness was practically zero;
- The applied layer of TiAlN had a microhardness of about 2500 HV compared to the basic substrate, which had a microhardness of about 1800 HV.

From this it can be concluded that the technology of magnetron coating using the HIPIMs method will guarantee the required excellent homogeneity in the distribution of individual elements in the coating of the milling drill, but it will not guarantee a uniform thickness of the coating for more complex shapes. It will be necessary to change the position of the milling drill more when coating, so that even the shaded places are covered and a more even coating is obtained.

**Author Contributions:** Conceptualization, Š.M., I.H. and J.N.; Data curation, Š.M., I.H., J.N., L.M. and V.Š.; Methodology, V.Š.; Resources, Š.M., I.H., L.M. and V.Š.; Supervision, Š.M. and J.N.; Visualization, I.H.; Writing—original draft, Š.M.; Writing—review & editing, I.H. and J.N. All authors have read and agreed to the published version of the manuscript.

**Funding:** This work was funded by with the support of OP VVV Project Development of new nano and micro coatings on the surface of selected metallic materials–NANOTECH ITI II. Reg. No. CZ.02.1.01/0.0/0.0/18_069/0010045.

**Institutional Review Board Statement:** Not applicable.

**Informed Consent Statement:** Not applicable.

**Data Availability Statement:** Data is contained within the article.

**Conflicts of Interest:** The authors declare no conflict of interest. The funders had no role in the design of the study; in the collection, analyses, or interpretation of data; in the writing of the manuscript, or in the decision to publish the results.

## References


1. Hovsepian, P.E.; Lewis, D.B.; Münz, W.-D. Recent progress in large scale manufacturing of multilayer/superlattice hard coatings. *Surf. Coat. Technol.* **2000**, *133–134*, 166–175. [CrossRef]
2. Tiron, V.; Velicu, I.; Mihăilă, I.; Popa, G. Deposition rate enhancement in HiPIMS through the control of magnetic field and pulse configuration. *Surf. Coat. Technol.* **2018**, *337*, 484–491. [CrossRef]
3. Sarakinos, K.; Alami, J.; Konstantinidis, S. High power pulsed magnetron sputtering: A review on scientific and engineering state of the art. *Surf. Coat. Technol.* **2010**, *204*, 1661–1684. [CrossRef]
4. Santiago, J.A.; Fernández-Martínez, I.; Kozák, T.; Capek, J.; Wennberg, A.; Molina-Aldareguia, J.M.; Bellido-González, V.; González-Arrabal, R.; Monclús, M.A. The influence of positive pulses on HiPIMS deposition of hard DLC coatings. *Surf. Coat. Technol.* **2019**, *358*, 43–49. [CrossRef]
5. Breidenstein, B.; Denkena, B. Significance of residual stress in PVD-coated carbide cutting tools. *CIRP Ann.* **2013**, *62*, 67–70. [CrossRef]
6. Tillmann, W.; Grisales, D.; Stangier, D.; Butzke, T. Tribomechanical behaviour of TiAlN and CrAlN coatings deposited onto AISI H11 with different pre-treatments. *Coatings* **2019**, *9*, 519–539. [CrossRef]
7. Gudmundsson, J.T.; Alami, J.; Helmersson, U. Evolution of the electron energy distribution and plasma parameters in a pulsed magnetron discharge. *Appl. Phys. Lett.* **2001**, *78*, 3427–3429. [CrossRef]
8. Broitman, E.; Czigány, Z.; Greczynski, G.; Böhlmark, J.; Cremer, R.; Hultman, L. Industrial-scale deposition of highly adherent CNx films on steel substrates. *Surf. Coat. Technol.* **2010**, *204*, 3349–3357. [CrossRef]
9. Lattemann, M.; Ehiasarian, A.P.; Bohlmark, J.; Persson, P.Å.; Helmersson, U. Investigation of high power impulse magnetron sputtering pretreated interfaces for adhesion enhancement of hard coatings on steel. *Surf. Coat. Technol.* **2006**, *200*, 6495–6499. [CrossRef]
10. Lattemann, M.; Moafi, A.; Bilek, M.; McCulloch, D.; McKenzie, D. Energetic deposition of carbon clusters with preferred orientation using a new mixed mode cathodic arc—Sputtering process. *Carbon* **2010**, *48*, 918–921. [CrossRef]
11. Jindal, P.C.; Santhanam, A.T.; Schleinkofer, U.; Shuster, A.F. Performance of PVD TiN, TiCN, and TiAlN coated cemented carbide tools in turning. *Int. J. Refract. Met. Hard Mater.* **1999**, *17*, 163–170. [CrossRef]

12. Bakoglidis, K.D.; Schmidt, S.; Greczynski, G.; Hultman, L. Improved adhesion of carbon nitride coatings on steel substrates using metal HiPIMS pretreatments. *Surf. Coat. Technol.* **2016**, *302*, 454–462. [CrossRef]
13. Vitelaru, C.; Parau, A.C.; Constantin, L.R.; Kiss, A.E.; Vladescu, A.; Sobetkii, A.; Kubart, T. A strategy for alleviating micro arcing during HIPIMs deposition of DLC coatings. *Materials* **2020**, *13*, 1038. [CrossRef]
14. Galvan, D.; Pei, Y.T.; De Hosson, J.T. TEM characterization of a Cr/Ti/TiC graded interlayer for magnetron-sputtered TiC/a-C:H nanocomposite coatings. *Acta Mater.* **2005**, *53*, 3925–3934. [CrossRef]
15. Shimizu, T.; Komiya, H.; Watanabe, T.; Teranishi, Y.; Nagasaka, H.; Morikawa, K.; Yang, M. HIPIMS deposition of TiAlN films on inner wall of micro-dies and its applicability in micro-sheet metal forming. *Surf. Coat. Technol.* **2014**, *250*, 44–51. [CrossRef]
16. Raaif, M. Constructing and characterizing TiAlN thin film by DC. Pulsed magnetron sputtering at different nitrogen/argon gas ratios. *J. Adv. Phys.* **2018**, *14*, 5638–5652. [CrossRef]
17. Mattox, D. *The Foundations of Vacuum Coating Technology*; Noyes Publications; William Andrew Publishing, Inc.: Albuquerque, NM, USA, 2003; ISBN 0-8155-1495-6.
18. CemeCon. Coating Technology. Available online: http://www.cemecon.cz/technologie-povlakovani (accessed on 19 May 2021).
19. Monteverde, F.; Medri, V.; Bellosi, A. Microstructure ofhot-pressed Ti(C.N)-based cermets. *J. Eur. Soc.* **2002**, *22*, 2587–2593.
20. *CSN EN ISO 6507-2:2005. Metallic Materials. Vickers Hardness Test. Part 2: Verification and Calibration of Testing Machines*; ISO: Geneva, Switzerland.
21. Anders, A. A review comparing cathodic arcs and high power impulse magnetron sputtering (HiPIMS). *Surf. Coat. Technol.* **2014**, *257*, 308–325. [CrossRef]
22. Hovsepian, P.E.; Ehiasarian, A.P.; Petrov, I. Structure evolution and properties of TiAlCN/VCN coatings deposited by reactive HIPIMS. *Surf. Coat. Technol.* **2014**, *257*, 38–47. [CrossRef]
23. Wagner, J.; Edlmayr, V.; Penoy, M.; Michotte, C.; Mitterer, C.; Kathrein, M. Deposition of Ti–Al–N coatings by thermal CVD. *Refract. Met. Hard Mater.* **2008**, *26*, 563–568. [CrossRef]
24. Jiang, W.; Malshe, A.P.; Goforth, R.C. Cubic Boron Nitride (CBN) based nanocomposite coatings on cutting inserts with chip breakers for hard turning applications. *Surf. Coat. Technol.* **2005**, *5*, 1848–1854. [CrossRef]
25. Panjan, P.; Čekada, M.; Panjan, M.; Kek-Merl, D. Growth defects in PVD hard coatings. *Vacuum* **2010**, *84*, 209–214. [CrossRef]
26. Surmenev, R.A. A review of plasma-assisted methods for calcium phosphate-based coatings fabrication. *Surf. Coat. Technol.* **2012**, *206*, 2035–2056. [CrossRef]
27. Lohberger, B.; Eck, N.; Glaenzer, D.; Kaltenegger, H.; Leithner, A. Surface Modifications of Titanium Aluminium Vanadium Improve Biocompatibility and Osteogenic Differentiation Potential. *Materials* **2021**, *14*, 1574. [CrossRef]
28. Ruffino, F.; Grimaldi, M.G.; Giannazzo, F.; Roccaforte, F.; Raineri, V. Atomic Force Microscopy Study of the Kinetic Roughening in Nanostructured Gold Films on $SiO_2$. *Nanoscale Res. Lett.* **2009**, *4*, 262. [CrossRef]
29. Tillmann, W.; Grisales, D.; Tovar, C.M.; Contreras, E.; Apel, D.; Nienhaus, A.; Stangier, D.; Dias, N.F.L. Tribological behaviour of low carbon-containing TiAlCN coatings deposited by hybrid (DCMS/HiPIMS) technique. *Tribol. Int.* **2020**, *151*, 106528. [CrossRef]
30. Xian, G.; Xiong, J.; Zhao, H.; Fan, H.; Li, Z.; Du, H. Evaluation of the structure and properties of the hard TiAlN-(TiAlN/CrAlSiN)-TiAlN multiple coatings deposited on different substrate materials. *Int. J. Refract. Met. Hard Mater.* **2019**, *85*, 105056. [CrossRef]
31. Ibrahim, R.N.; Rahmat, M.A.; Oskouei, R.H.; Singh Raman, R.S. Monolayer TiAlN and multilayer TiAlN/CrN PVD coatings as surface modifiers to mitigate treating fatigue of AISI P20 steel. *Eng. Fract. Mech.* **2015**, *137*, 64–78. [CrossRef]
32. Wei, Y.; Gong, C. Effects of pulsed bias duty ratio on microstructure and mechanical properties of TiN/TiAlN multilayer coatings. *Appl. Surf. Sci.* **2011**, *257*, 7881–7886. [CrossRef]

*Article*

# Fabrication and Characterization of Sn-Based Babbitt Alloy Nanocomposite Reinforced with $Al_2O_3$ Nanoparticles/Carbon Steel Bimetallic Material

**Mohamed Ramadan [1,2,*], Abdulaziz S. Alghamdi [1], Tayyab Subhani [1,*] and K. S. Abdel Halim [1,2]**

1 College of Engineering, University of Ha'il, Ha'il P.O. Box 2440, Saudi Arabia; a.alghamdi@uoh.edu.sa (A.S.A.); k.abdulhalem@uoh.edu.sa (K.S.A.H.)
2 Central Metallurgical Research and Development Institute (CMRDI), Cairo 11421, Egypt
* Correspondence: mrnais3@yahoo.com (M.R.); ta.subhani@uoh.edu.sa (T.S.)

Received: 19 May 2020; Accepted: 15 June 2020; Published: 18 June 2020

**Abstract:** Sn-based Babbitt alloy was reinforced with alumina nanoparticles to prepare a novel class of nanocomposites. The route of liquid metallurgy in combination with stirring mechanism was chosen to prepare nanocomposites with three different loadings of alumina nanoparticles, i.e., 0.25 wt%, 0.50 wt% and 1.0 wt%. The molten mixture of metallic matrix and nanoparticles was poured over carbon steel substrate for solidification to manufacture a bimetallic material for bearing applications. The underlying aim was to understand the effect of nanoparticle addition on microstructural variation of Sn-based Babbitt alloy as well as bimetallic microstructural interface. The addition of 0.25 wt% and 0.50 wt% alumina nanoparticles significantly affected both the morphology and distribution of $Cu_6Sn_5$ hard phase in solid solution, which changed from needle and asterisk shape to spherical morphology. Nanocomposites containing up to 0.50 wt% nanoparticles showed more improvement in tensile strength than the one containing 1.0 wt% nanoparticles, due to nanoparticle-agglomeration and micro-cracks at the interface. The addition of 0.5 wt% nanoparticles significantly improved the wear resistance of Sn-based Babbitt alloy.

**Keywords:** nanocomposite; nanoparticle; microstructure; mechanical; Babbitt; alumina

## 1. Introduction

In any engine, bearings are used to support moving mechanical parts, which protect them from frictional degradation. The importance of stronger bearing materials becomes many fold in powerful engines. Similarly, the significance of bearings in other mechanical systems such as pumps and turbines cannot be underestimated, and both the improved wear resistance and reliability of frictional components in terms of their mechanical strength become the prime requisites [1]. White metal is one of the bearing alloys that offers desired bearing surface properties in plain bearings; the common alloys have tin and lead as leading metals and are widely used in friction assemblies such as compressors, turbines, vehicles, and many other frictional units [2]. Babbitt alloys have two important lead-based and tin-based families. Generally poured bearing and insert bearings are made of Babbitt alloys; insert bearings have a backing of steel with inner layer made of Babbitt alloy. Lead-based Babbitt alloys have an outstanding low-load bearing characteristics. However, lead is a weak material and therefore tin, copper and antimony are added for improved strength [3].

One of the techniques to improve the mechanical and frictional properties of bearing materials is the incorporation of second phase as reinforcement. The resultant hybrid material named as composite provides tailored properties especially suitable for bearing applications [4]. After the arrival of nanotechnology, a novel class of composites emerged as nanocomposites containing reinforcements with one of their dimensions in nanometer scale [5,6]. Nano-reinforcements of zero, single and two-dimensions have been incorporated to prepare nanocomposites [7]. In addition to solitary reinforcement [8], dual reinforcements have also been incorporated at similar [9] and multiscale [10] levels. Among the three types of nanocomposites containing polymeric [11], ceramic [12] and metallic [13] matrices, comparatively little attention has been paid to nanocomposites containing ceramic and metallic matrices and still lesser to metallic matrices. In this context, the combination of traditional bearing materials with nano-reinforcements can lead to a novel class of nanocomposites ideally suitable for high performance bearing applications.

The two common processes appropriate for developing advanced metallic materials are: (a) new alloy formation and (b) novel heat-treatment process. The manufacturing of nanocomposites offers another route to prepare metal matrix composites containing nano-reinforcement of high strength and stiffness: typically ceramic materials. The nanoreinforcement of size down to nanometer-scale provides an opportunity to interact with dislocations and hence improve the mechanical properties of matrix material through novel strengthening mechanisms and thus remarkable improvement in mechanical properties [14].

A range of ceramic nano-reinforcements including alumina, silicon carbide, silicon dioxide, silicon nitride and zirconium dioxide have been incorporated into different metallic matrices such as aluminum, copper, magnesium, titanium and iron. Both liquid and solid-state processing routes have been utilized to prepare nanocomposites. Liquid processing routes offer better densification while solid-state routes provide better distribution of nano-reinforcement. To achieve optimum properties of nanocomposites, both near-theoretical density and uniform dispersion of nano-reinforcements are fundamental requirements. Liquid metallurgy in combination with stirring technique offer better dispersion while solid-state processing primarily powder metallurgy in combination with pressure-assisted techniques such as hot-pressing and spark plasma sintering provides better densification. The low wettability of ceramic nano-reinforcements with metallic matrices leads to their poor dispersion, which can be improved by chemical and mechanical routes. In addition to the ex-situ incorporation of nano-reinforcements, in-situ formation of nano-reinforcements has also been investigated. Finally, friction stir processing is another solid-state processing route to prepare nanocomposites with improved mechanical performance.

Among a variety of metallic matrices, aluminum and its alloys are widely used for preparing nanocomposites, and scarce attention has been paid to improve the mechanical properties of bearing materials by adopting the technique of composite preparation. Recently, carbon nanotubes have been added in white metal (Babbitt), which increased its wear resistance to almost an order of magnitude [15].

In the present work, alumina nanoparticles have been incorporated in Sn-based Babbitt alloy through stir casting technique. The developed nanocomposite was bonded with carbon steel substrate to prepare a bimetallic material for bearing applications. The underlying aim was to investigate the effect of ceramic nanoparticles addition upon the microstructural evolution and resultant mechanical property improvement especially tribology. Another objective was to observe the microstructural interface of nanocomposites with carbon steel substrate.

## 2. Materials and Methods

Alumina nanoparticles of spherical morphology and average particle size of ~50nm were used as nano-reinforcement, as procured from Metkon Technology. Tin-based Babbitt alloy (Rotometals, San Leandro, CA, USA) was used as the matrix material to prepare nanocomposites. To prepare the bimetallic material, carbon steel was used for bonding with nanocomposites. The chemical compositions of Sn-based Babbitt alloy and carbon steel (local manufacturing Co., Riyadh, Saudi Arabia) are given in Table 1. Chemical composition of Sn- Babbitt alloy shown in Table 1 as per conducting by manufacturing company (Rotometals). For carbon steel substrate, ASCert metal scan analyzers (Arun Technology, Crawley, UK) was used for chemical analysis.

**Table 1.** Chemical compositions of tin-based Babbitt alloy and carbon steel substrate.

| Materials | Chemical Compositions (wt.%) | | | | | | | | |
|---|---|---|---|---|---|---|---|---|---|
| | Sb | Cu | Pb | C | Si | Mn | Cr | Sn | Fe |
| Tin Babbitt Alloy | 6.5–7.5 | 2.5–3.5 | 0.35 | - | - | - | - | Bal. | |
| Carbon Steel | - | 0.20 | - | 0.14 | 0.30 | 0.48 | 0.14 | - | Bal. |

A charge of 400 g of tin-based Babbitt alloy was placed in a graphite crucible and heated in electrical furnace to a temperature of 375 °C (heating rate = 9.4 °C/min.). Subsequently alumina nanoparticles were added after lowering the temperature to 330°C (cooling rate = 5.6 °C/min.) in a semi-solid state of alloy. Alumina nanoparticles were wrapped in aluminum foil and were directly immersed in small packages. The stirring operation was performed continuously during the addition of nanoparticles. The semi-solid slurry was then heated to 420 °C so that a fully liquid state could be achieved before pouring. The stirring in semi-solid state was performed at a speed of 800 rpm for 10 min, which was followed by stirring at 200 rpm until pouring.

The stirring was carried out mechanically using a two-blade impeller. After mixing and stirring stage, the molten metal was poured on a tinned carbon steel substrate, which was inserted into a pre-heated metallic mold at 150 °C. The carbon steel substrate had a thickness of 4mm and diameter 59 mm. The schematic of the process is presented in Figures 1 and 2 along with the images of the furnace, stirring mechanism, metallic mold and carbon steel substrate. Tinning process of carbon steel substrate involves the deposition of metallic tin and flux mixture on the surface of steel, as explained elsewhere [16].

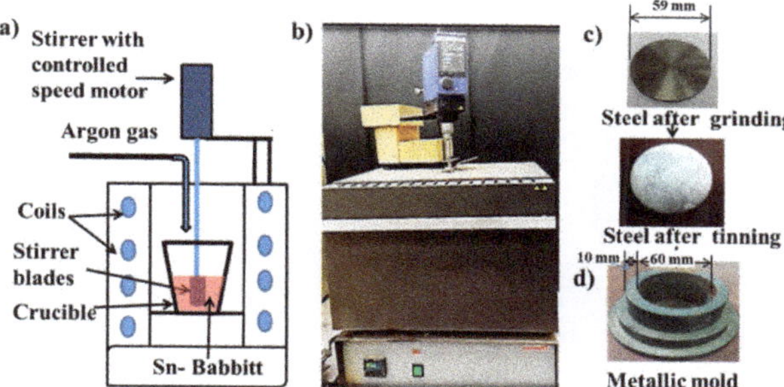


**Figure 1.** Furnace and mold used for preparing nanocomposite containing tin-based Babbitt alloy reinforced with alumina nanoparticles and carbon steel specimen before and after tinning, (**a**) Schematic drawing for furnace and stirrer, (**b**) photocopy of furnace and stirrer, (**c**) carbon steel substrate and (**d**) metallic mold.

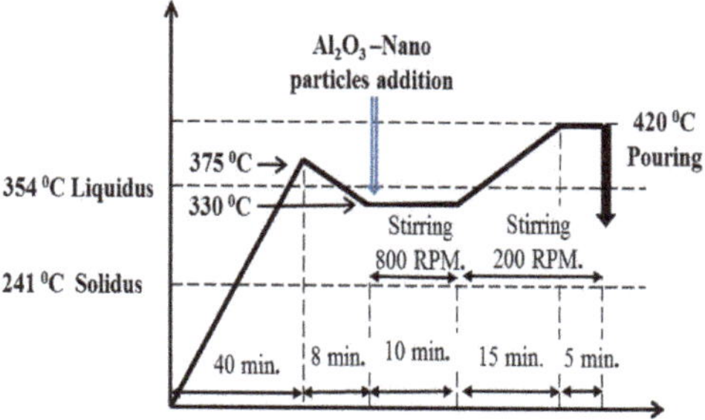


**Figure 2.** Sequence of preparing nanocomposite of tin-based Babbitt alloy reinforced with alumina nanoparticles by stir-casting process.

Nanocomposites of three types were prepared after incorporating three loading fractions of alumina nanoparticles in tin-based Babbitt alloy, i.e., 0.25 wt%, 0.50 wt% and 1.0 wt%. A reference sample without the addition of alumina nanoparticles was also prepared. For preparing bimetallic sample, the same type of carbon steel was used.

Optical microscopy was performed to observe the microstructure of bimetallic specimens containing nanocomposite and carbon steel alongwith their interface. The specimens were cut, ground, polished and etched with a solution of 4% nital (96% ethical alcohol and 4% nitric acid $HNO_3$). Optical microscope fitted with digital camera (Olympus GX51, Tokyo, Japan) was used for microstructural investigations at magnifications of 200× and 500×). For electron micrsocopy, field emission gun scanning electron microscope (FEG-SEM, FEI, Eindhoven, The Netherlands) was utilized by a Quanta 250 FEG, (Eindhoven, The Netherlands) with energy-dispersive X-ray spectroscopy (EDS) was used. Interface between the nanocomposites and substrate steel was observed in electron microscopy and compositional analysis was performed using EDS. Moreover, SEM and EDS was also performed on the worn surfaces after wear tests, as discussed further below.

The hardness of nanocomposites was tested using Vickers microhardness testing machine (VLS 3853, Shimadzu, Japan). The hardness testing was performed at the load of 1kgf for the dwell time of 5 sec. The specimens were polished until the surface finish of 1 µm. At least five readings of each of the nanocomposite were taken for the average value of hardness.

The tensile and shear testing until complete failure of the bond were conducted using tensile testing machine Instron 5969 characterized by a maximum load of 50 KN. The interfacial shear strength was measured by a tensile shear method. The dimensions of tensile shear specimen are reported in a previous work [17].

To evaluate the wear resistance of nanocomposites, pin-on-ring wear test was performed. The fixed samples of nanocomposites were tested against the rotating steel wheel at a constant speed of 260 rpm for 10 min under a constant loading of 28N.The wear resistance of the samples was measured in terms of the weight-loss.

## 3. Results

### *3.1. Microstructure of Reinforced Sn-based Babbitt Alloy and Bimetal Interface*

Microstructures of both of the materials used in the fabrication of tin-based Babbitt alloy/carbon steel bimetallic composite are shown in Figure 3 while Figure 4 shows SEM microstructures (Figure 4a,c) and EDS analysis (Figure 4b,d) of $Cu_6Sn_5$ hard phase and matrix of tin-based Babbitt alloy of bimetallic material. Figure 3a shows an optical microstructure of Sn-based Babbitt alloy revealing $Cu_6Sn_5$ and SnSb phases reinforced in solid solution of SnSbCu matrix. $Cu_6Sn_5$ phase has the shape of needles and asterisks. The microstructure of a matrix of solid solution with hard embedded phases of $Cu_6Sn_5$ and SnSb make a metal matrix composite possessing fatigue resistant properties [17–19]. Previous studies [20–22] reveal that the variation in the chemical composition of tin-based Babbitt alloy modifies the microstructure. Alloys containing less than 8% Sb are characterized by a solid solution with distributed needles of $Cu_6Sn_5$, copper-rich constituents and fine precipitates of SbSn. On the contrary, the alloys containing greater than 8% Sb exhibit a primary cuboid phase of SbSn. The microstructure of Sn-based Babbitt alloy acquired in the present study is in good agreement with those available in literature [20–22]. The microstructure of low carbon steel substrate (Figure 3b) shows the two very common phases of ferrite and pearlite; the mutual percentage of the presence of two phases verifies the carbon content in the steel.

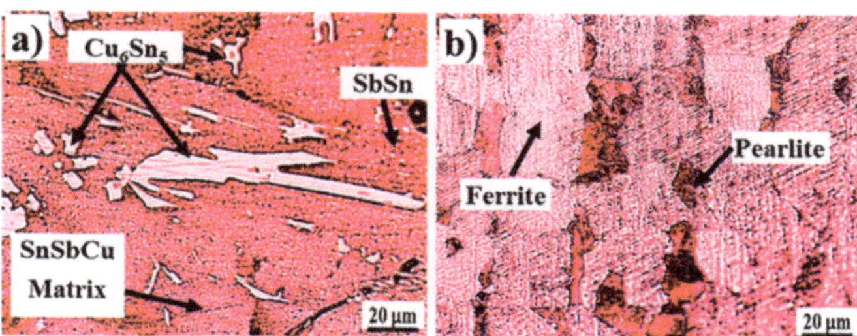


**Figure 3.** Microstructures of (**a**) tin-based Babbitt alloy, and (**b**) steel substrate, used for preparing bimetallic casting.

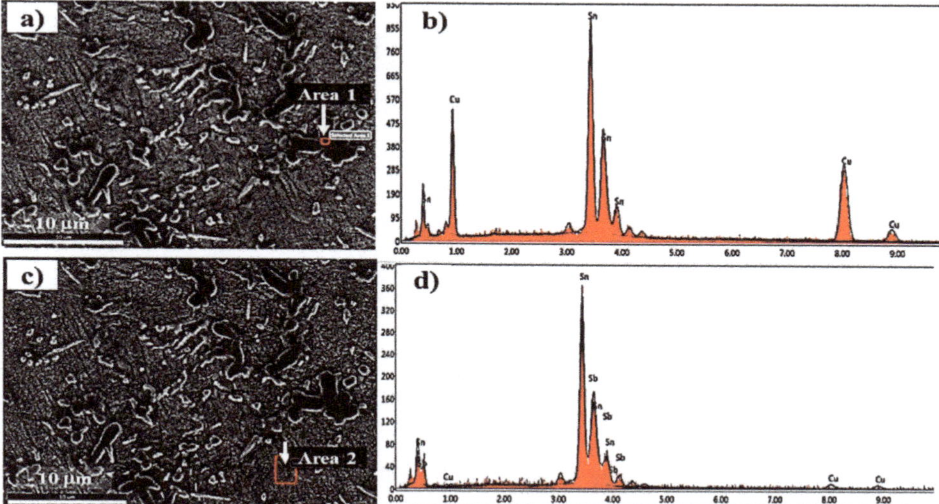


**Figure 4.** (**a**,**c**) SEM images of tin-based Babbitt alloy and (**b**,**d**) graphical presentation of EDS microanalysis of precipitates (Area 1) and matrix (Area 2) in Sn-based Babbitt alloy.

Figure 5 shows the microstructures of Sn-based Babbitt alloy without (Figure 5a) and with three different loadings of alumina nanoparticles, i.e., 0.25 wt%, 0.50 wt%, and 1.0 wt%, correspondingly (Figure 5b–d). The addition of alumina nanoparticles significantly affected both the morphology and distribution of $Cu_6Sn_5$ hard phase in solid solution. It is evident that $Cu_6Sn_5$ phase changed from needle and asterisk shape to spherical morphology after the addition of alumina nanoparticles. Pervious study [23] reported that addition of $Al_2O_3$ nanocomposites to hyper-eutectic Al-Si alloy, change the morphology of primary Si to a typical polyhedral shape. A possible reason of the change in the morphology of $Cu_6Sn_5$ phase from needle to spherical shape may be the hindrance in the growth due to the presence of alumina nanoparticles. However, a trend of $Cu_6Sn_5$ phase agglomeration was observed as the alumina nanoparticle content increased to 1.0 wt% loading where clearly identifiable regions of $Cu_6Sn_5$-rich and $Cu_6Sn_5$-depleted phases can be seen (Figure 5d). It may be due to the reason that at 1.0 wt% content, alumina nanoparticles could not be dispersed uniformly and at alumina-rich regions, the growth of $Cu_6Sn_5$ phase was very low and vice versa.

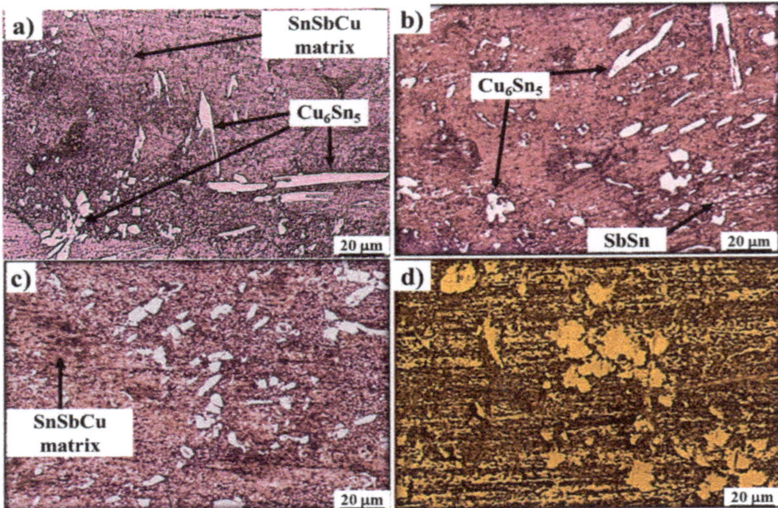


**Figure 5.** Microstructures of (**a**) tin-based Babbitt alloy, and tin-based Babbitt nanocomposites containing (**b**) 0.25 wt%, (**c**) 0.50 wt%, and (**d**) 1.0 wt% alumina nanoparticles correspondingly.

The microstructures of interfaces of bimetallic samples containing tin-based Babbitt alloy with and without the addition of alumina nanoparticles and substrate steel are shown in Figure 6. A visual observation of the interfacial microstructures of bimetallic interface shows a good bonding between the two metallic parts, i.e., nanocomposites of tin-based Babbitt alloy matrix containing alumina nanoparticles and carbon steel substrate. A uniform interfacial bonding is observed, which shows that the molten nanocomposite solidified on the surface of steel substrate in such way that no interfacial defects were noted. Tin-based Babbitt alloy with 0.25 wt% and 0.50 wt% additions of alumina nanoparticles revealed a good quality of interfacial bonding compared with 1 wt% addition.

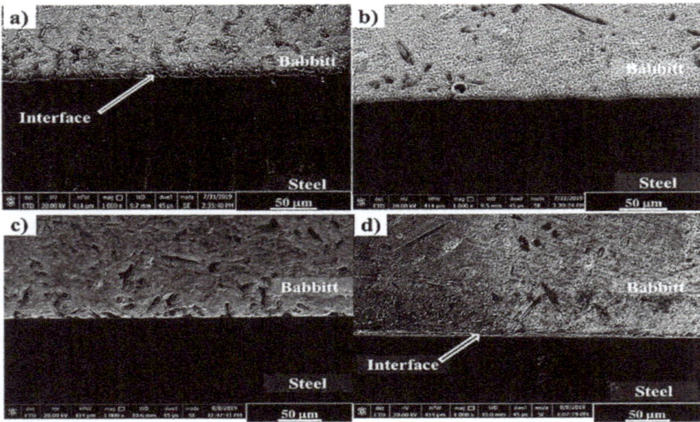


**Figure 6.** SEM images of interfaces in tin-based Babbitt/steel bimetallic cast specimens with different loading fraction of alumina nanoparticles in tin-based Babbitt alloy: (**a**) 0 wt %, (**b**) 0.25 wt%, (**c**) 0.50 wt% and (**d**) 1.0 wt%.

Interfacial micro-voids of tin-based Babbitt alloy with and without alumina nanoparticles were observed, as shown in Figure 7. EDS analysis of the micro-voids show the presence of all elements as found in both metals (Figure 7b). It was reported [24] that hot tearing resistance of the A206/1 wt% $Al_2O_3$ nanocomposite was significantly better than that of the pure A206 alloy. The nanocomposite containing 1.0 wt% alumina nanoparticles showed micro-crack (Figure 7d), which is another indication of the presence of alumina nanoparticle agglomerates. In case the nanoparticles were not completely wetted by the surrounding matrix material, these acted as defects and raised the localized stress level thus initiating cracks, as observed in Figure 7d. However, no such indication was noted without the addition of nanoparticles (Figure 7a) and with the incorporation of 0.50 wt% nanoparticles (Figure 7c).

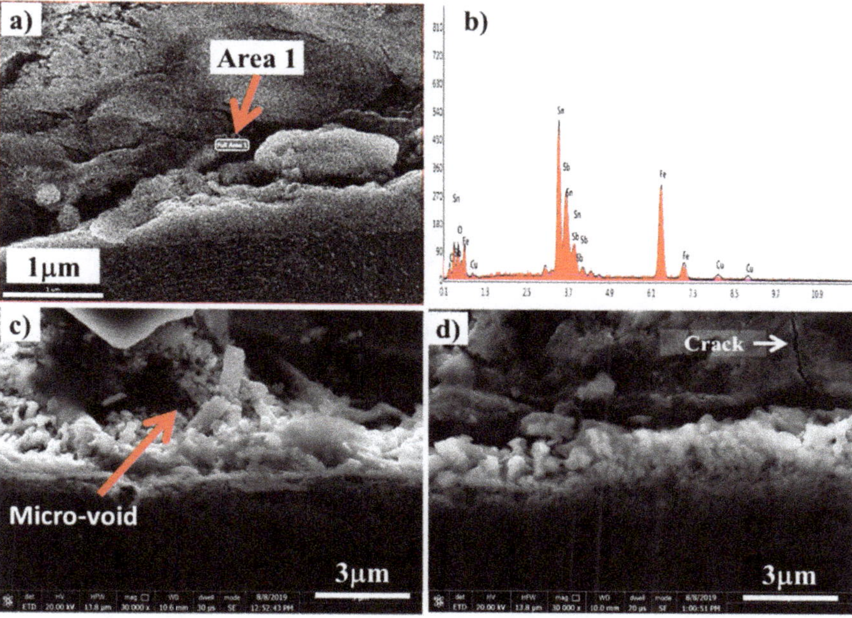


**Figure 7.** SEM images of interfaces in tin-based Babbitt/steel bimetallic specimens with different loading fractions of alumina nanoparticles in tin-based Babbitt alloy: (**a**) 0 wt%, (**c**) 0.50 wt% and (**d**) 1.0 wt%. (**b**) EDS of tin-based Babbitt interface with 0 wt%.

Figure 8 shows the SEM image and EDS analysis of nanocomposite containing 0.50 wt% alumina nanoparticles. The EDS area analysis shows the presence of Al and O atoms in both the $Cu_6Sn_5$ phase and Sn-based Babbitt matrix. It is clear that the concentration of Al and O in Babbitt matrix is relatively higher than that found in $Cu_6Sn_5$ phase. The EDS area analysis shows that the presence of $Al_2O_3$ nanoparticles in Babbitt matrix is relatively high compared to that found in $Cu_6Sn_5$ phase. The modification in $Cu_6Sn_5$ phase may be attributed to the result of the effect of pushed $\gamma$-$Al_2O_3$ on restricting the growth of the particles during solidification into remaining liquid.

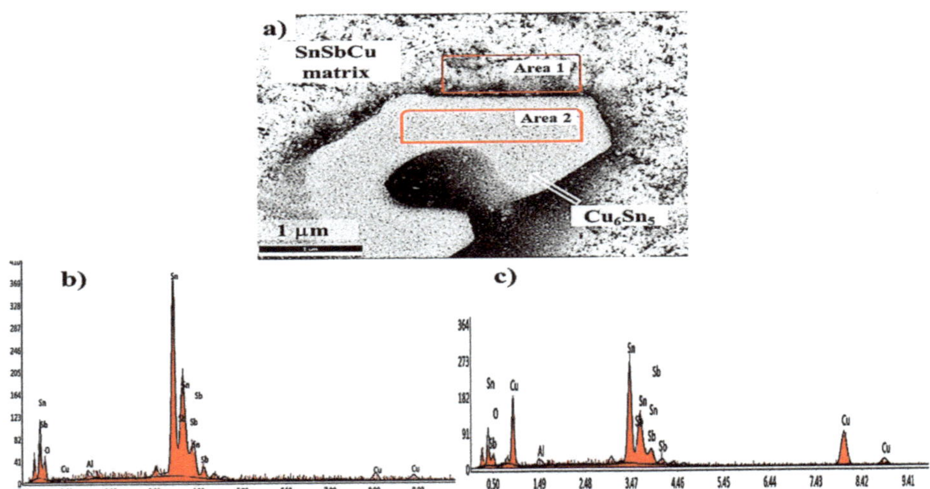


**Figure 8.** (**a**) SEM image of tin-based Babbitt alloy nanocomposite containing 0.5 wt% alumina nanoparticles, (**b**) and (**c**) graphical presentation of EDS results at two selected areas of matrix and $Cu_6Sn_5$ Phase consequently.

### 3.2. *Mechanical Properties of Reinforced Sn-based Babbitt Alloy and Bimetal Interface*

The hardness values of three nanocomposites and the reference sample of Sn-based Babbitt alloy are shown in Figure 9 and Table 2. It can be seen that the hardness of Sn-based Babbitt material was improved after the addition of alumina nanoparticles up to 0.5 wt%. In contrast, a minor decrease in hardness was noticed with rising content of nanoparticles up to 1.0 wt%. The modification in the morphology of hard $Cu_6Sn_5$ phase and its uniform distribution is the reason of high hardness up to 0.5 wt%. Otherwise, it can be inferred that the rising effect in hardness in nanocomposites was nullified by the gradual agglomeration of nanoparticles in their increasing content. A secondary effect may be due to the change in the morphology of hard $Cu_6Sn_5$ phase along with the presence of $Cu_6Sn_5$ phase agglomerates [23]. Nevertheless no significant decline in hardness profile was observed, which could otherwise be improved by increasing the quality of nanoparticle dispersion keeping in view the large difference in the hardness values of metallic tin-based Babbitt alloy and alumina.

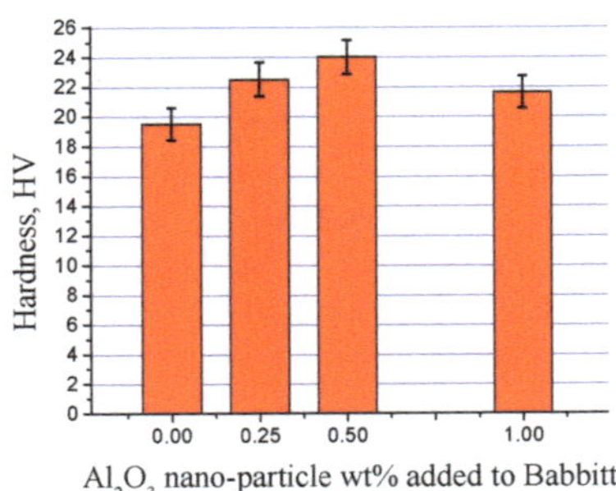


**Figure 9.** Hardness values of tin-based Babbitt alloy nanocomposite containing alumina nanoparticles.

**Table 2.** Tensile strength, hardness and shear strength of tin-based Babbitt/carbon steel bimetal with without $Al_2O_3$ nanoparticles additions.

| S/N | Composition | Sn-Babbitt Zone Tensile Strength, MPa | Sn-Babbitt Zone Hardness, HV | Interface Zone Shear Strength, MPa |
|---|---|---|---|---|
| 1 | Sn-Babbitt/steel bimetal | 56 ± 5.6 | 19.5 ± 1.08 | 27.2 ± 1.36 |
| 2 | Sn-Babbitt + 0.25% $Al_2O_3$ nanoparticles/steel bimetal | 65 ± 6.5 | 22.5 ± 1.13 | 29.9 ± 1.50 |
| 3 | Sn-Babbitt + 0.50% $Al_2O_3$ nanoparticles/steel bimetal | 68 ± 6.8 | 24.0 ± 1.13 | 30.6 ± 1.53 |
| 4 | Sn-Babbitt + 1.00% $Al_2O_3$ nanoparticles/steel bimetal | 63 ± 6.3 | 21.6 ± 1.08 | 27.7 ± 1.39 |

The average tensile strength values of nanocomposites containing three different nanoparticle loadings and corresponding shear strength values of bimetallic interfaces are shown in Figure 10 and Table 2. The addition of nanoparticles increased the strength up to 68 MPa with 0.50 wt% loading showing a rise of 21% in comparison to reference as-cast matrix. However, a further rise in nanoparticle addition up to 1.0 wt% reversed the rising trend and a decrease in strength was witnessed. A similar trend was observed in interfacial shear strength of the three nanocomposites after preparing bimetallic materials with carbon steel, i.e., maximum strength was observed at 0.50 wt% loading which decreased by further addition.

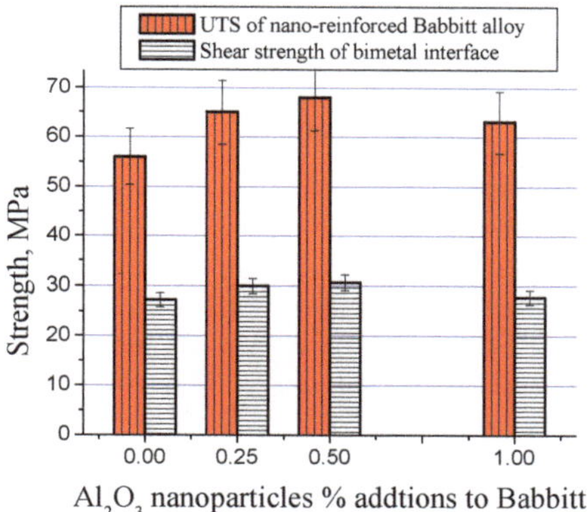


**Figure 10.** The average tensile strength values of nanocomposites containing alumina nanoparticles and average shear strength values of corresponding bimetallic interfaces.

### *3.3. Wear Properties of Reinforced Sn-based Babbitt Alloy*

Figure 11 shows the variation in weight-loss against the loading fraction of alumina nanoparticles in Sn-based Babbitt alloy. A value of 2.1 ± 0.1 mg was noted in unloaded sample of tin-based Babbitt alloy, which dramatically decreased to 1.30 ± 0.07 mg after adding 0.25 wt% alumina nanoparticles, which decreased further to 1 ± 0.05 mg at 0.5 wt% alumina nanoparticles. However, a regain in the value was noted by increasing the nanoparticle content to 1.0 wt%, i.e., 1.82 ± 0.09 mg. It can be seen that the addition of nanoparticles up to the values of 0.50 wt% decreased the weight-loss but further rise in the content of nanoparticles led to their agglomeration, which increased the weight-loss. Indeed poor dispersion of nanoparticles produces defects, which are actually the unfilled matrix areas between nanoparticles. These unfilled matrix areas or the space between particles is in fact the inability of the matrix to encapsulate the nano-reinforcement. Each nanocomposite manufacturing technique has a limitation in the loading fraction of the nano-reinforcement. In the present work, the stirring parameters together with the associated manufacturing setup provided a maximum nanoparticle dispersion of 0.5 wt%. As a result, the wear properties were also found optimum at the same loading fraction of nanoparticles.

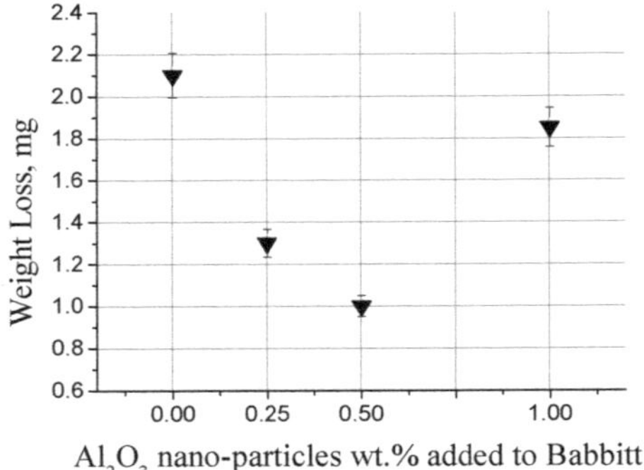


**Figure 11.** The average wear results of tin-based Babbitt alloy reinforced with alumina nanoparticles.

Figure 12 shows the SEM images of the worn surfaces of four types of specimens with and without nanoparticle addition along with the EDS results of unloaded sample at two points. The results of EDS analysis shown in Figure 4e,f confirmed that the proper peaks of elements coming from the SnSbCu matrix are clearer for two selected areas of the worn surfaces of Sn-based Babbitt alloy without $Al_2O_3$ nanoparticles addition. A lower content of Sb and Cu, as well as other additives nanoparticles in matrix, directly influences the hardness, as well as tribological properties of this area of Sn- based Babbitt structure. No cracks were observed on the wear tracks formed on the surfaces of the specimens. The grooves can be seen parallel to the sliding direction. The maximum wear was observed in the unloaded specimen, which may be due to large plastic deformation owing to low hardness value. With the systematic increase in hard nanoparticles in the matrix, the mechanism of wear shifted from adhesive to abrasive wear. As the contacting surface is steel during the wear test, therefore, an adhesion of tin-based Babbitt alloy was seen in the specimen without alumina nanoparticles, which shifted to abrasion in the presence of nanoparticles in specimens up to 1.0 wt%. The worn surfaces also became gradually smoother in the presence of nanoparticles. However, in specimen containing 1.0 wt% nanoparticles, the presence of defects resulted in poor density and, hence, the increased rate of material removal though the evidence of abrasive wear is evident.

The decrease in weight-loss in the wear test may be attributed to the change in the morphology of $Cu_6Sn_5$ phase from asterisk and needle shape to a spherical shape. A spherical shape is always smoother than the asterisk and needle shape and may have the ability to wear less in comparison to the other two shapes. During the wear test, the contacting surface may slip over the spherical shape in comparison to breaking the edges of needle and asterisk shapes. As evidence, tin-based Babbitt alloy exhibited the maximum weight-loss, which decreased with the addition of alumina nanoparticles, which changes the shape of $Cu_6Sn_5$ phase from asterisk and needle shape to spherical shape. However, the effect of the change in the morphology of $Cu_6Sn_5$ phase was nullified due to their agglomeration at 1 wt% alumina nanoparticle loading. The $Cu_6Sn_5$-rich and $Cu_6Sn_5$-depleted regions clearly developed identifiable soft and hard regions. Soft regions may have shown more wear and hard regions may have exhibited more pull-out phenomenon. At the same time, the loading of 1 wt% alumina nanoparticles may also have developed nanoparticle-agglomerates, which also adversely affected the wear process. Nevertheless, the minimum wear and weight-loss was observed at the loading of 0.50 wt% alumina nanoparticles, which is highly desirable for bearing applications.

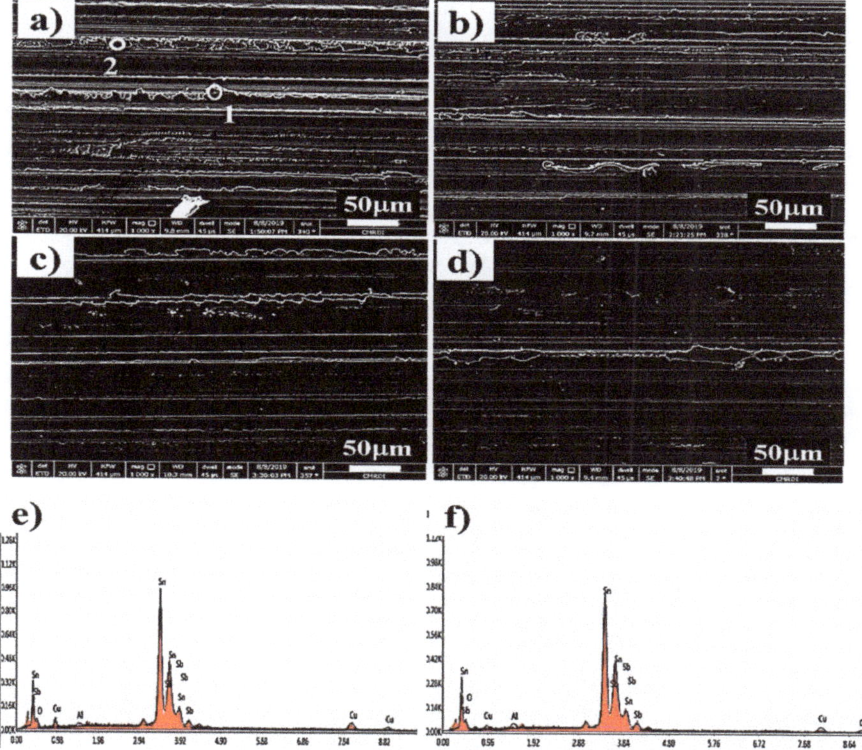


**Figure 12.** SEM images of worn surfaces of tin-based Babbitt alloy nanocomposites containing (**a**) 0 wt%, (**b**) 0.25 wt%, (**c**) 0.5 wt% and (**d**) 1.0 wt% alumina nanoparticles, and (**e**,**f**) EDS results of two selected areas of worn surfaces Sn-based Babbitt alloy without $Al_2O_3$ nanoparticles addition.

## 4. Discussion

The acquired results show that alumina nanoparticle loadings of 0.25 wt% and 0.50 wt% increase the hardness, tensile strength, and wear resistance of Sn-based Babbitt alloy, which decrease by further addition in nanoparticles. A possible reason may be the misfit effect in the lattice parameters of the matrix phase, which may be due to the result of cooling induced changes caused by the difference between the coefficients of thermal expansion between the $Cu_6Sn_5$ phase and the nanoparticles, thus resulting in an increase in the hardness [25]. Moreover, enhanced dislocation generation and reduced $Cu_6Sn_5$ phase owing to the presence of the nanoparticles may be contributing to the increased hardness of the Sn-based Babbitt matrix. In addition to mismatch effect in thermal expansion, the increase in the hardness is also mismatch in elastic moduli of the nanoparticles.

It has been shown [26] that the addition of the alumina nanoparticles results in a constraint in the plastic flow of matrix. This matrix can flow only with the movement of nanoparticles or over the particles during plastic deformation. The matrix shows a constrained plastic deformation because of smaller inter-particle distance and results in improvement in the flow stress. Finally, the strengthening mechanisms that are well-documented for coarse-grained metals and alloys can also operate in nano-crystalline materials after being modified by nano-scale particles and non-equilibrium microstructures [27]. It is suggested that the alumina nano-particles contributed to Orowan hardening. Hence, nano-dispersed cast structures carry both the nature of composites and refined Sn-based Babbitt structures. It is therefore important at this stage to understand the active strengthening mechanisms in nanoparticle-reinforced Sn-based Babbitt materials, as this could be the key to developing novel microstructures with improved properties.

## 5. Conclusions

A novel class of nanocomposites containing alumina nanoparticles in Sn-based Babbitt alloy was prepared with nanoparticle additions of 0.25 wt%, 0.50 wt%, and 1.0 wt%. It was found that the addition of nanoparticles changed the morphology of $Cu_6Sn_5$ phase from needle and asterisk shape to a spherical form. A parallel trend of $Cu_6Sn_5$ phase agglomeration was also observed while increasing nanoparticle content leading to identifiable $Cu_6Sn_5$-rich and $Cu_6Sn_5$-depleted areas. Sn-based Babbitt alloy with 0.25 wt% and 0.50 wt% nanoparticles revealed improvements in the tensile strength and quality of interfacial bonding, while the one containing 1.0 wt% nanoparticles showed lower tensile and shear strength values due to nanoparticles agglomerations in the matrix and micro-cracks at the interface. A substantial decrease in weight-loss was observed in the wear test by adding 0.5 wt% nanoparticles, which otherwise increased at 1.0 wt% loading. The developed bimetallic material containing 0.50 wt% alumina nanoparticles in Sn-based Babbitt alloy bonded with carbon steel substrate may be considered as a promising material for high-performance bearing applications.

**Author Contributions:** Conceptualization, M.R. and A.S.A.; methodology, M.R. and K.S.A.H.; validation, M.R., T.S. and K.S.A.H.; investigation, M.R.; resources, T.S. and M.R.; data curation, M.R. and T.S.; writing—original draft preparation, M.R. and T.S.; writing—review and editing, M.R. and A.S.A.; supervision, K.S.A.H. and A.S.A.; project administration, A.S.A. and M.R. All authors have read and agreed to the published version of the manuscript.

**Funding:** The present research work has been undertaken within the funded research projects of University of Ha'il, Kingdom of Saudi Arabia (Deanship of Scientific Research RG-191193). This is gratefully acknowledged.

**Conflicts of Interest:** The authors declare no conflict of interest.

## References

1. Gebretsadik, D.W.; Hardell, J.; Prakash, B. Tribological performance of tin-based overlay plated engine bearing materials. *Tribol. Int.* **2015**, *92*, 281–289. [CrossRef]
2. Ünlü, B. Determination of the tribological and mechanical properties of SnPbCuSb (white metal) bearings. *Mater. Sci.* **2011**, *46*, 478–485. [CrossRef]
3. Leszczyńska-Madej, B.; Madej, M. The Tribological Properties and the Microstructure Investigations of Tin Babbit with Pb Addition after Heat Treatment. *Arch. Metall. Mater.* **2016**, *61*. [CrossRef]
4. Khan, M.; Zulfaqar, M.; Ali, F.; Subhani, T. Microstructural and mechanical characterization of hybrid aluminum matrix composite containing boron carbide and Al-Cu-Fe quasicrystals. *Met. Mater. Int.* **2017**, *23*, 813–822. [CrossRef]
5. Khan, M.; Amjad, M.; Khan, A.; Ud-Din, R.; Ahmad, I.; Subhani, T. Microstructural evolution, mechanical profile, and fracture morphology of aluminum matrix composites containing graphene nanoplatelets. *J. Mater. Res.* **2017**, *32*, 2055–2066. [CrossRef]
6. Alghamdi, A.; Ramadan, M.; Abdel-Halim, K.; Fathy, N. Microscopical Characterization of Cast Hypereutectic Al-Si Alloys Reinforced with Graphene Nanosheets. *Eng. Technol. Appl. Sci. Res.* **2018**, *8*, 2514–2519.
7. Khurram, A.A.; Khan, A.; Gul, I.H.; Subhani, T. Glass Fiber Epoxy Matrix Composites Containing Zero and Two Dimensional Carbonaceous Nanoreinforcements. *Polym. Compos.* **2018**, *39*, E2056–E2063. [CrossRef]
8. Asghar, Z.; Latif, M.A.; Rafi ud, D.; Nazar, Z.; Ali, F.; Basit, A.; Badshah, S.; Subhani, T. Effect of distribution of B4C on the mechanical behaviour of Al-6061/B4C composite. *Powder Metall.* **2018**, *61*, 293–300. [CrossRef]
9. Subhani, T.; Latif, M.; Ahmad, I.; Rakha, S.A.; Ali, N.; Khurram, A.A. Mechanical performance of epoxy matrix hybrid nanocomposites containing carbon nanotubes and nanodiamonds. *Mater. Des.* **2015**, *87*, 436–444. [CrossRef]
10. Subhani, T.; Shaukat, B.; Ali, N.; Khurram, A.A. Toward improved mechanical performance of multiscale carbon fiber and carbon nanotube epoxy composites. *Polym. Compos.* **2017**, *38*, 1519–1528. [CrossRef]
11. Ahmad, M.S.; Subhani, T. Thermal and ablative properties of binary carbon nanotube and nanodiamond reinforced carbon fibre epoxy matrix composites. *Plast. Rubber Compos.* **2015**, *44*, 397–404. [CrossRef]

12. Ahmad, I.; Subhani, T.; Wang, N.; Zhu, Y. Thermophysical Properties of High-Frequency Induction Heat Sintered Graphene Nanoplatelets/Alumina Ceramic Functional Nanocomposites. *J. Mater. Eng. Perform.* **2018**, *27*, 2949–2959. [CrossRef]
13. Khan, M.; Zulfaqar, M.; Ali, F.; Subhani, T. Hybrid aluminium matrix composites containing boron carbide and quasicrystals: Manufacturing and characterisation. *Mater. Sci. Technol.* **2017**, *33*, 1955–1963. [CrossRef]
14. Khan, M.; Rehman, A.; Aziz, T.; Shahzad, M.; Naveed, K.; Subhani, T. Effect of inter-cavity spacing in friction stir processed Al 5083 composites containing carbon nanotubes and boron carbide particles. *J. Mater. Process. Technol.* **2018**, *253*, 72–85. [CrossRef]
15. Aleshin, N.; Kobernik, N.; Mikheev, R.; Vaganov, V.; Reshetniak, V.; Aborkin, A. Plasma–powder application of antifrictional babbitt coatings modified by carbon nanotubes. *Russ. Eng. Res.* **2016**, *36*, 46–52. [CrossRef]
16. Ramadan, M.; Ayadi, B.; Rajhi, W.M.A.; Alghamdi, A. Influence of Tinning Material on Interfacial Microstructures and Mechanical Properties of Al-Sn /Carbon Steel Bimetallic Castings for Bearing Applications. *Key Eng. Mater.* **2020**, *835*, 108–114. [CrossRef]
17. Koutský, J.; Veselá, J. Evaluation of white metal adhesion (conventional casting and thermal wire arc spraying) by ultrasonic non-destructive method. *J. Mater. Process. Technol.* **2004**, *157–158*, 724–728.
18. Barykin, N.; Sadykov, F.; Aslanian, I. Wear and failure of babbit bushes in steam turbine sliding bearings. *J. Mater. Eng. Perform.* **2000**, *9*, 110–115. [CrossRef]
19. Sadykov, F.A.; Barykin, N.P.; Valeev, I.S.; Danilenko, V.N. Influence of the structural state on mechanical behavior of tin babbit. *J. Mater. Eng. Perform.* **2003**, *12*, 29–36. [CrossRef]
20. Dean, R.R.; Evans, C.J. Plain bearing materials: The role of tin. *Tribol. Int.* **1976**, *9*, 101–108. [CrossRef]
21. Pratt, G.C. Materials for plain bearings. *Int. Metall. Rev.* **1973**, *18*, 62–88. [CrossRef]
22. Moazami Goudarzi, M.; Jenabali Jahromi, S.A.; Nazarboland, A. Investigation of characteristics of tin-based white metals as a bearing material. *Mater. Des.* **2009**, *30*, 2283–2288. [CrossRef]
23. Choi, H.; Konishi, H.; Li, X. Al2O3 nanoparticles induced simultaneous refinement and modification of primary and eutectic Si particles in hypereutectic Al–20Si alloy. *Mater. Sci. Eng. A* **2012**, *541*, 159–165. [CrossRef]
24. Algarín, D.F.; Marrero, R.; Li, X.; Choi, H.; Suárez, O.M. Strengthening of aluminum wires treated with A206/alumina nanocomposites. *Materials* **2018**, *11*, 413. [CrossRef] [PubMed]
25. Dieringa, H. Properties of magnesium alloys reinforced with nanoparticles and carbon nanotubes: A review. *J. Mater. Sci.* **2010**, *46*, 289–306. [CrossRef]
26. El Mahallawi, I.; Shash, A.; Amer, A. Nanoreinforced cast Al-Si alloys with $Al_2O_3$, $TiO_2$ and $ZrO_2$ nanoparticles. *J. Met.* **2015**, *2015*, 802–821. [CrossRef]
27. Scattergood, R.O.; Koch, C.C.; Murty, K.L.; Brenner, D. Strengthening mechanisms in nanocrystalline alloys. *Mater. Sci. Eng. A* **2008**, *493*, 3–11. [CrossRef]

*Article*

# Effect of Chromium and Molybdenum Addition on the Microstructure of In Situ TiC-Reinforced Composite Surface Layers Fabricated on Ductile Cast Iron by Laser Alloying

**Damian Janicki**

Department of Welding, Silesian University of Technology, Konarskiego 18A, 44-100 Gliwice, Poland; damian.janicki@polsl.pl

Received: 9 November 2020; Accepted: 11 December 2020; Published: 16 December 2020

**Abstract:** In situ TiC-reinforced composite surface layers (TRLs) were produced on a ductile cast iron substrate by laser surface alloying (LA) using pure Ti powder and mixtures of Ti-Cr and Ti-Mo powders. During LA with pure Ti, the intensity of fluid flow in the molten pool, which determines the TRL's compositional uniformity, and thus Ti content in the alloyed zone, was directly affected by the fraction of synthesized TiC particles in the melt—with increasing the TiC fraction, the convection was gradually reduced. The introduction of additional Cr or Mo powders into the molten pool, due to their beneficial effect on the intensity of the molten pool convection, elevated the Ti concentration in the melt, and, thus, the TiC fraction in the TRL. It was found that the melt enrichment of Cr, in conjunction with non-equilibrium cooling conditions, suppressed the martensitic transformation of the matrix, which lowered the total hardness of the TRL. Moreover, the presence of Cr in the melt (~3 wt%) altered the growth morphology of the synthesized primary TiC precipitates compared with that obtained using pure Ti. The addition of Mo in the melt produced (Ti, Mo)C primary precipitates that exhibited a nonuniform Mo distribution (coring structure). The dissolution of Mo in the primary TiC precipitates did not affect its growth morphology.

**Keywords:** laser surface alloying; ductile cast iron; in situ composite; titanium carbide

## 1. Introduction

Recently, there has been considerable interest in improving the wear properties of machine parts' working surfaces made from ductile cast iron (DCI) using laser surface treatment methods [1–7]. A highly effective approach to enhance the wear properties of metallic substrates is the formation of metal matrix composite (MMC) surface layers via laser alloying (LA) [8–10]. The wear performance of MMC materials is strongly affected by the morphology, fraction, and types of reinforcing particles, as well as the microstructure of the matrix [11–13]. Thus, the processing conditions and the chemical composition of the alloying material are extremely important to obtain the desired wear properties [14]. Depending on the chemistry of the alloying material, LA can produce MMC surface layers on cast iron substrates using both ex-situ and in situ routes [15–17]. The in situ MMCs offer the advantage of stronger interfacial bonding between reinforcements and matrix materials than ex-situ MMCs. Moreover, the in situ synthesized reinforcing phases exhibit higher chemical and thermal stability (long-term stability) [18,19]. High C contents in cast iron substrates provide excellent conditions for the in situ formation of MMC surface layers via LA, with elements that tend to form carbides. Several studies on the LA of cast iron substrates have focused on the in situ synthesis of TiC-reinforced layers (TRLs) by introducing Ti into the molten pool [16,20–26]. From the point of view of the wear properties, the microstructure of the synthesized TRLs should contain a martensitic matrix with homogeneously-dispersed TiC precipitates [24,26]. Moreover, published data on the wear performance of Fe-based MMCs reinforced with TiC phase suggest a significant increase in the wear resistance of TRLs with increasing the TiC fraction [24,26–29]. However, reported works have shown that during the in situ synthesis of TRLs on cast iron substrates via LA, it is difficult to maintain a homogeneous dispersion of the reinforcing phase upon increasing the Ti concentration in the melt [23]. As a consequence, control over the fraction and growth morphology of the TiC particles, and also the phase composition of the matrix, is highly limited.

Previous studies [23–25] were undertaken to determine the feasibility of the in situ synthesis of TRLs on the DCI substrate via LA with pure Ti powder. The aim of that investigation was to establish the processing conditions that allowed the use of the entire C content in the processed substrate to optimize the TRL's microstructure, i.e., to achieve the highest possible TiC fraction and a fully martensitic matrix. However, the results indicated that a limited amount of Ti can be introduced into the molten pool while maintaining a homogeneous dispersion of the reinforcing phase in the matrix. This limitation was a consequence of a gradual decrease in the intensity of the fluid flow in the molten pool with an increasing TiC fraction in the melt.

In general, the current work extends the concept outlined in earlier works, with an emphasis on assessing the possibility of influencing both the microstructural characteristics of TRL and the intensity of the fluid flow in the molten pool by properly selecting the alloying material chemistry. The primary objective was to determine the effect of the presence of additional Cr and Mo in the molten pool, during the LA of the DCI substrate with Ti, on the growth morphology of TiC particles and also the microstructure of the matrix.

## 2. Materials and Methods

50 × 50 × 10 mm plates of DCI-grade EN-GJS-700-2 were used as a substrate material (SM). The SM composition, as analyzed by glow discharge optical emission spectrometry, is given in Table 1. The alloying materials were prepared using commercially available powders of pure Ti (99% purity, the particle size range of 45 to 70 µm, H.C. Starck), pure Cr (99% purity, the particle size range of 45 to 70 µm, abcr GmbH), and pure Mo (99.95% purity, the particle size range of 5 to 10 µm, abcr GmbH). Based on unpublished work [30], Ti-Mo and Ti-Cr powder mixtures were prepared in weight percentage ratios of 80–20. The main purpose of the additional introduction of Cr and Mo powders into the molten pool during the synthesis of TRLs was to affect the morphology and composition of the reinforcing phase (TiC precipitates) and also the phase composition of the matrix material. For comparative purposes, the alloying process was also conducted with pure Ti powder.

**Table 1.** Chemical composition of the used ductile cast iron (DCI) grade EN-GJS-700-2 (wt%).

| C | Si | Cu | Mn | Cr | Ni | Ti | Mo | S | P | Fe |
|---|---|---|---|---|---|---|---|---|---|---|
| 3.60 | 2.51 | 0.78 | 0.25 | 0.02 | 0.04 | 0.02 | 0.02 | 0.005 | 0.016 | balance |

The LA trials were conducted using an experimental setup with a diode laser (DL020, Rofin-Sinar Laser GmbH, Hamburg, Germany). A detailed description of the experimental setup and the laser processing conditions are given elsewhere [24]. A laser beam spot size (rectangular in shape) was 1.5 × 6.6 mm. During all trials, the laser spot was located on the top surface of SM, and the fast-axis of the laser beam was set parallel to the scanning direction. Based on previous work [23–25], the laser power and traverse speed were held constant at 1500 W and 1.25 mm/s, respectively. The powder feed rates depended on the type of the alloying material and are listed in Table 2.

**Table 2.** Effect of the chemical composition of the alloying material and powder feed rate on the fusion area of the single-alloyed beads (SAB) and the concentration of alloying elements in the bead.

| Processing Condition No./SAB No. | Alloying Material | Powder Feed Rate [2] (mg/mm) | Fusion Area of the SAB ($mm^2$) | Average Ti Content, (wt%) | Average Cr Content, (wt%) | Average Mo Content, (wt%) | Quality [3] |
|---|---|---|---|---|---|---|---|
| T1 | Ti | 8.0 | 7.34 ± 0.61 | 7.0 ± 0.8 | – | – | U |
| T2 | | 10.0 | 7.05 ± 0.58 | 8.8 ± 1.1 | – | – | U |
| T3 | | 11.0 | 6.80 ± 0.50 | – | – | – | N |
| TC1 | Ti-Cr [1] | 9.0 | 7.73 ± 0.63 | 8.4 ± 0.29 | 1.6 ± 0.11 | – | U |
| TC2 | | 11.0 | 7.89 ± 0.65 | 10.3 ± 0.38 | 1.9 ± 0.15 | – | U |
| TC3 | | 12.0 | 8.50 ± 0.67 | 11.6 ± 0.58 | 2.3 ± 0.24 | – | U |
| TC4 | | 13.0 | 8.42 ± 0.71 | – | – | – | N |
| TM1 | Ti-Mo [1] | 10.0 | 8.20 ± 0.63 | 7.0 ± 0.38 | – | 1.8 ± 0.16 | U |
| TM2 | | 12.5 | 8.09 ± 0.62 | 8.4 ± 0.46 | – | 2.3 ± 0.23 | U |
| TM3 | | 13.5 | 8.10 ± 0.64 | 9.9 ± 0.69 | – | 2.8 ± 0.32 | U |
| TM4 | | 14.0 | 8.07 ± 0.69 | – | – | – | N |

[1] Powder mixtures prepared with a weight ratio of 80–20; [2] defined as the amount of the powder provided per unit length of the SAB; [3] assessed in terms of the compositional uniformity (Ti concentration): U—uniform, N—non-uniform.

To investigate the influence of the alloying material chemistry on the maximal possible Ti concentration in the molten pool and the compositional homogeneity of the alloyed zone, single-alloyed beads (SAB) were produced (Table 2). For microstructure analysis of the TRLs synthesized at different melt compositions, surface alloyed layers (SAL) were produced via a multi-pass overlapping alloying process with an overlap ratio of 30% (Tables 3 and 4). Abbreviations used in the manuscript are also listed in Appendix A.

**Table 3.** Effect of the chemical composition of the alloying material on the microstructural parameters of the surface alloyed layers (SALs).

| TRL No. | Processing Condition No. (Table 2) [1] | α-Fe (Martensite) Fraction (wt%) [2] | Retained Austenite Fraction (wt%) [2] | Cementite Fraction (vol%) | TiC Fraction (vol%) |
|---|---|---|---|---|---|
| TR | T2 | 66.4 ± 1.1 | 15.3 ± 1.4 | 8.1 ± 2.9 | 15.4 ± 2.1 |
| TRC | TC3 | 52.9 ± 1.6 | 23.7 ± 1.2 | 2.1 ± 0.6 | 20.8 ± 1.1 |
| TRM | TM3 | 69.1 ± 1.4 | 5.7 ± 1.4 | 1.9 ± 0.6 | 21.1 ± 1.3 |

[1] Overlap ratio: 30%; [2] based on the Rietveld refinement of X-ray diffraction (XRD) data.

**Table 4.** Chemical compositions of the SALs.

| TRL No. | Element (wt%) | | | | | | | | | |
|---|---|---|---|---|---|---|---|---|---|---|
| | C | Ti | Cr | Mo | Mn | Si | Cu | S | P | Fe |
| TR | 3.15 | 9.5 | 0.05 | 0.02 | 0.13 | 1.99 | 0.79 | 0.148 | 0.023 | balance |
| TRC | 3.10 | 12.4 | 2.87 | 0.04 | 0.13 | 1.90 | 0.81 | 0.181 | 0.025 | balance |
| TRM | 3.07 | 10.5 | 0.06 | 3.15 | 0.14 | 1.85 | 0.76 | 0.177 | 0.023 | balance |

Geometrical parameters of SABs and SALs were measured with an optical microscope (Eclipse MA100, Nikon Corporation, Tokyo, Japan) and image processing software Nikon NIS-EBR (ver. 3.13, Nikon Corporation, Tokyo, Japan).

The microstructure of TRLs was characterized by scanning electron microscopy (SEM) with energy-dispersive spectroscopy (EDS) analysis and X-ray diffraction (XRD). EDS measurements were performed at an accelerating voltage of 15 kV. EDAX Genesis software (ver. 6.0, EDAX Inc., Mahwah, NJ, USA) was used for the collection and analysis of EDS data. The quantitative EDS analysis was made to estimate the concentration of the alloying material in the matrix and also to confirm the compositional homogeneity of the alloyed layers. Additionally, chemical analysis of the SM and SALs was performed on ground surface specimens using a LECO GDS 500A optical emission spectrometer (LECO Corporation, St. Joseph, Michigan, USA). The C concentration in the SM was determined using a LECO CS125 Carbon Analyzer (LECO Corporation, St. Joseph, Michigan, USA). The procedures and equipment used during microstructural characterization are described in greater detail elsewhere [24,25].

Hardness measurements were made on cross-sections of SALs using a Vickers hardness tester (401 MVD, Wilson Wolpert Instruments, Aachen, Germany). Hardness profiles were determined with a 200 g load. The matrix material hardness was measured with a 25 g load applied over 10 s.

## 3. Results and Discussion

### 3.1. Macrostructure Analysis

Figures 1 and 2 present a series of cross-sectional optical macrographs of the SABs produced using Ti-Cr and Ti-Mo powder mixtures respectively, at different powder feed rates. The corresponding cross-sectional areas of the SAB and concentrations of alloying elements in the bead are summarized in Table 2. A detailed description of the effect of the powder feed rate of pure Ti on the shape of the alloyed zone can be found in previous works [23–25]. It has been reported that, during synthesis with pure Ti powder, the presence of TiC precipitates in the melt leads to a decrease in the intensity of the fluid flow in the molten pool, which reduces the compositional homogeneity of the alloyed zone [24,25]. Table 2 shows that for SABs produced using a pure Ti, upon increasing the Ti content, i.e., upon increasing the fraction of TiC precipitates in the melt, the fusion area of the SAB decreased. The reduction of the fusion area is, to a certain extent, associated with the fact that the energy of the laser beam melts the alloying powder. On the other hand, the TiC precipitates formed due to an exothermic reaction between Ti and C in the molten pool. Consequently, the higher fraction of TiC precipitates (associated with a higher Ti concentration in the melt) leads to a larger amount of heat generated directly in the molten pool. However, the calculation of the heat generated during the reaction between Ti and C for the TiC content of 15 vol% (SAB no. T2, Table 2) indicated that its value is negligible (~18 J/mm) compared with the heat input of the process (1200 J/mm—the ratio of the laser power and the traverse speed). The above calculations considered the value of enthalpy ($\Delta H = -203$ kJ/mol [31]) of the formation of TiC at 2000 °C (the temperature established during examination of the thermal conditions in the molten pool [32]).

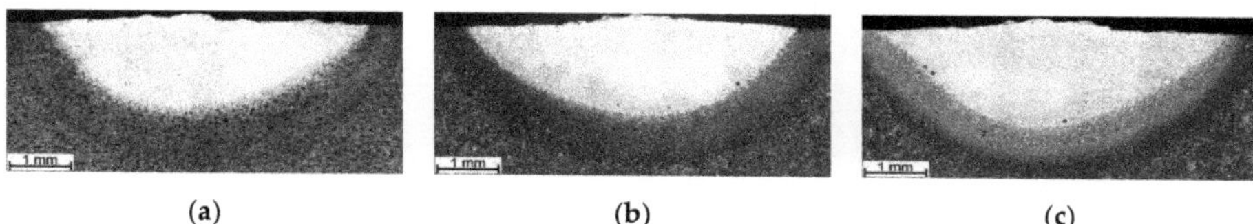


(a) (b) (c)

**Figure 1.** Optical macrographs of the SABs produced using a powder mixture of 80Ti-20Cr (weight ratio); SAB no. (Table 2): (**a**) TC1, (**b**) TC2, (**c**) TC3.

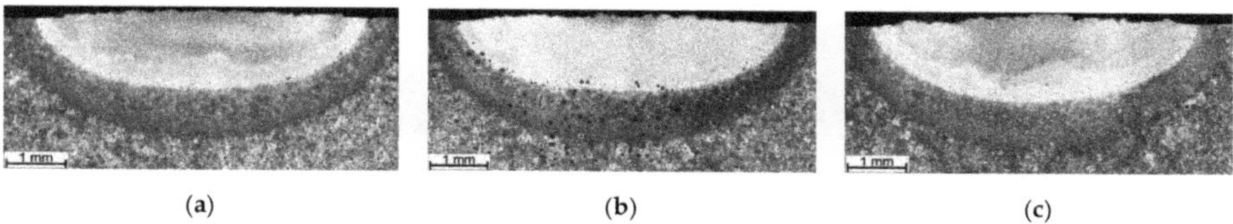


(a) (b) (c)

**Figure 2.** Optical macrographs of the SABs produced using a powder mixture of 80Ti-20Mo (weight ratio); SAB no. (Table 2): (**a**) TM1, (**b**) TM2, (**c**) TM3.

Interestingly, the use of a Ti-Cr powder mixture increased the fusion zone upon increasing the amount of alloying material introduced into the molten pool in the optimal powder feed rate range (Table 2). The fusion zone shape of SABs produced with Cr addition indicates that the coefficient of the surface tension temperature on the molten pool surface was positive, producing fluid flow inward along the molten pool surface [33–35]. The increase in the powder feed rate increased the fluid flow intensity, leading to a deeper molten pool. In the case of the Ti-Mo mixture, despite an increased powder feed rate, the cross-sectional area of the SAB was generally constant within the optimal powder feed rate range.

From the point of view of shaping the TRL's microstructure, the use of a complex powder mixture increased the amount of alloying materials in the molten pool at a given heat input (Table 2). This, in turn, significantly increased the Ti content and, thus, the TiC fraction in the uniformly alloyed bead compared with that processed with pure Ti (Table 3). Additionally, the homogeneity of the TiC dispersion throughout the alloyed zone was also improved (Figure 3). Note that, based on the thermodynamic calculations (for 3.6 wt% of C in the DCI substrate), the Ti concentration leading to the formation of a microstructure composed of TiC precipitates embedded in a martensitic matrix was estimated to be in the range of 12 to 13 wt% [24,25]. In the case of SABs made with an 80Ti-20Cr powder mixture, the Ti content reached 11.6 wt% (SAB no. TC3, Table 2). The multi-pass overlapping alloying process under the above processing conditions using an overlap ratio of 30% ensured the production of homogenous SALs (TRC, Table 3) with average Ti and Cr contents of 12.4 and 2.9 wt% respectively, and a uniform thickness of ~1.9 mm (Figure 4a). In turn, the use of the 80Ti-20Mo powder mixture produced a uniform SAB containing up to 9.9 wt% Ti and 2.8 wt% Mo (SAB no. TM3, Table 2). The corresponding SALs (TRM, Table 3) produced at an overlap ratio of 30% contained about 10.5 wt% Ti and 3.1 wt% Mo. The thickness of this SAL was about 1.6 mm (Figure 4b).

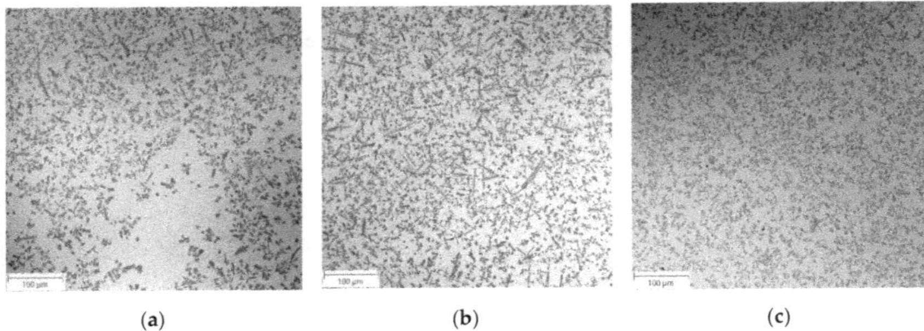

(**a**) (**b**) (**c**)

**Figure 3.** Low-magnification BSE SEM micrographs taken at the mid-section of SAB no. (**a**) T2, (**b**) TC3, (**c**) TM3 (Table 2).

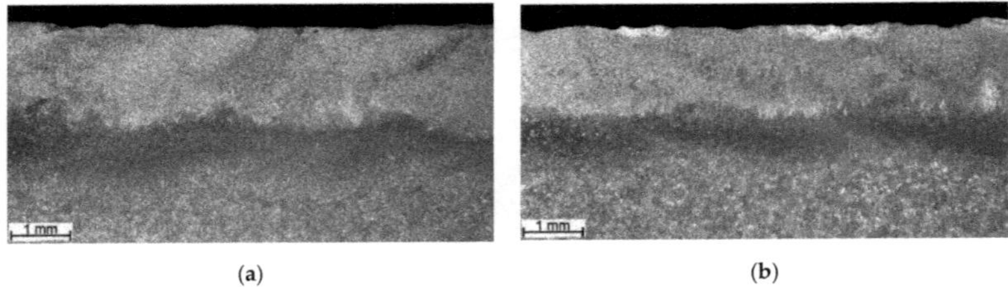


(**a**) (**b**)

**Figure 4.** Optical macrographs of (**a**) TRC and (**b**) TRM (Table 3).

Regardless of the used alloying material, both SABs and SALs contained cracks induced due to a lack of preheating the substrate. The chemistry of alloying material had little effect on the cracking tendency of the alloyed zone. All types of SALs exhibited minor porosity and negligible tendency to a formation of microvoids.

It should be pointed out that the research program adopted in the presented work was not directly aimed at fully understanding the above effect of Cr and Mo on the heat and fluid flow phenomena in the molten pool during the investigated LA process. The complexity of the heat and mass transfer phenomena in the molten pool, especially comprising the significant changes in the surface and bulk chemical composition of the molten pool, requires a detailed examination using both experimental and numerical approaches. However, the results of the performed research revealed a possible factor that may affect the observed changes in the intensity of the fluid flow in the molten pool.

It is well known that mass transport in the molten pool can be substantially altered by the presence of surface-active trace elements such as sulfur and oxygen. Additionally, it is well documented [33,34] that in the case of Fe-based alloys, surface-active trace elements affect the molten pool shape by altering surface tension gradients, leading to changes in both the magnitude of fluid flow and its direction in the molten pool. The chemical analysis presented in Table 4 revealed a significantly elevated S content in all SALs compared with that of the as-received DCI (Table 1). The S content in the as-received DCI was approximately 0.005 wt%. In the case of the SALs produced with added pure Ti powder, the S content gradually increased upon increasing the Ti concentration [25], i.e., with increasing the powder feed rate, and reached 0.148 wt%. The SALs produced using powder mixtures of 80Ti-20Cr and 80Ti-20Mo under the optimal processing parameters exhibited essentially the same sulfur content of 0.180 wt%. Thus, the elevated S content in all types of SALs was associated with the purity of the alloying powders. Based upon investigations on the heat transport in the molten pool [35], which indicate that small concentrations of surface-active elements can significantly alter the magnitude of surface tension, it is suggested that the above differences in the S content may affect the intensity of fluid flow in the molten pool.

Summarizing the above results, the introduction of additional Cr or Mo powder into the molten pool during the LA of the DCI with Ti powder enables significantly higher Ti concentrations in the uniform SAL relative to that achieved using a pure Ti powder by influencing the intensity of fluid flow in the molten pool. However, a full understanding of this phenomenon requires additional examinations and presents a challenge for further research.

### 3.2. Microstructural Analysis

BSE SEM micrographs taken from different locations of the TRC and TRM are presented in Figures 5 and 6, respectively. Representative XRD patterns for TRC and TRM are presented in Figure 7a,b, respectively. A detailed microstructural characteristic of TRL synthesized during LA with a pure Ti powder is presented in previous works [23–25]. The chemical compositions and microstructural parameters of the TRLs produced using all types of alloying materials are outlined in Tables 3 and 4, respectively.

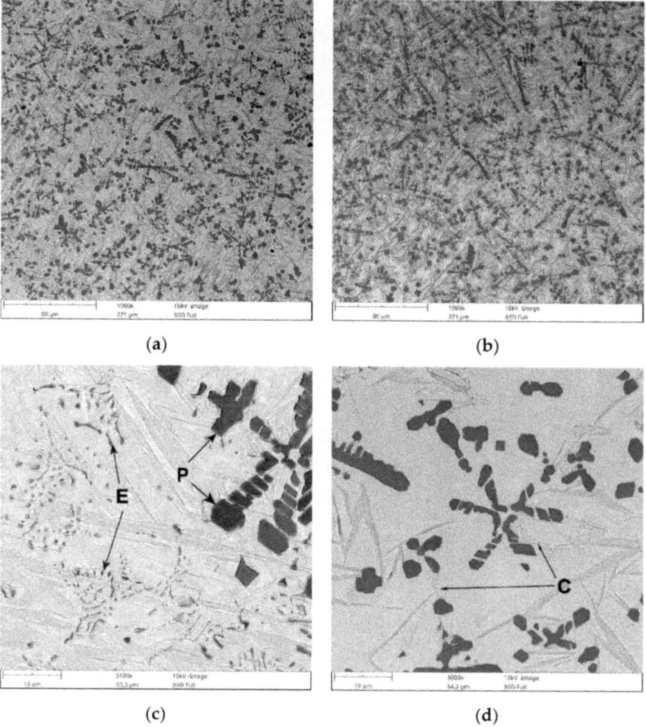


**Figure 5.** BSE SEM images showing the microstructure of TRC (Table 3): (**a**) mid-section of the layer, (**b**) the overlap boundary of the layer, (**c**) a detail from (**a**) showing the primary (*P*) and eutectic (*E*) TiC precipitates, and (**d**) a detail from (**a**) showing the distribution of eutectic-like structures formed during the last stages of solidification (**c**).

The micrograph analysis showed that the TRC and TRM exhibit a highly uniform distribution of TiC precipitates throughout the matrix and have similar fractions of about 21 vol%. This TiC fraction is close to that predicted by the thermodynamic calculations for the processed DCI substrate [24,25]. As a result, both the TRC and TRM contained minor fractions of the cementite phase (~2 vol%). Moreover, in contrast to the TRLs formed with pure Ti [25], there is no evidence of graphitization in the heat-affected zone (HAZ) between consecutive beads in SALs (Figures 5b and 6d). As can be seen from Figures 5c and 6c, the obtained TRC and TRM compositions also led to the formation of fine eutectic precipitates of the TiC phase with a plate-like morphology.

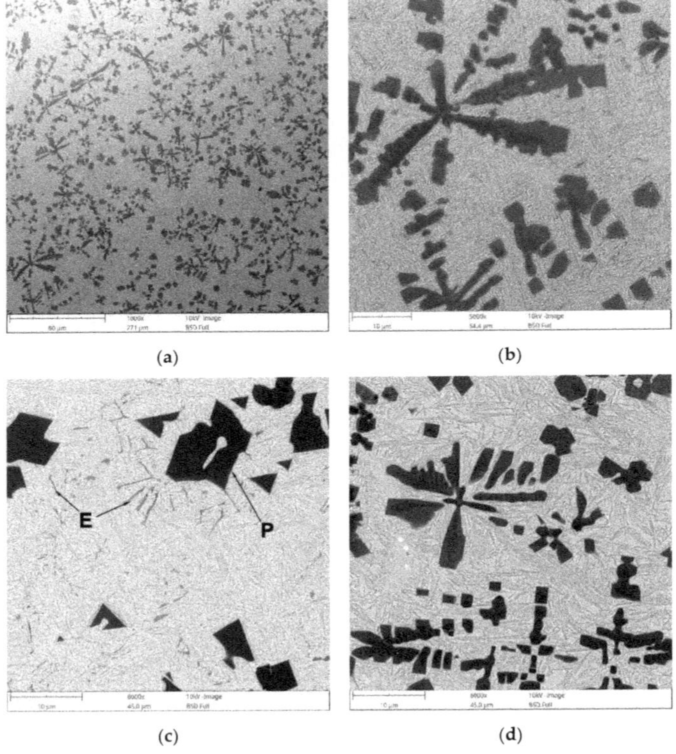


**Figure 6.** BSE SEM images showing the microstructure of the TRM (Table 3): (**a**) mid-section of the layer, (**b**) a detail from (**a**) showing typical morphology of primary TiC precipitates, (**c**) a detail from (**a**) showing a distribution of primary (*P*) and eutectic (*E*) TiC precipitates, and (**d**) image taken from the overlap boundary of the layer.

The microstructure of the TRC also contained a minor amount of fine eutectic-like regions that exhibited two discrete structures (Figure 8a). The corresponding SEM elemental mapping (Figure 8c–g) indicated C enrichment in both structures, wherein that containing remarkably larger lamellar precipitates was also significantly enriched in Cr. In contrast, the structures composed of fibrous-like precipitates (Figure 8b) were depleted in Cr and enriched in Si. Moreover, the elemental maps of C and Ti clearly suggest the presence of TiC precipitates within this eutectic-like region. The above eutectic-like regions formed as a result of the last solidification stages in the Fe-C-Ti-Cr alloy system taking place under non-equilibrium conditions. It is reasonable to assume that the lamellar precipitates enriched in Cr were $(Fe, Cr)_3C$ phase. Quantitative analysis of the micrographs indicated an average volume fraction of (Fe, Cr)$_3$C precipitates in the TRC of 2.1% ± 0.6%. (Fe, Cr)$_3$C phase was not detected by the XRD due to the low fraction of this phase.

The typical eutectic-like structure formed during the last stages of solidification in TRM is presented in Figure 9. The fraction of these eutectic regions in the layer was estimated to be 1.9 ± 0.6 vol%. As in the case of (Fe, Cr)$_3$C phase in TRC, the XRD did not detect the presence of these eutectic precipitates. However, considering the morphology of the above eutectic-like structure and the chemistry of the layer, one can assume that it contained mainly the cementite phase.

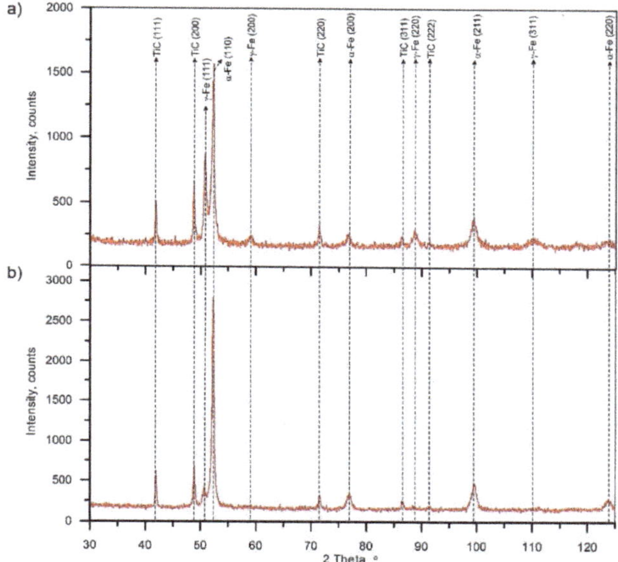


**Figure 7.** X-ray diffraction (XRD) patterns of (**a**) TRC and (**b**) TRM (Table 3).

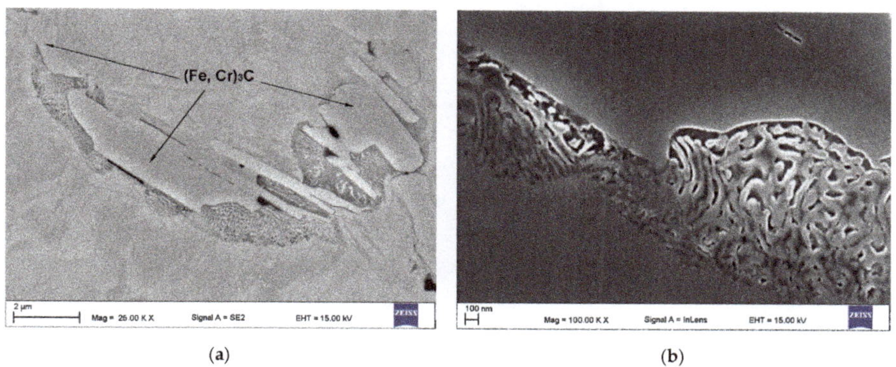


(**a**) (**b**)

**Figure 8.** *Cont.*

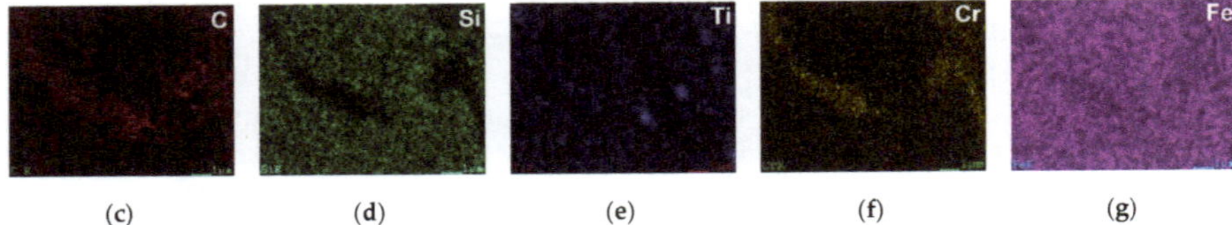


(c) (d) (e) (f) (g)

**Figure 8.** (**a**,**b**) SEM micrographs taken from the mid-section of the TRC (Table 3) showing eutectic-like structures due to the last solidification stages, (**b**) detail from (**a**), and (**c**–**g**) EDS maps of C, Si, Ti, Cr, and Fe distributions respectively, from (**a**).

The metallographic data indicate that the addition of Cr significantly elevated the retained austenite fraction in the matrix material (Table 3). The fraction of retained austenite in the TRC was estimated to be 24 wt%, whereas the TRM had austenite fractions of about 5.7 wt%. The overall high retained austenite fraction in the TRC was directly attributed to the combined effect of non-equilibrium cooling conditions in the molten pool and enrichment of primary austenite grains in Cr, which inhibited the martensitic transformation. Based on the SEM/EDS analysis, the Cr content in the martensitic/austenitic matrix was approximately 2.4 wt%. The suppressed martensitic transformation in the laser-processed SALs on the DCI substrate by relatively low Cr additions has been reported in the previous work [36].

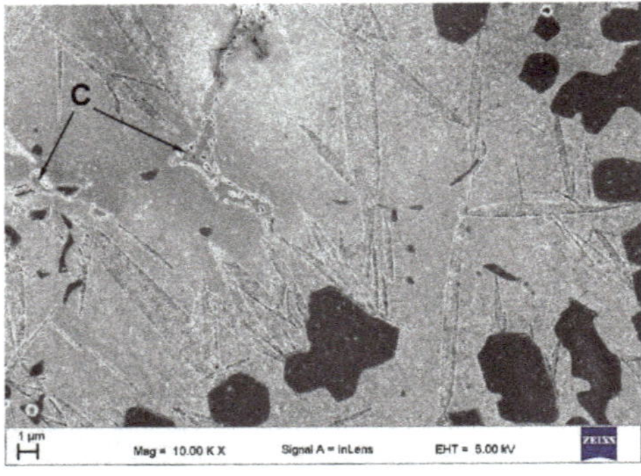


**Figure 9.** In-lens SEM micrograph taken from the mid-section of the TRM (Table 3) showing the eutectic-like structures (C) as a result of the last solidification stages.

The TRC exhibited a non-uniform distribution of retained austenite throughout the layer because of the overlap reheating during multi-pass alloying (Figure 5a,b). Metallographic data suggest that, in the HAZ between consecutive beads in the TRC, a significant retained austenite fraction was converted to martensite due to reheating in this region. This is supported by the hardness distribution in the layer discussed in the next section. In turn, the microstructural characteristics of the TRM clearly show a Mo concentration of about 3.1 wt% in the molten pool during the LA, which Ti provides a completely martensitic matrix, as desired. Furthermore, minor microstructural changes in the overlap boundary (not a significantly elevated austenite fraction) indicates a notable improvement in the thermal stability of the TRM matrix compared with that of the TRC.

Note that in the case of the cubic MC carbide, Mo is a reactive element, i.e., Mo atoms can substitute for Ti atoms to form (Ti, Mo)C phase [37]. Thus, the addition of Mo into the melt provided an opportunity to change the composition of the synthesized cubic MC (TiC) carbide and led to the formation of primary and eutectic (Ti, Mo)C precipitates (Figure 10). It should be pointed out the SEM/EDS analysis detected negligible Mo content in the TRM matrix material. This indicates that the entire amount of Mo introduced into the molten pool dissolved in the TiC phase. Thus, considering the total concentration of MC carbide-forming elements in the melt (~10.5 wt% of Ti and ~3.1 wt% of Mo), it can be concluded that the TRM composition was optimal from a thermodynamics perspective, i.e., required to achieve the highest possible TiC fraction and a fully martensitic matrix [24,25]. Primary (Ti, Mo)C precipitates were observed in the compositional gradient between the core and outer layer (coring). The SEM/EDS elemental mapping, presented in Figure 10, suggests that the center of the primary dendritic (Ti, Mo)C precipitates was depleted in Mo, whereas the outer regions solidified with progressively higher Mo concentrations. A detailed characterization of the primary (Ti, Mo)C precipitates is presented elsewhere [38].

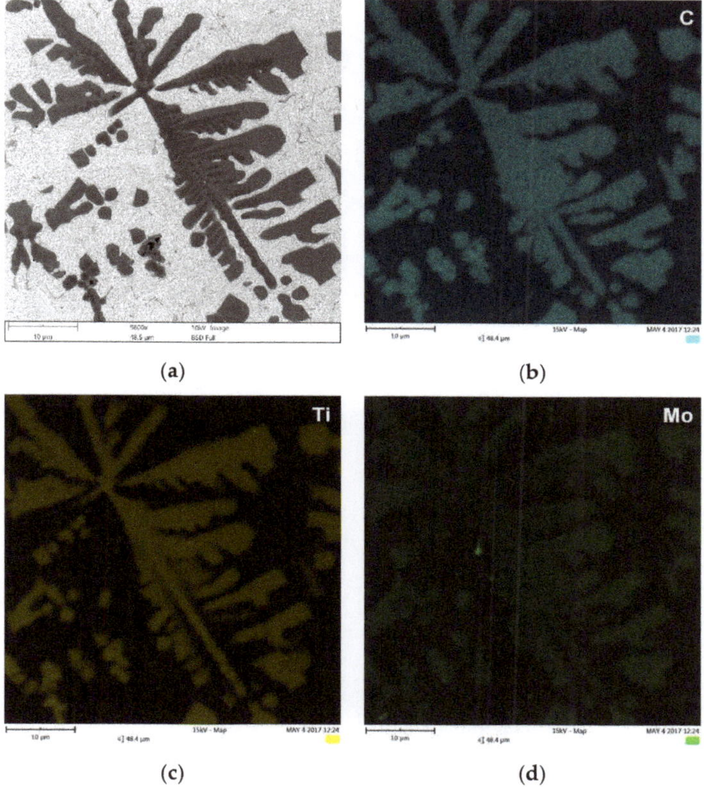


**Figure 10.** (**a**) BSE SEM micrograph taken from the mid-section of the TRM (Table 3), showing the primary (Ti, Mo)C precipitates in dendritic form and eutectic plate-like (Ti, Mo) C precipitates, and (**b–d**) corresponding distribution maps of C, Ti, and Mo, respectively.

As expected, based on reported data [39,40], Cr addition did not affect the composition of the synthesized TiC precipitates. The EDS elemental mappings presented in Figure 11 indicted homogenous compositions of the TiC precipitates in TRC. Figure 11 also shows a distribution of the previously mentioned Cr-rich eutectic regions in the martensitic/austenitic matrix of TRC.

Quantitative metallography revealed that, despite the same fractions of primary TiC/(Ti, Mo)C precipitates, TRC exhibited a slightly lower mean free path (MFP) than TRM. The MFP of TRC and TRM were 15.1 ± 2.4 and 17.6 ± 3.4 μm respectively, which was directly attributed to differences in the morphology between dendritic precipitates of primary TiC and (Ti, Mo)C phase. It is well-established that the MFP of composites depends on both the fraction and morphology of the reinforcing particles [41,42]. Comparing the primary TiC precipitates in TRC (Figures 5, 11a and 12a) and (Ti, Mo)C in TRM (Figures 6, 10a and 12b) shows that the difference in the composition of the molten pool affected the growth morphology of the primary TiC/(Ti, Mo)C phase. The TiC precipitates in TRC have markedly thinner primary and secondary arms compared with (Ti, Mo)C in TRM. For comparative purposes, Figure 13 presents an SEM image taken of the TRL processed with pure Ti, which was non-uniformly alloyed, but the analyzed region contained approximately 12.5 wt% Ti (the same Ti content as in the TRC). This micrograph generally showed the same growth morphology of the primary TiC phase as (Ti, Mo)C, proving that Mo addition did not affect the morphology of the primary (Ti, Mo)C phase in the TRM. Thus, it is reasonable to suggest that Cr addition during the investigated LA process influenced the nucleation or growth mechanism of crystals of the primary TiC phase. It was reported in several works [43–48] that the growth morphology of TiC crystals during the melt reaction method strongly depends on the composition of the melt environment.

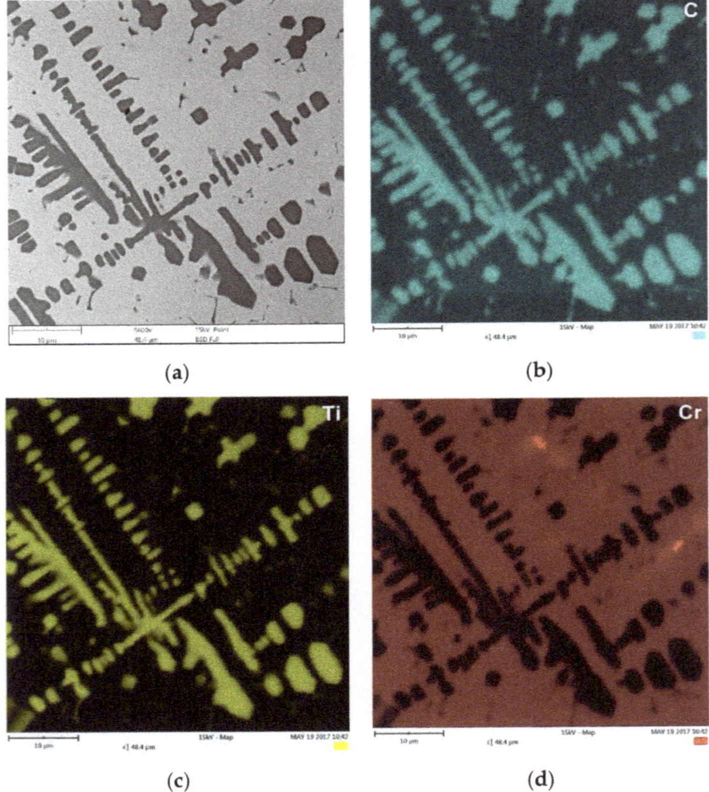


**Figure 11.** (**a**) BSE SEM micrograph taken from the mid-section of TRC (Table 3) showing the primary TiC precipitates in dendritic form and eutectic plate-like TiC precipitates, and (**b–d**) corresponding maps of C, Ti, and Cr distribution, respectively.

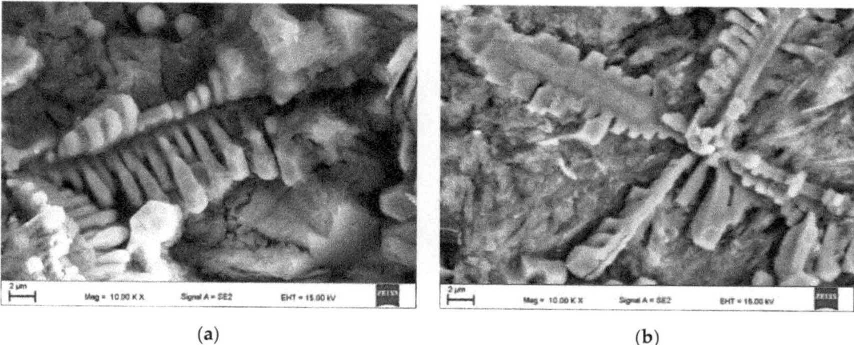

(**a**) (**b**)

**Figure 12.** SEM micrographs of deep-etched (**a**) TRC and (**b**) TRM showing primary dendritic precipitates of TiC and (Ti, Mo)C phases, respectively (Table 3).

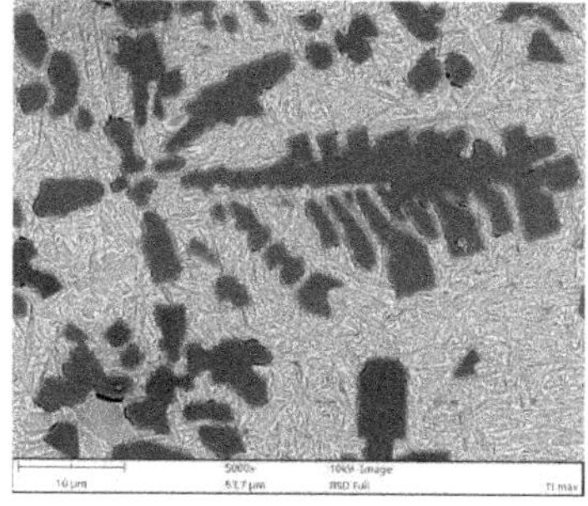

**Figure 13.** BSE SEM micrograph taken from the TRL region processed with pure Ti containing approximately 12.5 wt% Ti.

### 3.3. Hardness Analysis

Figure 14a,b shows typical hardness depth profiles of TRC and TRM, respectively. The corresponding hardness profiles across the overlap boundary in these layers are presented in Figure 14c,d. In general, due to the minor fraction of retained austenite, the TRM exhibited a higher overall hardness (610 ± 20 HV0.2) than the TRC (545 ± 25 HV0.2). Moreover, the Mo addition during the investigated LA process provided an almost homogeneous hardness distribution in the layer, even at the overlap boundary region (Figure 14d). The hardness uniformity of the TRM confirms that the addition of Mo produced a matrix material with high thermal stability. In the case of the TRC, as previously mentioned in the microstructure section, Cr addition leads to a relatively high fraction of retained austenite that, in turn, undergoes martensitic transformation in the HAZ of the overlap boundary, resulting in large hardness variations in the layer. The average hardness of the matrix in the TRM was estimated to be 580 ± 34 HV0.025. In the case of TRC, the matrix hardness ranged from 480 ± 32 to 550 ± 41 HV0.025 near the center of a single processed bead and in the HAZ of the overlap boundary, respectively.

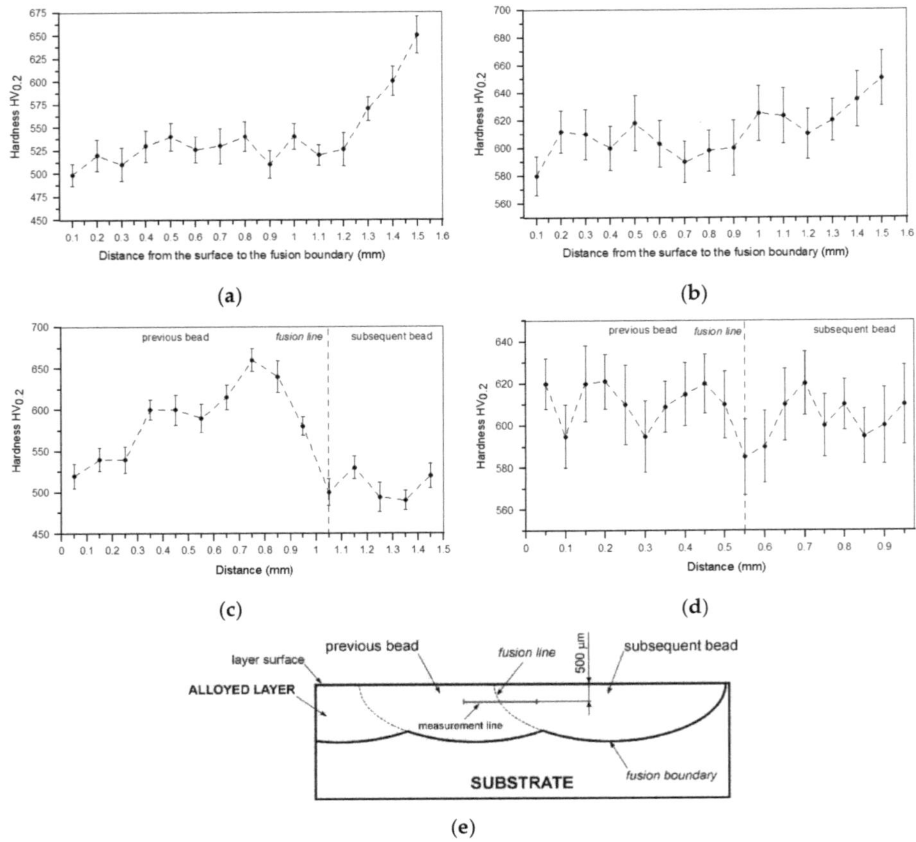


**Figure 14.** (**a**,**b**) Hardness depth profiles for TRC and TRM (Table 3), respectively. (**c**,**d**) Hardness profiles across the overlap boundary between consecutive beads in TRC and TRM, respectively. (**e**) Diagram showing a location of the measurement line in the overlap boundary.

## 4. Conclusions

This manuscript presented a novel approach to the in situ synthesis of TiC-reinforced composite surface layers on DCI substrates via LA that, by properly selecting the molten pool composition, changed the characteristics of TiC precipitates, the phase composition of the matrix material, and also the dispersion uniformity of TiC phase. In contrast to alloying with pure Ti, using 80Ti-20Cr and 80Ti-20Mo (weight ratio) powder mixtures produced the optimal TiC fraction from a thermodynamics perspective, i.e., the highest possible TiC fraction for the C content in the processed DCI [24]. As a result, the developed approach makes it possible to convert an as-cast structure into a composite structure containing TiC precipitates uniformly dispersed in the completely metallic matrix (eutectic cementite was limited to ~2 vol%). However, the presence of additional Cr in the molten pool, in conjunction with non-equilibrium cooling conditions, suppressed the martensitic transformation of the matrix material, leading to an undesired elevated amount of retained austenite (~24 wt%), and, consequently, a low matrix hardness. Additionally, the elevated retained austenite fraction in the matrix significantly changed the hardness in the overlap boundary in the SAL. In contrast, the addition of Mo to the melt led to the formation of an almost fully martensitic matrix (~5.7 wt% of retained austenite fraction), ensuring a uniform hardness distribution throughout the SAL.

It was found that the introduction of Mo into the molten pool led to the synthesis of primary and eutectic precipitates of (Ti, Mo)C phase. The primary (Ti, Mo)C precipitates exhibited a nonuniform Mo distribution (coring structure). The growth morphology of the primary (Ti, Mo)C precipitates was similar to that of TiC precipitates synthesized during processing using pure Ti. The presence of additional Cr in the molten pool did not affect the composition of the primary TiC phase. However, the melt enrichment of Cr (~3 wt%) refined the primary dendritic precipitates of TiC phase compared with that formed during alloying with a pure Ti and Ti-Mo powder mixture. As a result, at a constant fraction of TiC precipitates, the mean free path between the reinforcing precipitates was reduced.

**Funding:** Publication supported under the Rector's grant in the area of research and development. The Silesian University of Technology, 10/050/RGP_20/0078.

**Conflicts of Interest:** The authors declare no conflict of interest.

## Appendix A

| | |
|---|---|
| TRL | TiC-reinforced composite surface layer |
| LA | Laser surface alloying |
| DCI | Ductile cast iron |
| MMC | Metal matrix composite |
| SM | Substrate material |
| SAB | Single-alloyed bead |
| SAL | Surface alloyed layer (the layer produced via a multi-pass overlapping alloying process) |
| TRC | SAL produced using a powder mixture of Ti-Cr |
| TRM | SAL produced using a powder mixture of Ti-Mo |
| TR | SAL produced using a pure Ti powder |
| HAZ | Heat-affected zone |
| MFP | Mean free path |
| BSE | Back-Scattered Electron |

## References

1. Paczkowska, M.; Makuch, N.; Kulka, M. The influence of various cooling rates during laser alloying on nodular iron surface layer. *Opt. Laser Technol.* **2018**, *102*, 60–67. [CrossRef]
2. Jeyaprakash, N.; Yang, C.; Duraiselvam, M.; Prabu, G.; Tseng, S.; Kumar, D.R. Investigation of high temperature wear performance on laser processed nodular iron using optimization technique. *Results Phys.* **2019**, *15*, 102585. [CrossRef]
3. Janicki, D. Direct diode laser surface melting of nodular cast iron. *Appl. Mech. Mater.* **2015**, *809–810*, 423–428. [CrossRef]
4. Ceschini, L.; Campana, G.; Pagano, N.; Angelin, V. Effect of laser surface treatment on the dry sliding behaviour of the EN-GJS400-12 ductile cast iron. *Tribol. Int.* **2016**, *104*, 342–351. [CrossRef]
5. Cui, W.; Liu, J.; Zhu, H.; Shi, Y. Investigation on wear resistance of nodular cast iron by laser surface treatment. *Mater. Res. Express* **2019**, *6*, 8. [CrossRef]
6. Pagano, N.; Angelini, V.; Ceschini, L.; Campana, G. Laser remelting for enhancing tribological preferences of a ductile iron. *Procedia CIRP* **2016**, *41*, 987–991. [CrossRef]
7. Janicki, D.; Górka, J.; Kwaśny, W.; Pakieła, W.; Matus, K. Influence of solidification conditions on the microstructure of laser-surface-melted ductile cast iron. *Materials* **2020**, *13*, 1174. [CrossRef]
8. Jeyaprakash, N.; Yang, C.; Duraiselvam, M.; Prabu, G. Microstructure and tribological evolution during laser alloying WC-12%Co and $Cr_3C_2$−25%NiCr powders on nodular iron surface. *Results Phys.* **2019**, *12*, 1610–1620. [CrossRef]
9. Tański, T.; Pakieła, W.; Janicki, D.; Tomiczek, B.; Król, M. Properties of the aluminium alloy EN AC-51100 after laser surface treatment. *Arch. Metall. Mater.* **2016**, *61*, 199–204. [CrossRef]
10. Lisiecki, A. Titanium matrix composite Ti/TiN produced by diode laser gas nitriding. *Metals* **2015**, *5*, 54–69. [CrossRef]

11. Szala, M.; Łatka, L.; Walczak, M.; Winnicki, M. Comparative study on the cavitation erosion and sliding wear of cold-sprayed Al/$Al_2O_3$ and Cu/$Al_2O_3$ coatings, and stainless steel, aluminium alloy, copper and brass. *Metals* **2020**, *10*, 856. [CrossRef]
12. Rasool, G.; Mridha, S.; Stack, M.M. Mapping wear mechanisms of TiC/Ti composite coatings. *Wear* **2015**, *328–329*, 498–508. [CrossRef]
13. Szala, M.; Hejwowski, T. Cavitation erosion resistance and wear mechanism model of flame-sprayed Al2O3-40% $TiO_2$/NiMoAl cermet coatings. *Coatings* **2018**, *8*, 254. [CrossRef]
14. Yan, H.; Wang, A.; Xiong, Z.; Xu, K.; Huang, Z. Microstructure and wear resistance of composite layers on a ductile iron with multicarbide by laser surface alloying. *Appl. Surf. Sci.* **2010**, *256*, 7001–7009. [CrossRef]
15. Man, H.C.; Zhang, S.; Cheng, F.T.; Yue, T.M. In situ synthesis of TiC reinforced surface MMC on Al6061 by laser surface alloying. *Scr. Mater.* **2002**, *46*, 229–234. [CrossRef]
16. Park, H.I.; Nakata, K.; Tomida, S. In situ formation of TiC particulate composite layer on cast iron by laser alloying of thermal sprayed titanium coating. *J. Mater. Sci.* **2000**, *35*, 747–755. [CrossRef]
17. Janicki, D.; Górka, J.; Kwaśny, W.; Gołombek, K.; Kondracki, M.; Żuk, M. Diode laser surface alloying of armor steel with tungsten carbide. *Arch. Metall. Mater.* **2017**, *62*, 473–481. [CrossRef]
18. Aikin, R.M. The mechanical properties of in-situ composites. *JOM* **1997**, *49*, 35. [CrossRef]
19. Parashivamurthy, K.I.; Kumar, R.K.; Seetharamu, S.; Chandrasekharaiah, M.N. Review on TiC reinforced steel composites. *J. Mater. Sci.* **2001**, *36*, 4519–4530. [CrossRef]
20. Janicki, D. In-situ synthesis of titanium carbide particles in an iron matrix during diode-laser surface alloying of ductile cast iron. *Mater. Technol.* **2016**, *51*, 317–321. [CrossRef]
21. Gilev, V.G.; Morozov, E.A. Laser melt injection of austenitic cast iron $Ch_{16}D_7$GKh with titanium. *Russ. J. Non-Ferrous Met.* **2016**, *57*, 625–632. [CrossRef]
22. Liu, Y.; Qu, W.; Su, Y. TiC reinforcement composite coating produced using graphite of the cast iron by laser cladding. *Materials* **2016**, *9*, 815. [CrossRef] [PubMed]
23. Janicki, D. Microstructural evolution during laser surface alloying of ductile cast iron with titanium. *Arch. Metall. Mater.* **2017**, *62*, 2425–2431. [CrossRef]
24. Janicki, D. Microstructure and sliding wear behaviour of in-situ TiC-reinforced composite surface layers fabricated on ductile cast iron by laser alloying. *Materials* **2018**, *11*, 75. [CrossRef]
25. Janicki, D. Fabrication of TiC-reinforced surface layers on ductile cast iron substrate by laser surface alloying. *Solid State Phenom.* **2020**, *308*, 76–99. [CrossRef]
26. Ding, J.; Liu, Y.; Zhu, L.; Zhou, L.; Li, Y. Microstructure and properties of a titanium carbide reinforced coating on grey iron applied with laser cladding. *Mater. Technol.* **2019**, *53*, 377–381. [CrossRef]
27. Dogan, O.N.; Hawk, J.A.; Tylczak, J.H.; Wilson, R.D.; Govier, R.D. Wear of titanium carbide reinforced metal matrix composite. *Wear* **1999**, *225–229*, 758–769. [CrossRef]
28. Olejnik, E.; Szymański, Ł.; Kurtyka, P.; Tokarski, T.; Maziarz, W.; Grabowska, B.; Czapla, P. Locally reinforcement TiC-Fe type produced in situ in castings. *Arch. Foundry Eng.* **2016**, *16*, 77–82. [CrossRef]
29. Zhu, H.; Hua, B.; Huang, J.; Li, J.; Wang, G.; Xie, Z. Dry sliding wear behaviour of TiC/Fe composites fabricated by exothermic dispersion of Fe–Ti–C powders. *Tribol. Mater. Surf. Interfaces* **2014**, *8*, 165–171. [CrossRef]
30. Janicki, D. *The Effect of Melt Composition on Microstructure of In-Situ Tic-Reinforced Fe-Based Composite Surface Layers Synthesized via a Laser Surface Alloying*; Department of Welding, Silesian University of Technology: Gliwice, Poland, 2017; unpublished work.
31. Barin, I. *Thermochemical Data of Pure Substances*; VCH: Weinheim, Germany, 1995.
32. Janicki, D. The effect of cooling rate on microstructure of in-situ TiC-reinforced Fe-based composite surface layers synthesized on the ductile cast iron via a laser surface alloying. *Surf. Coat. Technol.* **2021**, in preparation.
33. Heiple, C.R.; Roper, J.R. Mechanism for minor element effect on GTA fusion zone geometry. *Weld. J.* **1982**, *61*, 97–102.
34. Heiple, C.R.; Roper, J.R.; Stagner, R.T.; Aden, R.J. Surface active element effects on the shape of GTA, laser, and electron beam welds. *Weld. J.* **1983**, *62*, 72–77.
35. Heiple, C.R.; Burgardt, P. Effects of $SO_2$ shielding gas additions on GTA weld shape. *Weld. J.* **1985**, *64*, 159–162.
36. Janicki, D. Fabrication of high chromium white iron surface layers on ductile cast iron substrate by laser surface alloying. *Stroj. Vestn.* **2017**, *63*, 705–714. [CrossRef]

37. Jang, J.H.; Lee, C.H.; Heo, Y.U.; Suh, D.W. Stability of (Ti, M)C (M = Nb, V, Mo and W) carbide in steals using first-principle calculations. *Acta Mater.* **2012**, *60*, 208–217. [CrossRef]
38. Janicki, D. The friction and wear behaviour of in-situ titanium carbide reinforced composite layers manufactured on ductile cast iron by laser surface alloying. *Surf. Coat. Technol.* **2020**, 126634, in press. [CrossRef]
39. Pierson, H.O. *Handbook of Refractory Carbides and Nitrides: Properties Characteristic, Processing and Application*; Noyes Publications: Westwood, NJ, USA, 1996.
40. Storms, E.K. *The Refractory Carbides*; Academic Press Inc.: New York, NY, USA, 1967.
41. Zum Ghar, K.-H. *Microstructure and Wear of Materials*; Tribology Series 10; Elsevier Science Publishers, B.V.: Amsterdam, The Netherlands, 1987.
42. Thakare, M.R.; Wharton, J.A.; Wood, R.J.K.; Menger, C. Effect of abrasive particle size and the influence of microstructure on the wear mechanisms in wear-resistant materials. *Wear* **2012**, *276–277*, 16–28. [CrossRef]
43. Strzeciwilk, D.; Wokulski, Z.; Tkacz, P. Growth and TEM and HREM characterisation of TiC crystals grown from high-temperature solutions. *Cryst. Res. Technol.* **2003**, *38*, 283–287. [CrossRef]
44. Nie, J.; Liu, X.; Ma, X. Influence of trace boron on the morphology of titanium carbide in an Al–Ti–C–B master alloy. *J. Alloys Compd.* **2010**, *491*, 113–117. [CrossRef]
45. Strzeciwilk, D.; Wokulski, Z.; Tkacz, P. Microstructure of TiC crystals obtained from high temperature nickel solution. *J. Alloys Compd.* **2003**, *350*, 256–263. [CrossRef]
46. Liu, Z.; Tian, J.; Li, B.; Zhao, L. Microstructure and mechanical behaviors of in situ TiC particulates reinforced Ni matrix composites. *Mater. Sci. Eng. A* **2010**, *527*, 3898–3903. [CrossRef]
47. Gopagoni, S.; Hwang, J.Y.; Singh, A.R.P.; Mensah, B.A.; Bunce, N.; Tiley, J.; Scharf, T.W.; Banerjee, R. Microstructural evolution in laser deposited nickel–titanium–carbon in situ metal matrix composites. *J. Alloys Compd.* **2011**, *509*, 1255–1260. [CrossRef]
48. Nie, J.; Wu, Y.; Li, P.; Li, H.; Liu, X. Morphological evolution of TiC from octahedron to cube induced by elemental nickel. *CrystEngComm* **2012**, *14*, 2213. [CrossRef]

*Article*

# Development and Optimization of Tin/Flux Mixture for Direct Tinning and Interfacial Bonding in Aluminum/Steel Bimetallic Compound Casting

**Mohamed Ramadan [1,2,*], Abdulaziz S. Alghamdi [1], K. M. Hafez [2], Tayyab Subhani [1,*] and K. S. Abdel Halim [1,2]**

1 College of Engineering, University of Ha'il, Ha'il P.O. Box 2440, Saudi Arabia; a.alghamdi@uoh.edu.sa (A.S.A.); k.abdulhalem@uoh.edu.sa (K.S.A.H.)

2 Central Metallurgical Research and Development Institute (CMRDI), P.O. Box 87, Helwan 11421, Egypt; khalidhafez@yahoo.com

* Correspondence: mrnais3@yahoo.com (M.R.); ta.subhani@uoh.edu.sa (T.S.)

Received: 8 November 2020; Accepted: 8 December 2020; Published: 10 December 2020

**Abstract:** Interfacial bonding highly affects the quality of bimetallic bearing materials, which primarily depend upon the surface quality of a solid metal substrate in liquid–solid compound casting. In many cases, an intermediate thin metallic layer is deposited on the solid substrate before depositing the liquid metal, which improves the interfacial bonding of the opposing materials. The present work aims to develop and optimize the tinning process of a solid carbon steel substrate after incorporating flux constituents with the tin powder. Five ratios of tin-to-flux—i.e., 1:1, 1:5, 1:10, 1:15, and 1:20—were used for tinning process of carbon steel solid substrate. Furthermore, the effect of volume ratios of liquid Al-based bearing alloy to solid steel substrate were also varied—i.e., 5:1, 6.5:1 and 8.5:1—to optimize the microstructural and mechanical performance, which were evaluated by interfacial microstructural investigation, bonding area determination, hardness and interfacial strength measurements. It was found that a tin-to-flux ratio of 1:10 offered the optimum performance in AlSn12Si4Cu1/steel bimetallic materials, showing a homogenous and continuous interfacial layer structure, while tinned steels using other percentages showed discontinuous and thin layers, as in 1:5 and 1:15, respectively. Furthermore, bimetallic interfacial bonding area and hardness increased by increasing the volume ratio of liquid Al alloy to solid steel substrate. A complete interface bonding area was achieved by using the volume ratio of liquid Al alloy to solid steel substrate of $\geq$8.5.

**Keywords:** nanocomposite; nanoparticle; bimetallic; mechanical; interfacial; compound casting

---

## 1. Introduction

Although the development of bimetallic bearings is progressing unceasingly, the stringent requirements of preparing ideal bearings have yet not been achieved. Simultaneously, the demand for outstanding quality bearings is rising continuously with the emergence of advanced technologies in turbines and jet engines. In gas and steam turbines, bearings are used for supporting and positioning the rotating components while journal or roller bearings provide radial support, and axial positioning is acquired by thrust bearings. Ball or roller bearings are generally used for radial support in aircraft jet engines. The desired attributes while designing the bearings are long service life, high reliability, and economic efficiency. Moreover, load, speed, lubrication, temperature, shaft arrangement, mounting/dismounting, noise and environmental conditions are other influencing factors that are considered by design engineers to meet the above specifications.

In most cases, main bearings, such as journal and thrust bearings, are manufactured from bimetallic materials. Bimetallic materials can be fabricated by bonding similar and dissimilar materials. Because of their unique physical, wear and mechanical properties, bimetallic materials were classified as advanced functional materials [1]. The quality of bearings and their lifetime are governed by their wear behavior—i.e., working layer, and bonding strength between the pair metals. The wear behavior of the working layer depends primarily on the bearing material and its structure. The bond between the two metals often depends on their physical, thermal and chemical properties, as well as their fabrication techniques [2–4].

Various techniques have been used to fabricate bimetallic bearings, such as casting, welding cast-rolling, cladding and powder metallurgy [1,5–7]. Liquid–solid casting (compound casting) is a unique technique for the fabrication of bimetallic bearings, wherein the working bearing materials are in a liquid state (Al-Sn alloy, Sn-Babbitt alloys, and Cu-Sn alloys) and the supported substrate (mild steel, cast iron, copper alloys) are in a solid state. In spite of the development of high-performance materials for bearing applications, there is a continuous demand for increasingly improved tribological properties and higher bonding strength of bimetallic bearing alloys. Although serious efforts have been made worldwide to improve the tribological properties of working layer bimetallic bearing materials, limited work was carried out to improve the bonding strength of the interface layer in bimetallic materials.

In liquid–solid bimetallic castings, metallic interlayers between the two metals are commonly used [6,8,9]. These metallic interlayers are deposited on the solid substrate before pouring the paired liquid metal. The major problems related to the fabrication of bimetallic joints without interlayers are the occurrence of brittle intermetallic phases in the bonding zone that significantly affect the bonding strength of the bimetal elements. In addition, in bimetallic bearings, both metals have different melting temperatures that result in poor wettability between liquid metal and solid substrate [6,8]. Reaction between liquid and solid in liquid–solid casting improves wettability and one of their characteristics is that the wetting angle changes with time as the reaction progresses [10]. If both of the metals of bimetal casting have relatively high melting points, the reaction time will increase, resulting in improved wettability. Otherwise, for low melting point metals, the reaction time will be much lower, resulting in lower wettability. A wise approach is to introduce an additional metal or alloy between the coupled metals and control the structure of the interfacial bonded zone to prevent the formation of intermetallic phases and improve the bimetallic wettability [6,9–11]. Recently, Ramadan et al. [9] have improved the shear strength of the Sn interlayer in bimetallic bearings by the addition of Cu. Therefore, Sn, Zn and Sn-Pb alloy are the proposed metals that can be used in the fabrication of interlayers in liquid–solid bimetallic journal and thrust bearings [6,9,11].

Among the available bi-metallic bearing fabrication techniques, liquid–solid static casting technology is considered to be the most economical for bearing layer elements. The improvement of the interfacial bond is a critical factor in the production of bimetallic bearings using a liquid–solid technique wherein the substrate should be considered carefully to achieve higher bond strength, durability and performance. For most liquid–solid bimetallic fabrication techniques, low carbon steel, stainless steel or copper alloys are the nominate materials used as solid substrates [1,9,12–14].

It has been reported [11] that Zn interlayer for bimetallic Mg/Al joints significantly increased the bonding strength. Sn and Sn-Pb alloys [6,8,15] are commonly used as interlayers for the fabrication of bimetallic materials using liquid–solid compound casting technique. In our previous work, the Sn metallic interlayer was reinforced by 3 wt.% Cu to improve the bonding strength of Al-Sn/mild steel bimetallic materials [9]. The results showed significant improvement in the shear strength by ~59% compared to the pure Sn interlayer. However, in spite of all the previous works [6,8,9,15] that used the interlayer to improve the bonding strength of bimetallic bearing materials, the bonding strength of bimetallic bearing with interlayer is still low and needs more improvements to increase the performance and lifetime of the bearings.

In view of the above, the current research was designed to develop and optimize a novel tinning process including Sn powder and flux mixture for direct tinning of the solid substrate. The proposed direct tinning process is considered as an alternative new tinning process designed for fast, easy and low-cost surface improvement of solid substrate used in liquid–solid compound casting. This newly designed tinning process is more suitable for flat and horizontal solid substrates like bimetallic thrust bearing applications.

## 2. Materials and Methods

Pure metals, including Sn, Cu, Al and Al-25%Si master alloy, were used for preparing Al-12Sn-4Si-1Cu bearing material. The charge of 1.5 kg of Al-alloy constituents was melted using an electrical furnace in a graphite crucible. Low carbon steel was used for bonding with Al-based bearing alloy to prepare the bimetallic material. The chemical compositions of Al-based bearing alloy and low carbon steel substrate, as manufactured by a local manufacturing Company, Riyadh, Saudi Arabia, are given in Table 1.

**Table 1.** Chemical compositions of Al-based bearing alloy and low carbon steel substrate (wt.%).

| | C | Si | Mn | Cu | Sn | Cr | Ni | Fe | Al |
|---|---|---|---|---|---|---|---|---|---|
| Al-based bearing alloy | - | 4 | - | 1 | 12 | - | - | - | Bal. |
| Steel substrate | 0.14 | 0.30 | 0.48 | 0.20 | - | 0.14 | 0.09 | Bal. | - |
| (Equivalent standard of steel substrate, Japanese Industrial Standards, JIS, G4051) | 0.13 0.18 | 0.15 0.35 | 0.30 0.60 | ≤0.3 | - | ≤0.2 | ≤0.2 | Bal. | - |

Solid steel cylindrical specimens of 59 mm diameter and 4 mm thickness were grinded with emery papers up to 400 grade. Flux was prepared by mixing zinc chloride, sodium chloride, ammonium chloride and hydrochloric acid in water, as reported in detail in our previous work [9,16]. Five different ratios of tin powder to flux were used for tinning process for bonding liquid Al-based bearing alloy with solid steel substrate—i.e., 1:1, 1:5, 1:10, 1:15, and 1:20. Ratios of tin powder to flux of 1:1 and 1:20 were not characterized due to reasons discussed further below. For a 1:1 ratio, a good distribution of tin/flux mixture on substrate surface was not achieved. For a 1:20 ratio, the amount of Sn was not enough to deposit a considerable Sn-interlayer on the steel substrate surface. For the preparation of bimetallic specimens, the mixture of different ratios of tin and flux were individually spread on a designed area of 2734 $mm^2$ of pre-grinded steel specimen. Subsequently the steel specimens with five different ratios of tin/flux mixture were heated on a hot plate for 2 min at 350 °C. Later, steel specimens were washed with warm water and cleaned using cotton cloth to remove the remaining flux from the surface of substrates whereupon the tin layers were produced. Finally, molten Al-based bearing alloy was poured after heating at a temperature of 750 °C into a metallic mold (Figure 1) that contained pre-heated (350 °C) tinned steel specimens.

After solidification, the specimens were removed from the metallic mold for microstructural and mechanical property characterization. For microstructural investigation, the specimens were cut, ground, polished and etched with a solution consisting of 0.5 mL $HNO_3$, 0.3 mL HCl, 0.2 mL HF and 19 mL $H_2O$. Optical microscope (OM) (Olympus GX51, Tokyo, Japan) and scanning electron microscope (SEM) (FEG-SEM, FEI, Eindhoven, The Netherlands) were used to investigate bimetallic materials and their interfacial structure. The chemical compositions of the bimetallic interfaces were measured by an energy-dispersive X-ray spectroscopy (EDS, Quanta 250 FEG, Eindhoven, The Netherlands).

A Rockwell hardness testing machine was used for measuring the hardness of bimetallic castings. The hardness testing (HRF) was performed at a load of 60 kgf using a 1/16 inch ball indenter for the dwell time of 5 sec. At least five readings of each of the bimetallic specimens were taken for the average value of hardness. The shear strength value of the interface of the bimetallic specimen was performed using a tensile testing machine (Instron 5969, Instron, Norwood, MA, USA), as discussed in a previous work along with the dimensions of the tensile-shear specimen [17]. The dimensions of the tensile shear specimens, having two notches of 4 × 8 × 4 mm on the Al alloy side and of 4 × 8 × 4 on the steel substrate side, were machined to examine the completely exposed bonding interface. The shear test was carried out at a constant strain rate of $\dot{\varepsilon} = 3.33 \times 10^{-4}\left(\dot{\varepsilon} = V/l_0\right)$ and tensile test velocity $V$ of 1 mm/min, where $l_0$ is the distance between shoulders. At least three samples for each Sn-to-flux ratio were used in order to minimize errors.

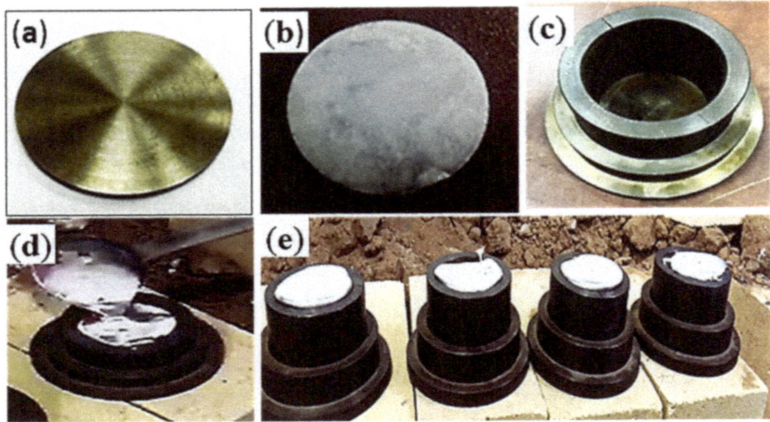


**Figure 1.** Solid steel substrate (**a**), tinned steel substrate (**b**), metallic mold (**c**), pouring molten Al-based bearing alloy onto tinned steel substrate (**d**), and bimetallic castings after pouring (**e**).

## 3. Results

Microstructures of Al-based bearing alloy and low carbon steel solid substrate used in the fabrication of bimetallic materials are shown in Figure 2. Microstructure of Al-Sn-Si-Cu bearing alloy shows α-Al matrix with $Cu_6Sn_5$, Sn and Si phases (Figure 2a) while the microstructure of low carbon steel has a ferritic–pearlitic microstructure (Figure 2b). The microstructure derives from alloying elements and the treating process parameters, as given in Table 1.

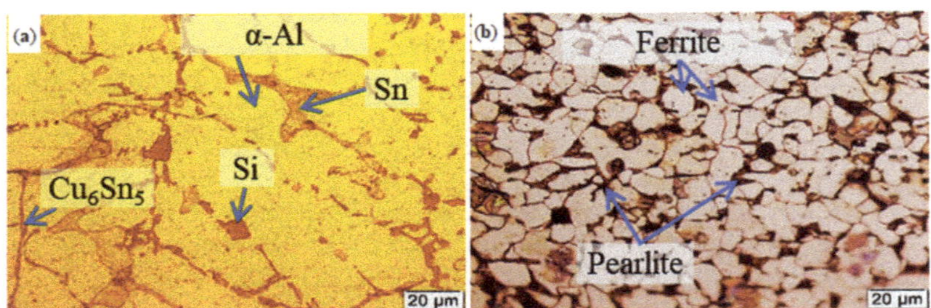


**Figure 2.** Microstructures of (**a**) Al-based bearing alloy and (**b**) steel substrate used for preparing bimetallic materials.

Figure 3 shows SEM microstructures (Figure 3a,b), EDS analysis (Figure 3c) and elemental mapping (Figure 3d–f) of precipitates, phases and matrix of Al-based bearing alloy in bimetallic material. It is clear that the Al-based bearing alloy has fine and well-dispersed precipitates in the α-Al matrix. It was reported [18] that the presence of large particles initiate fatigue, resulting in a low lifetime of Al-based bearing material.

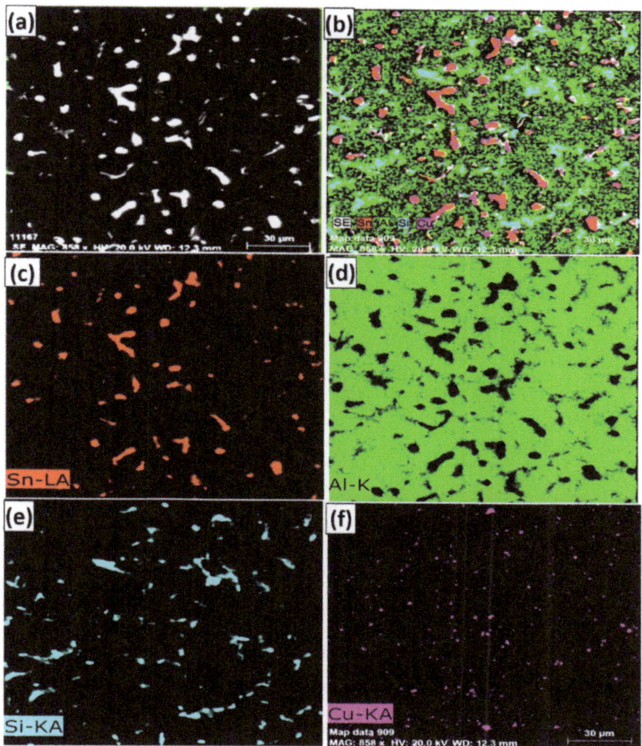


**Figure 3.** (**a**,**b**) SEM images, (**c**) EDS and (**d–f**) mapping analysis of Al-based bearing alloy.

### 3.1. *Effect of Sn-to-Flux Percentage on Structure and Properties of Bimetallic Interface*

Figure 4 shows the interfacial microstructures of three specimens containing tin-to-flux ratios of 1:5, 1:10 and 1:15. Figure 4a shows an irregular interface wherein the disconnected interface material is also visible at the interface, which may be due to an insufficient flux ratio in the tin/flux mixture (1:5). Using an insufficient flux ratio resulted in decreased wettability of the tin with the steel substrate. The specimen containing 1:10 ratio of tin-to-flux shows a smooth interfacial surface without the formation of isolated regions or areas. In contrast, a continuous layer of tin seems to be developed at the interface. An irregular and relatively thin interface can also be seen in the specimen containing a tin-to-flux ratio of 1:15 but no detachment at the interface was noted in that observed with the specimen of a 1:5 ratio. The development of a continuous tin phase and the emergence of intermetallic phases and/or oxides have their separate effects upon the mechanical performance of the interface, which are discussed further below.

Figure 5 shows the interfacial SEM microstructures of the three bimetallic specimens containing different tin-to-flux ratios along with the elemental analysis along a line across the interface. The interfacial SEM images (Figure 5a,c,e) replicate the images acquired by optical microscopy and Figure 5a shows the presence of thick interface of specimen containing 1:5 tin: flux ratio indicating the presence of intermetallic phases [19] (Fe-Al intermetallic phases) due to the reaction of aluminum with iron, as revealed in Figure 5b. EDS and line scan confirm the presence of composition of bimetal and the formation of the Sn interlayer. Figure 5c indicates a comparatively continuous and smooth interphase of tin at the interface of adjoining materials of specimen containing 1:10 ratio, which was verified in the graph of EDS line analysis of the same interface indicating the absence of intermetallic phases (Fe-Al intermetallic phases) and the presence of tin at the interface (Figure 5d). Figure 5e shows the interfacial image of the specimen containing a tin-to-flux ratio of 1:15; wherein the thickness of the interphase is discontinuing and relatively thin compared to the previous two samples and the presence of both the tin layer and intermetallic phases (Fe-Al intermetallic phases) can be detected (Figure 5f).

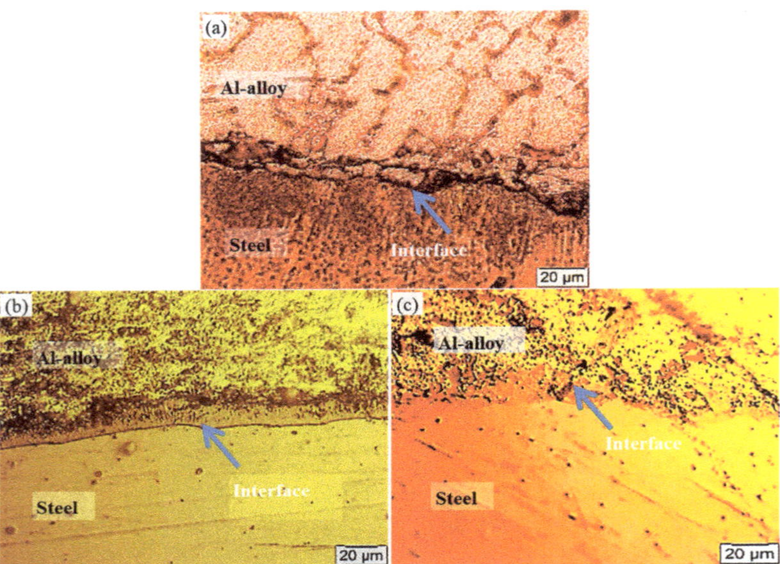


**Figure 4.** Optical interfacial microstructures of bimetallic materials with tin-to-flux ratios of (**a**) 1:5, (**b**) 1:10 and (**c**) 1:15.

EDS patterns in interface layers adjacent to Al-bearing alloy zones in bimetallic casting containing a tin-to-flux ratio of 1:5, 1:10 and 1:15 are shown in Figure 6. Higher percentagesa of Fe and Al elements are detected in bimetallic casting containing a tin-to-flux ratio of 1:5 and 1:15 (Figure 6a,c). Figure 6b shows the lowest Fe percentage in bimetal casting containing a tin-to-flux ratio of 1:10. Line scans of Fe and Al intersect across the interface of bimetallic materials (Figure 5a,c) and EDS patterns of points analysis (Table 2) in interface layers adjacent to Al-bearing alloy zones in bimetallic casting (Figure 6a,c), containing a tin-to-flux ratio of 1:5 and 1:15, confirm the formation of Fe-Al intermetallic at the interlayer. Fe-Al intermetallic is, otherwise, absent at the interface of bimetallic casting (Figure 5b) containing tin-to-flux ratio of 1:10 and EDS pattern in interface layers adjacent to the Al-bearing alloy zones in bimetallic casting (Figure 6b) containing a tin-to-flux ratio of 1:10. It shows that the continuous and smooth Sn interlayer can significantly decrease or prevent the formation of Fe-Al intermetallic phases at the interface of Al/Fe.

Figure 7a shows the SEM image of bimetallic specimen containing tin-to-flux ratio of 1:15 along with the elemental mapping analysis (Figure 7b–g) of the interface. This specimen was specially chosen for EDS as it contains the mutual effect of the other two specimens containing tin-to-flux ratios of 1:5 and 1:10—i.e., presence of intermetallic phases and the indication of the presence of tin film at the interface. Image in Figure 7b indicates that tin has moved to aluminum and a very low presence of tin was detected in steel. Figure 7c shows the distribution of aluminum in aluminum specimen, while Figure 7g shows the presence of iron in the steel specimen. This shows the absence of migration of aluminum atoms in opposing material while a very low fraction of iron atoms moved to aluminum, as also observed elsewhere [20]. Figure 7d shows the distribution of oxygen across the interface. Comparatively bimetallic casting fabricated using a tin-to-flux ratio of 1:5 has significantly higher oxygen content at the interface, which indicates the presence of tin oxide. The contents of silicon (Figure 7e) in aluminum and steel matches with their relative percentages in specimens, as shown in Table 1. The presence of copper in both the aluminum and steel was also observed (Figure 7f).

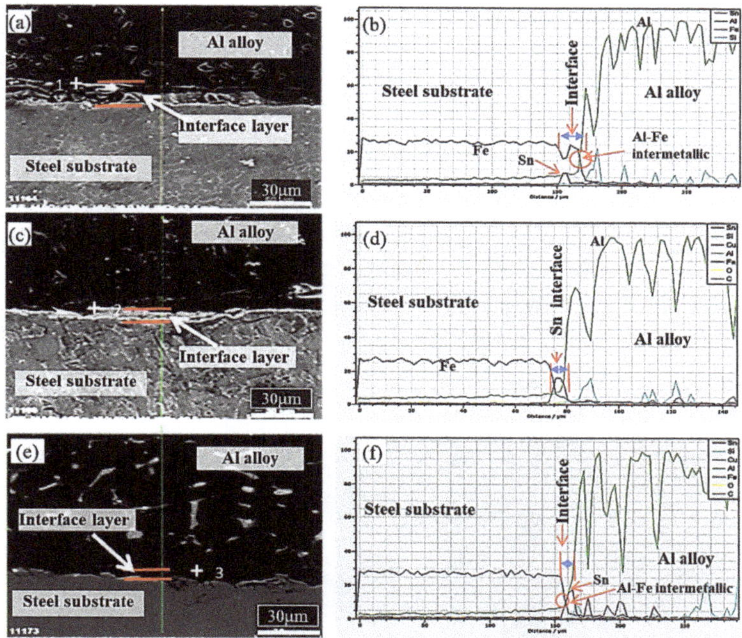


**Figure 5.** SEM (**a**,**c**,**e**) and EDS (**b**,**d**,**f**) analysis and line scan results across the interface of bimetallic materials using the tin:flux ratios of percentage of 1:5, 1:10 and 1:15.

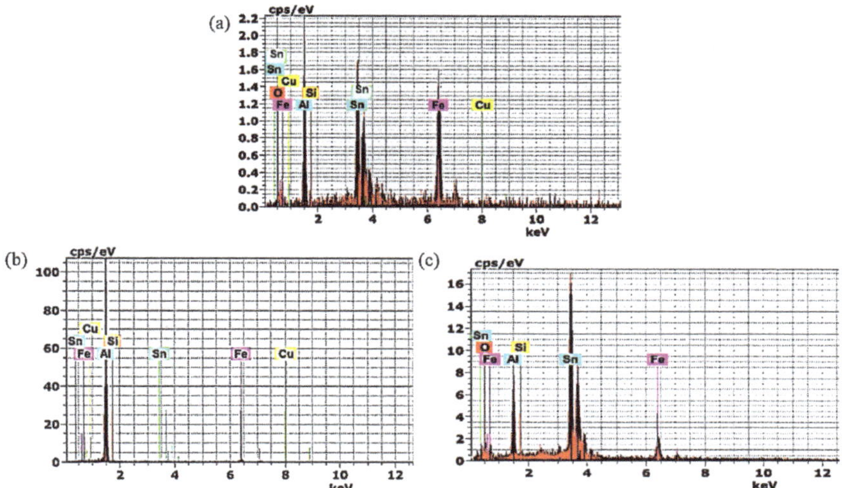


**Figure 6.** EDS patterns in interface layers adjacent to Al bearing alloy zones in bimetallic castings containing tin-to-flux ratio of (**a**) 1: 5, (**b**) 1:10, and (**c**) and 1:15 at points 1, 2 and 3 shown in Figure 5a–c, respectively.

**Table 2.** EDS analysis corresponding to the points 1, 2 and 3 indicated in Figure 5.

| Number | Element Compositions (at.%) | | | | | | Inference Adjacent to Al Bearing Alloy, Al, Fe-Al (IMC) |
|---|---|---|---|---|---|---|---|
| | Al | Fe | Sn | Si | O | Cu | |
| 1 | 49.6 | 22.94 | 14.71 | 1.16 | 10.68 | 0.91 | $Fe_2Al_5$ |
| 2 | 94.56 | 3.71 | 0.29 | 1.31 | 0 | 0.13 | Al |
| 3 | 30.32 | 10.44 | 33.02 | 0.92 | 25.31 | 0 | $FeAl_3$ |

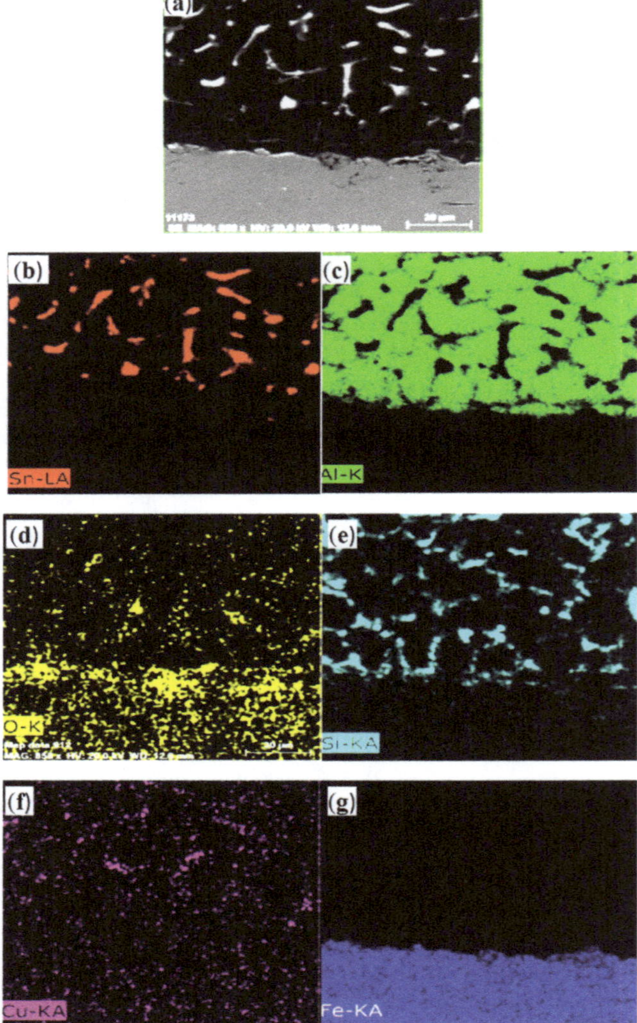


**Figure 7.** SEM image (**a**), and mapping analysis (**b–g**) of bimetallic materials containing a tin:flux ratio of 1:15.

Figure 8 shows the interfacial bonded areas of the three bimetallic specimens with tin-to-flux ratios of 1:5, 1:10 and 1:15. It can be seen that a tin-to-flux ratio of 1:5 has the lowest value, while the ratios of 1:10 and 1:15 have shown comparable results: 1:15 ratio has shown comparatively higher value, which is however not significantly higher than 1:10 ratio. The results of bonded areas immediately suggest that the two ratios of 1:10 and 1:15 are better than 1:5 but their quantitative characterization may further explore the preference of one over other, as performed in the interfacial strength measurement test discussed below.

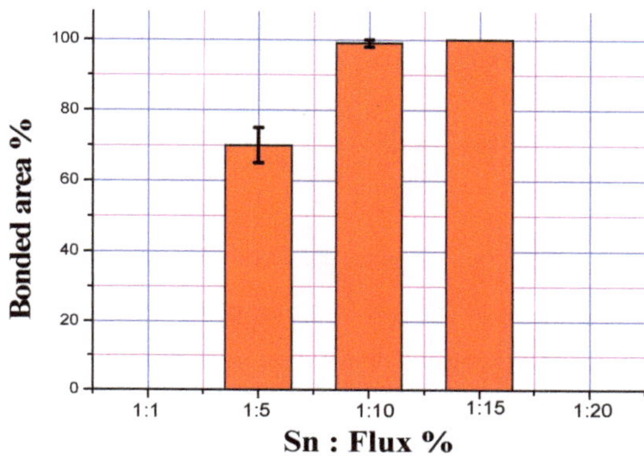


**Figure 8.** Bimetallic interfacial bonded areas in bimetallic specimens as a function of Sn-to-flux ratios.

Figure 9 shows the stress–strain curves of the three bimetallic specimens with different Sn-to-flux ratios of 1:5, 1:10 and 1:15. Matching with the results of bonded surface areas (Figure 8), the Sn-to-flux ratio of 1:5 demonstrates the minimum stress level (5.5 ± 0.16 MPa), while the stress levels of 1:10 and 1:15 are comparable (6.75 ± 0.30 MPa and 6.0 ± 0.12 MPa, respectively) though 1:10 exhibits a slightly better value. However, the strain value of 1:10 is significantly higher (1.26%) than 1:15 (0.76%), which is still lower than the 1:5 ratio (1.0%). The presence of intermetallic phases in 1:5 and 1:15 may be related to low fracture strain in comparison to the absence of intermetallic phases in 1:10 ratio specimen, as discussed above. In a separate study, Sn + 3%Cu was used along with flux, which significantly improved the shear strength up to 59% in comparison to pure Sn, which was related to the improvement of the interfacial bond structure and low tin oxide content [9]. Although the proposed direct tinning process shows a relatively low shear strength (Table 3), it can be considered as a promising new tinning process for steel for its simplicity, low-cost, easily application and capability to improve by using powder of low melting point solder alloys with a variety of reinforcements.

**Table 3.** Reports on the shear property of Al/Fe bimetallic composites for different interlay materials and deposition processes in the literature.

| Interlayer Material | Shear Stress, MPa | | |
|---|---|---|---|
| | Deposition Process | | |
| | Hot Dipping | Electroplating | Direct Tinning |
| Brass | - | 17.5 [21] | - |
| Al-7.2 wt.% Si | 8.5 [22] | - | - |
| Pure Zn | 16.0 [22] | 20.0 [17] | - |
| Pure Sn | - | - | 6.75 [This work] |

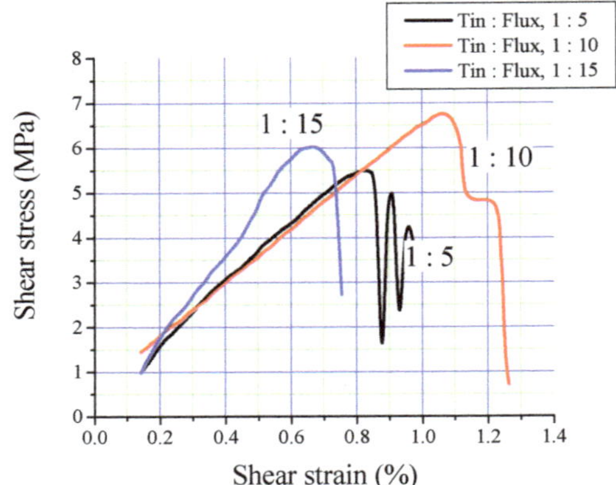


**Figure 9.** Interfacial shear strength values of bimetallic specimens containing different Sn-to-flux ratios.

### 3.2. *Effect of Volume Ratio of Liquid-to-Solid on the Interfacial Bonded Area and Hardness*

After optimizing the tin-to-flux ratio of 1:10 in bimetallic specimens of Al-based bearing alloys and carbon steel substrate, the optimization of the ratio of Al-based bearing alloy and carbon steel substrate was also performed (Figure 10). Figure 4b shows interfacial microstructure of bimetal using liquid-solid volume ratio 6.5:1; the other ratios show same interfacial microstructural morphology except that the thickness of interface layer decreases with increasing liquid-solid volume ratio. Three different ratios of aluminum-to-steel were prepared—i.e., 5:1, 6.5:1 and 8.5:1—and it was found that by increasing the content of molten aluminum, which was deposited on the solid substrate and solidified, the bonded area increased. It is to be mentioned here that, in the three aluminum-to-steel ratios, the tin-to-flux ratio was kept constant after optimization.

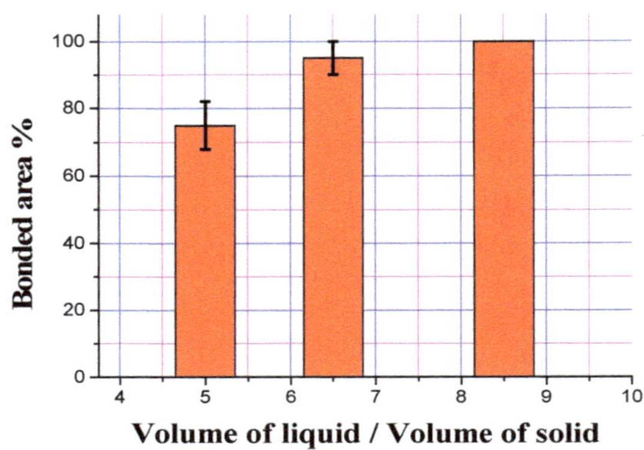


**Figure 10.** Effect of volume ratio of liquid Al alloy to solid steel substrate on bonded interface area of 1:10 Sn: flux tinned steel substrate.

The hardness characteristics of the three bimetallic specimens containing different ratios of aluminum-to-steel—i.e., 5:1, 6.5:1 and 8.5:1—were measured (Figure 11). Special focus was given on the hardness value determination of the interface, as the hardness of the two opposing materials was found the same in the three specimens. The hardness of the specimens increased continuously with the increase in the bonded area owing to increased molten aluminum deposited on the solid steel substrate.

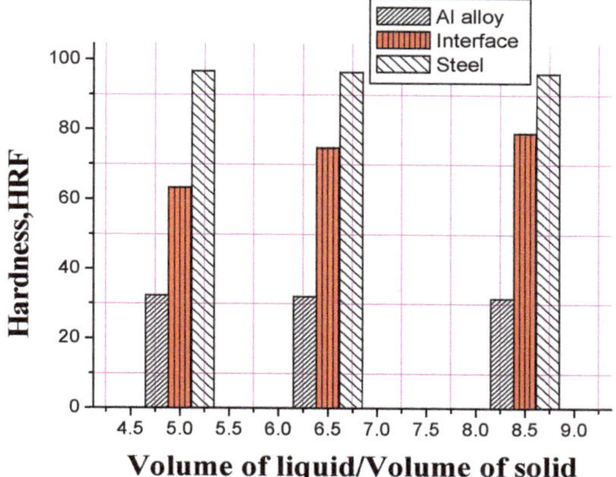


**Figure 11.** Effect of volume of liquid aluminum to solid steel substrate on interfacial hardness of 1:10 Sn: flux tinned steel substrate.

The achieved results show that the optimum interfacial structure of Al-based bearing alloy/carbon steel bimetal can be accomplished using direct tinning process with tin-to-flux ratio of 1:10. This specified ratio resulted in the formation of a continuous and smooth Sn interlayer between Al bearing alloy and low carbon steel in bimetallic specimens. Fluxes are defined as the chemical compounds that improve the bonding properties of a joint when applied uniformly on the surface to be jointed [23,24]. The 1:5 ratio of tin-to-flux resulted in an irregular interface wherein the disconnected interface material is also visible at the interface. Insufficient flux ratio in tin/flux mixture (1:5) led to the decreased wettability of tin interlayer with steel substrate. Decreasing the wettability of tin with steel substrate results in the improper distribution of tin on the surface of the steel substrate that leads to Al-Fe intermetallic formation at the interlayer between poured Al-alloy and the bare surface of the steel substrate. Otherwise, increased ratio of flux in tin/flux mixture (1:15) leads to decreasing percentage of tin on steel substrate forming a very thin tin interlayer on steel substrate. Before pouring liquid metal (Al-bearing alloy), the tinned steel substrate should be preheated using a hot plate. This preheating process usually partially melts the outer surface of the tin interlayer that leads to a thin layer formation of tin oxide before pouring [25]. In the presence of a thin interlayer of tin, as in the case of 1:15 tin-to-flux ratio, most of the tin interlayer is expected to oxidize. This thin tin oxide layer fails upon the impact of liquid metal pouring onto it due to its brittle nature, as well as its poor wettability with the steel substrate. A fraction of this oxide moves through liquid metal while the remaining stays at the steel substrate (Figure 7b,d).

Aside from the structure of interface and its influence on the properties and strength of bimetallic castings, the desired metallurgical bonding and high-quality bimetallic products cannot be achieved in the presence of unbonded regions and gaps at the interface [4]. It is reported that a good bonding of bimetallic specimen, free of unbonded regions, was fabricated in liquid–solid compound casting using relatively higher volume ratios of liquid to the volume of the solid substrate. In the present study, an Al/Fe bimetal with fully bonded area was achieved by increasing the volume of the liquid to the volume of the solid at 8.5:1. Many metallurgical factors influence a full bond while optimizing the ratio of the volume of the liquid to the volume of the solid in bimetallic specimens. The quantity of the two metals, their melting points, pouring temperatures and interlayer materials are critical factors that determine the optimum ratio of the volume of liquid to volume of solid. Xiong et al. [4] reported that a sound interface between high chromium cast iron and medium carbon steel bimetallic specimen was achieved without using the interlayer but using the liquid–solid volume ratios of ≥10:1. Fathy et al. [6] reported that liquid to solid volume ratio should be kept higher than 5:1 to successfully fabricate Babbitt-steel bimetallic composite at low pouring temperature with the assistance of Sn-Pb interlayer. Similarly, Liu et al. [26] found that the effective interfacial bonding could be achieved using a liquid-to-solid volume ratio of 8:1 that mainly depends on the imported heat energy of the poured liquid metal. One of the previous investigations [4] is in good agreement with the above findings that the liquid-to-solid volume ratios of 5:1 and 6.5:1 are not enough to partially remelt the tin + tin oxide layer (that formed during atmosphere preheating) on the surface of tinned solid steel, resulting in the appearance of an unboned interface area. In contrast, the liquid–solid volume ratio of 8.5:1 imported enough heat energy to achieve optimum interfacial bonding.

## 4. Conclusions

Bimetallic specimens of AlSn12Si4Cu1/carbon steel were successfully prepared by compound casting after depositing the molten aluminum alloy on the solid steel substrate. A novel tinning process comprising the tin powder in combination with varying ratios of flux was performed. Moreover, the optimization of the tinning process was performed after finalizing the tin-to-flux ratio for improved bonded area and corresponding rise in hardness and interfacial shear strength values. The tin-to-flux ratio of 1:10 showed the best combination of interfacial structure, bonded area and interfacial shear strength, which is due to the restriction in the formation of Al-Fe and Fe-Al intermetallic phases due to stable and continuous Sn interface layer formation. Finally, the ratio of the content of aluminum-to-steel was optimized at 8.5:1. Considering it a new process, the direct tinning offers encouraging results in interfacial shear strength characterization, though the scope of further improvement still exists. The developed tinning process, therefore, offers a promising route to manufacture bimetallic flat thrust bearings with improved interfacial strength for hydro, gas and steam turbines, coal pulverizing, and defense applications.

**Author Contributions:** Conceptualization, M.R.; methodology, M.R., K.M.H. and K.S.A.H.; validation, M.R., T.S. and K.S.A.H.; investigation, M.R. and K.M.H., resources, T.S. and M.R.; data curation, M.R., K.M.H. and T.S.; writing—original draft preparation, M.R. and T.S.; writing—review and editing, M.R., K.M.H. and A.S.A.; supervision, K.S.A.H. and A.S.A.; project administration, A.S.A. and M.R. All authors have read and agreed to the published version of the manuscript.

**Funding:** This research has been funded by Scientific Research Deanship at University of Ha'il, Saudi Arabia through project number RG-20074.

**Conflicts of Interest:** The authors declare no conflict of interest.

## References

1. Simsir, M.; Kumruoglu, L.C.; Ozer, A. An investigation into stainless-steel/structural-alloy-steel bimetal produced by shell mould casting. *Mater. Des.* **2009**, *30*, 264–270. [CrossRef]
2. Paramsothy, M.; Srikanth, N.; Gupta, M. Solidification processed Mg/Al bimetal macrocomposite: Microstructure and mechanical properties. *J. Alloy. Compd.* **2008**, *461*, 200–208. [CrossRef]
3. Kurt, B.; Calik, A. Interface structure of diffusion bonded duplex stainless steel and medium carbon steel couple. *Mater. Charact.* **2009**, *60*, 1035–1040. [CrossRef]
4. Xiong, B.; Cai, C.; Lu, B. Effect of volume ratio of liquid to solid on the interfacial microstructure and mechanical properties of high chromium cast iron and medium carbon steel bimetal. *J. Alloy. Compd.* **2011**, *509*, 6700–6704. [CrossRef]
5. Aleshin, N.P.; Kobernik, N.V.; Mikheev, R.S.; Vaganov, V.E.; Reshetnyak, V.V.; Aborkin, A.V. Plasma–Powder Application of Antifrictional Babbitt Coatings Modified by Carbon Nanotubes. *Russ. Eng. Res.* **2016**, *36*, 46–52. [CrossRef]
6. Fathy, N.; Ramadan, M. Influence of volume ratio of liquid to solid and low pouring temperature on interface structure of cast Babbitt-steel bimetal composite. *AIP Conf. Proc.* **2018**, *1966*, 020028.
7. Belov, N.A.; Akopyan, T.K.; Gershman, I.; Stolyarova, O.O.; Yakovleva, A.O. Effect of Si and Cu additions on the phase composition, microstructure and properties of Al-Sn alloys. *J. Alloy. Compd.* **2017**, *695*, 2730–2739. [CrossRef]
8. Diouf, P.; Jones, A. Investigation of Bond Strength in Centrifugal Lining of Babbitt on Cast Iron. *Metall. Mater. Trans. A* **2010**, *41*, 603–609. [CrossRef]

9. Ramadan, M.; Ayadi, B.; Rajhi, W.; Alghamdi, A.S. Influence of Tinning Material on Interfacial Microstructures and Mechanical Properties of Al12Sn4Si1Cu/Carbon Steel Bimetallic Castings for Bearing Applications. *Key Eng. Mater.* **2020**, *835*, 108–114. [CrossRef]
10. Soderhjelm, C. Multi-Material Metal Casting: Metallurgically Bonding Aluminum to Ferrous Inserts. Ph.D. Thesis, Worcester Polytechnic Institute, Worcester, MA, USA, April 2017.
11. Mola, R.; Bucki, T.; Dzik, M.G. The Effect of a Zinc Interlayer on the Microstructure and mechanical Properties of a Magnesium Alloy (AZ31)–Aluminum Alloy (6060) Joint Produced by Liquid–Solid Compound Casting. *JOM* **2019**, *71*, 2078–2086. [CrossRef]
12. Gawronski, J.; Szajnar, J.; Wróbel, P. Study on theoretical bases of receiving composite alloy layers on surface of cast steel castings. *J. Mater. Process. Technol.* **2004**, *157*, 679–682. [CrossRef]
13. Cholewa, M.; Wróbel, T.; Tenerowicz, S. Bimetallic layer castings. *J. Achiev. Mater. Manuf. Eng.* **2010**, *43*, 385–391.
14. Wrobel, T. Characterization of Bimetallic Castings with an Austenitic Working Surface Layer and an Unalloyed Cast Steel Base. *J. Mater. Eng. Perform.* **2014**, *23*, 1711–1717. [CrossRef]
15. Liaw, P.K.; Gungor, M.N.; Logsdon, W.A.; Ijiri, Y.; Taszarek, B.J.; Fröhlich, S. Effect of Phase Morphologies on the Mechanical Properties of Babbitt-Bronze Composite Interfaces. *Metall. Trans. A* **1990**, *21*, 529–538. [CrossRef]
16. Ramadan, M.; Alghamdi, A.S.; Subhani, T.; Halim, K.S.A. Fabrication and Characterization of Sn-Based Babbitt Alloy Nanocomposite Reinforced with $Al_2O_3$ Nanoparticles/Carbon Steel Bimetallic Material. *Materials* **2020**, *13*, 2759. [CrossRef] [PubMed]
17. Shin, J.; Kim, T.; Lim, K.; Cho, H.; Yang, D.; Jeong, C.; Yi, S. Effects of steel type and sandblasting pretreatment on the solid-liquid compound casting characteristics of zinc-coated steel/aluminum bimetals. *J. Alloy. Compd.* **2020**, *778*, 170–185. [CrossRef]
18. Mwanza, M.C.; Joyce, M.R.; Lee, K.K.; Syngellakis, S.; Reed, P.A.S. Microstructural characterisation of fatigue crack initiation in Al-based plain bearing alloys. *Int. J. Fatigue* **2003**, *25*, 1135–1145. [CrossRef]
19. Jiang, W.; Fana, Z.; Li, C. Improved steel/aluminum bonding in bimetallic castings by a compound casting process. *J. Mater. Process. Technol.* **2015**, *226*, 25–31. [CrossRef]
20. Jiang, W.; Fan, Z.; Li, G.; Liu, X.; Liu, F. Effects of hot dipping galvanizing and aluminizing on interfacial microstructures and mechanical properties of aluminum/iron bimetallic composites. *J. Alloy. Compd.* **2016**, *25*, 742–751. [CrossRef]
21. Raja, V.; Kavitha, M.; Chokkalingam, B.; Ashraya, T.S. Effect of Interlayers on Mechanical Properties of Aluminium Casting over Stainless Steel Pipe for Heat Exchanger Applications. *Trans. Indian Inst. Met.* **2020**, *73*, 1555–1560. [CrossRef]
22. Guo, Z.; Liu, M.; Bian, X.; Liu, M.; Li, J. An Al–7Si alloy/cast iron bimetallic composite with super-high shear strength. *J. Mater. Res. Technol.* **2019**, *8*, 3126–3136. [CrossRef]
23. Afolalu, S.A.; Akinlabi, S.A.; Ongbali, S.O.; Abioye, A.A. Investigation of morphology characterization andmechanical properties of nano-flux (MnO) powder for TIG welding. *Int. J. Mech. Prod. Eng. Res. Dev.* **2019**, *9*, 887–898.
24. Laurent, S.; Forge, D.; Port, M.; Roch, A.; Obic, C.; Vander, E.L. Magnetic iron oxide nanoparticles: Synthesis, stabilization, vectorization, physicochemical characterizations, andbiological applications. *Chem. Rev.* **2010**, *108*, 2064–2110. [CrossRef] [PubMed]
25. Cho, S.; Yu, J.; Kang, S.K.; Shih, D.-Y. Oxidation Study of Pure Tin and Its Alloys via Electrochemical Reduction Analysis. *J. Electron. Mater.* **2005**, *34*, 635–642. [CrossRef]
26. Liu, Y.H.; Liu, H.F.; Yu, S.R.; Guo, G.C. Special Cast. *Nonferr. Alloys* **2001**, *2*, 17–19. (In Chinese)

*Article*

# Non-Equilibrium Crystallization of Monotectic Zn-25%Bi Alloy under 600 *g*

Grzegorz Boczkal [1,*], Pawel Palka [1], Piotr Kokosz [1], Sonia Boczkal [2] and Grazyna Mrowka-Nowotnik [3]

1 Faculty of Non-Ferrous Metals, AGH University of Science and Technology, 30-059 Krakow, Poland; pawel.palka@agh.edu.pl (P.P.); piotrek.kokosz@gmail.com (P.K.)
2 Łukasiewicz Research Network—Institute of Non-Ferrous Metals, Division in Skawina, ul. Pilsudskiego 19, 32-050 Skawina, Poland; sboczkal@imn.skawina.pl
3 Department of Material Science, Rzeszów University of Technology, Al. Powstańców Warszawy 12, 35-959 Rzeszów, Poland; mrowka@prz.edu.pl
* Correspondence: gboczkal@agh.edu.pl; Tel.: +48-126172697

**Abstract:** This study investigated the influence of supergravity on the segregation of components in the Zn–Bi monotectic system and consequently, the creation of an interface of the separation zone of both phases. The observation showed that near the separation boundary, in a very narrow area of the order of several hundred microns, all types of structures characteristic for the concentration range from 0 to 100% bismuth occurred. An additional effect of crystallization in high gravity is a high degree of structural order and an almost perfectly flat separation boundary. This is the case for both the zinc-rich zone and the bismuth-rich zone. Texture analysis revealed the existence of two privileged orientations in the zinc zone. Gravitational segregation also resulted in a strong rearrangement of the heavier bismuth to the outer end of the sample, leaving only very fine precipitates in the zinc region. For comparison, the results obtained for the crystallization under normal gravity are given. The effect of high orderliness of the structure was then absent. Despite segregation, a significant part of bismuth remained in the form of precipitates in the zinc matrix, and the separation border was shaped like a lens. The described method can be used for the production of massive bimaterials with a directed orientation of both components and a flat interface between them, such as thermo-generator elements or bimetallic electric cell parts, where the parameters (thickness) of the junction can be precisely defined at the manufacturing stage.

**Keywords:** supergravity crystallization; gravitational segregation; texture; hexagonal alloys; monotectic transformation

**Citation:** Boczkal, G.; Palka, P.; Kokosz, P.; Boczkal, S.; Mrowka-Nowotnik, G. Non-Equilibrium Crystallization of Monotectic Zn-25%Bi Alloy under 600 *g*. *Materials* **2021**, *14*, 4341. https://doi.org/10.3390/ma14154341

Academic Editor: Tomasz Wróbel and Daolun Chen

Received: 29 June 2021
Accepted: 30 July 2021
Published: 3 August 2021



## 1. Introduction

Metal crystallization is a well-known, natural process that has been modelled under equilibrium conditions. However, moving away from equilibrium conditions leads to interesting effects, such as inhomogeneous chemical composition, the formation of metastable phases and the displacement of phase transition points. The research carried out by other authors so far includes the study of the effect of the crystallization rate in microgravity conditions on the morphology of the precipitates [1] as well as microstructure refinement in the magnetic field [2].

Hence, the idea to investigate the alloys previously tested under these conditions in supergravitation conditions has appeared. The presented experiment is of a cognitive nature. The processes taking place during the gravitationally oriented crystallization have a universal character and can also be applied to other alloys. The obtained knowledge will allow a broader understanding of the mechanisms responsible for a forming of the alloy microstructure in non-equilibrium conditions.

The direct transfer of the results of the experiment to the real technology requires further work, although a similar method is used in the centrifugal casting process. It should also be mentioned that the conditions described in the experiment, which are extreme on Earth habitats, might be normal on other places in space. The current technological expansion and solar system exploration require new information about material responses under various conditions, from microgravity in Earth orbit to overload effects, which are given elements of spacecraft.

The overload used in the experiment, resulting from rotational motion, is the equivalent of the gravitational force of gravity multiplied by 600×. One of the research studies concerning the influence of centrifugal force on the crystallized structure of alloy was carried out by Lőffler et al. and its results were published in 2002–2004 [3,4]. In the experiment, two magnesium (Mg)-based alloys and one aluminum (Al)-based alloy were melted and simultaneously rotated. The inertial acceleration was 60,000 $g$, where $g$ is gravitational acceleration. As a result of these unique conditions, the researchers obtained samples where the gravitational segregation of phases could be observed. The primary phases sedimented at the end of the samples, while binary and ternary eutectics were solidified in the central regions of the samples. The authors of the experiment hoped that their research would help to discover a new type of bulk metallic glasses. Centrifugal force can also be used to purify metal. Ying et al. [5] carried out research on a lead (Pb) alloy with 3% the weight of copper (Cu). The researchers managed to get 1000 $g$ which caused significant segregation of Cu in the upper part of the tested sample. The volume fraction of Cu changed from 15.57% in the top of the sample to 0% in the bottom. Another attempt to purify metal using supergravity was taken up by Lixin Zhao et al [6]. In this research, industrial Al was purified by gravitational segregation. The scientists obtained 400 $g$ and the concentration of iron (Fe) changed from 3.5 to 1.5 weight percent between the bottom and top of a sample, which was 1.5 cm long. The same group of scientists suggested that supergravity could be used for grain refining [7]. In the experiment, scientists crystallized industrial Al with different gravity coefficients (ratio of super-gravitational acceleration to normal gravitational acceleration). It was observed that the average grain size decreased exponentially and after reaching 250 $g$, the average grain size was constant.

Phase separation during the sedimentation process in Cu–Tin (Sn) alloy under supergravity was investigated in 2019 by Wierzba et al. [8]. Centrifugal casting is used in industry; the main application is the improvement of the filling of molds [9,10]. It is also used in the production of metallic glasses [11]. The described method engages changes in the parameters of nucleation kinetics caused by high overloads. By crystallizing a material under high overload, one can potentially control its structure. Partial amorphization mentioned in the paper [11], is a consequence of a decrease in thickness of the formed lamellar eutectic phases. At sufficiently small thickness of eutectic lamellas, certain chemical composition and appropriate cooling rate, the conditions for partial amorphizations in the microstructure can be fullfiled. On the other hand, after changing the parameters, it is possible to obtain highly ordered structures of quasi-crystalline character.

There are reports of the use of this method to produce highly ordered structures, not only based on metals [12]. As a consequence of the above considerations, the Zinc (Zn)-Bismuth (Bi) alloy was selected for research due to the simple equilibrium system and the low solidus temperature. It has also not been previously investigated in non-equilibrium crystallization conditions, especially in supergravity crystallization conditions.

### *1.1. Investigations of the Zn–Bi Alloys*

This article focuses on the effect of centrifugal force on the crystallization process of Zn-25%Bi alloy. Low mutual solubility of both metals should generate a microstructure with regular precipitates, convenient for analyses. The proportions of the components in the alloy were selected on the assumption that strong gravitational segregation would lead to the generation of a wide spectrum of microstructures in the material, characteristic for various compositions of the alloy.

Previous results obtained by the authors for other zinc alloys showed a tendency to grow a preferred orientation along $\langle 11\bar{2}0 \rangle$ direction during zone crystallization [13,14]. It was a concept that similar effect could occur in the supergravity solidification of hexagonal alloys.

Different aspects of Zn–Bi alloys were investigated before and described in literature.

One of the research concepts was crystallization under microgravity. In zinc-rich content alloys, Bi precipitates have had a drop shape and their sizes were larger than from diffusion-controlled growth. A concentration of more than 24% Bi in samples congregate in large (Bi) areas in the center of the sample and generate a (Bi) halo-layer around the sample [15].

Different aspects of Zn–Bi alloy research were presented in work [16]. The authors applied the squeeze casting method for the production of the alloy chips. Crystallization proceeds under pressure, 120 MPa, until complete solidification. The obtained material showed increases in density, tensile strength and hardness. Their electrical resistivity decreased [16].

Moreover, bismuth as a metal is a potential substitute for lead in low-melting solders and was described in a phase control method in the Zn–Bi alloy by tungsten nanoparticles [17]. The effect was refined for the microstructure and to avoid sedimentation.

Zinc–bismuth alloys are also interesting for their bactericidal applications [18]. This is allowed for improvement of typical anticorrosion zinc covers with bactericidal and antifungal effect.

These facts motivated the authors to expand their knowledge of this interesting alloy to include the effect of supergravity on crystallization. This aspect has not been previously studied for monotectic systems.

### *1.2. Zn–Bi System Characteristic*

The Zn–Bi phase system is shown in Figure 1. A characteristic feature of the system is the monotectic presence in the tested range of composition. Figure 1 shows the microstructure of various Zn–Bi alloys obtained under the conditions of equilibrium crystallization.

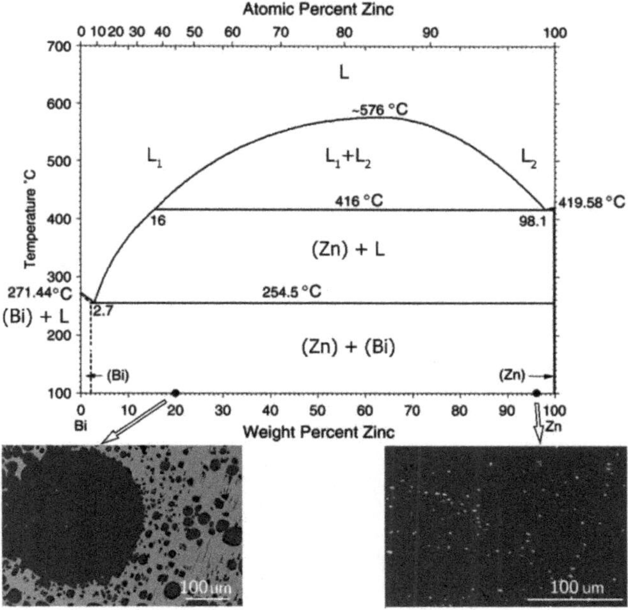


**Figure 1.** Bi–Zn phase diagram [19–21] and microstructures observed at different chemical compositions of investigated alloys.

The alloy components differ in density, i.e., ρBi = 9.79 g/cm$^3$, while ρZn = 7.14 g/cm$^3$. This promotes the occurrence of the effect of gravitational segregation. A significant difference in the melting point of the components was reflected in the surface tension of the liquid: at 426 °C (419 °C is the melting temperature of zinc), its values are γZn = 810 mJ/m$^2$ for Zn [22] and γBi = 360 mJ/m$^2$ for Bi [23].

Microstructures obtained with different zinc contents are shown in Figure 1. At 20 wt% of Zn, spherical zinc phase precipitates were noted to occur in a (Bi) matrix. A characteristic feature was the large scatter of the values of the diameter of individual precipitates. Primary precipitates with diameters in the order of 200–300 microns and small precipitates with dimensions of several microns were directly adjacent to each other. It was the result of crystallization from the monotectic phase with a large difference in the surface tension of both liquids. Zinc was crystallized first and had a high surface tension, allowing for the formation of large spherical precipitates during primary crystallization. Additionally, minor secondary precipitates of Zn-rich needle-shaped phases were visible.

The second type of structure presented on the right side of Figure 1 contains 4 wt% Bi. The observed microstructure consisted of (Zn) matrix and small, spherical bismuth particles of similar size. The monotectic transformation occurring in the Zn–Bi system favors the spherical shape of the precipitates regardless of their chemical composition [10,24].

## 2. Materials and Methods

Tests were carried out on a test stand designed and constructed for the purpose of this study. A furnace was made in a ring configuration, as is shown in Figure 2.

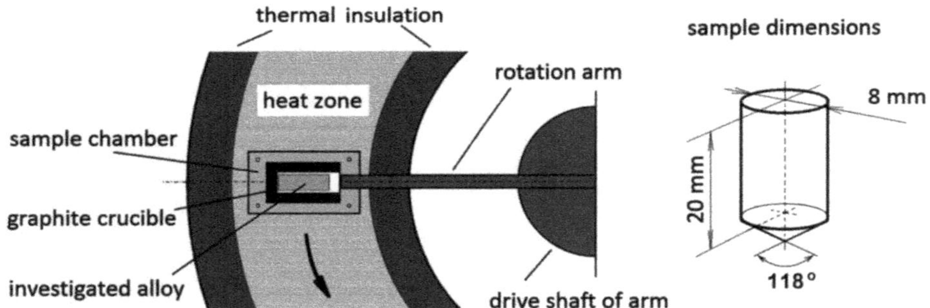


**Figure 2.** Scheme of the ring furnace and sample geometry. One experiment needs two identical samples in two chambers on both tips of arm.

The furnace with heating elements is stationary, only the symmetric arm with two chambers on both sides rotates. The arm system is balanced and has bearings on both sides. The obtained parameters, such as supergravity and temperature were limited mainly by the strength of the material used for the arm and crucible. At a temperature of 500 °C and a machine operation cycle of about 3 h, the material of the arm is exposed to creep processes and therefore, the arm has been made of high alloy steel. The crucible material also imposed some constraints. The tests were carried out using graphite chambers, but it became necessary to use disposable elements made in the form of a monolith with a cylindrical chamber of 8 mm diameter and a length of up to 20 mm.

The motor used in the device had a nominal rotational speed of 1350 RPM and therefore, the samples were rotating with frequency of 1350/60 = 22.5 Hz. The correlation between frequency and angular velocity is given by the formula:

$$\omega = 2\pi f \tag{1}$$

where: $f$—frequency s$^{-1}$.

It can be calculated that the angular velocity of the sample was: $\omega = 141.31\ \frac{rad}{s}$.

The radius of the circle in which the samples were moving was 300 mm. The radius was measured up to a central point of the sample, at a half-length. Using a simple formula for angular acceleration, it was possible to calculate it as equal to 5995 $m/s^2$. The measure of the load exerted on the sample will be expressed in the angular acceleration divided by gravitational acceleration of 9.81 $m/s^2$. The normal acceleration was 611.15 $g$. It was assumed that the arms had an angular velocity equal to the nominal speed of the motor used and the normal acceleration can be estimated at 600 $g$ for the whole length of the sample. Due to the finite length of the sample, a slight overload gradient will be present along the length of the sample, hence the efforts to obtain the maximum long arm to reduce this gradient.

To prepare samples, measured amounts of the alloy were cast into the crucibles, followed by the balancing of the system. In the first step of the experiment, the electric motor was turned on and the arm with samples started to rotate. When the rotation speed (overload) was stabilized, the furnace was turned on. The furnace was heated up to 500 °C, which took 30 min. The samples were held at 500 °C for 30 min to make sure that they melted down completely, then the furnace was turned off and the samples cooled down with the furnace to 200 °C. Cooling the furnace to 200 °C took another 30 min and after that time, the rotation of the samples was stopped and they were taken out of the device and further cooled in the air.

The samples thus obtained were cut lengthwise. The cutting was done using a Struers Secotom-10 circular saw with a corundum disc. Cutting parameters were: disc rotation—2000 rpm, movement—0.05 mm/s, with cooling by lubricant mixture. The sample surface was then polished with diamond pastes and finally, with 0.04 microns graded alumina suspension.

Structural examinations were carried out using a Hitachi S-3400N scanning electron microscope (SEM, Hitachi High-Technologies Corporation, Tokyo, Japan) with an EDS Thermo Noran System Six chemical composition analyzer in microregions and an Inspect F50 with EBSD TSL detector (Edax Ametek, Tokyo, Japan). The EDS line scan step resolution was 68 micrometers and the measurement was performed at the accelerating voltage 20 kV with the acquisition time of 60 s per each measurement point (50 measurement points in line). Another SEM (Hitachi SU-70) equipped with wavelength dispersive spectrometer (WDS) and Ultra Dry by Thermo Noran was used to analyze fine Bi precipitates.

Hardness measurements were carried out by the Vickers method with a load of 0.19807 N, using a Innovatest Nova 240 microhardness tester (Innovatest, Maastricht, The Netherlands).

In total, 15 indentations were made in the Bi-rich zone and 30 indentations in the Zn-rich zone.

The phase composition studies were carried out using the Bruker D8 Advanced/Discover device, equipped with a copper tube. For improving the monochromatic radiation of wavelength $\lambda CuK\alpha 1$ = 1.5406 Å, the Ni filter was used. The analysis of the phase composition of the tested materials was carried out using the PDF-2 crystallographic basis. The measurement was made with a scan step size of 0.02°. The following cards were used to analyze the phase composition: Zn: 00-004-0831 and Bi: 00-044-1246.

Additionally, the same diffractometer was used to determine of pole figures from the planes (0002) and $(10\overline{1}0)$ for zinc and $(01\overline{1}2)$, $(20\overline{2}2)$ for bismuth. The measurement was performed by averaging the results of a scan of the longitudinal section of the sample from a length of 12 mm, thereby obtaining mean value of macro texture. Orientation distribution function (ODF) was determined using M-TEX software in the Matlab environment.

## 3. Results

### 3.1. Microstructures Analysis

Figure 3 presents the Zn-25Bi alloy cast to mold at room temperature and crystallized under 1 $g$ normal gravity.

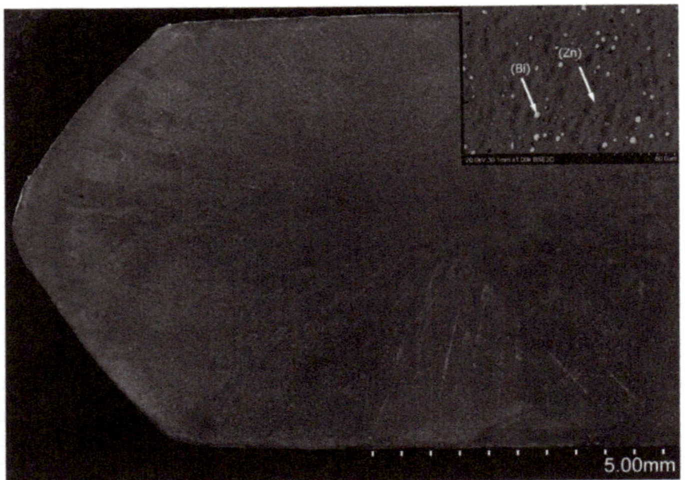


**Figure 3.** Microstructure of the Zn-25 wt% Bi alloy on longitudinal section after casting into the cold mold, with magnification of the representative microstructure containing (Bi) particles.

Casting molds were made from cold graphite and were the same as during the supergravity test. Cooling processes were done in air. The cross-sectional microstructure was typical of a volume crystallizing casting. Crystallization in accordance with the directions of heat dissipation were visible. Bismuth was evenly distributed in a sample volume. Gravity segregation was not observed.

Figure 4 presents the alloy casted and crystallized under 1 *g* normal gravity, then cooled with the furnace. Long time exposition in the liquid state caused natural gravity segregation of components. A bismuth-rich and zinc-rich areas are visible in the microstructure. The separation line between them is a curve that represents a shape of a crystallization front.

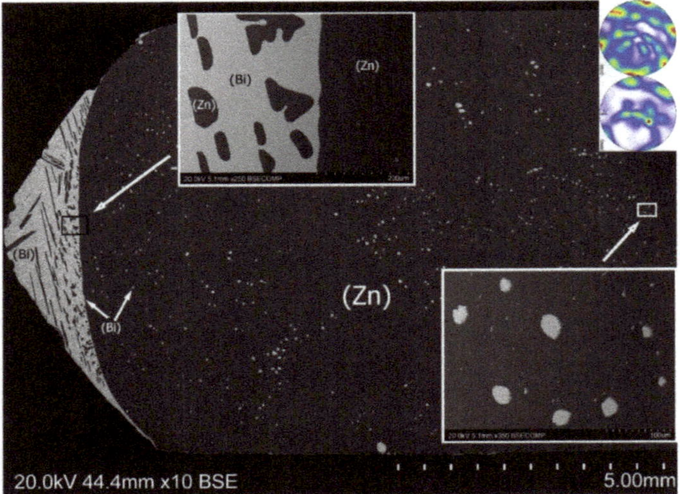


**Figure 4.** Microstructure of the Zn-25 wt% Bi alloy on longitudinal section, cooled with the furnace at normal gravity 1 *g*. Microstructures in two characteristic places are shown in higher magnification. Macro-textures from (Zn) phase are included in the right upper corner.

The radius of the boundary can be interpreted as a characteristic parameter in relation to the cooling rate and temperature field.

Two observed zones represent individual microstructures. In the zinc-rich area, many spheroidal particles of almost pure bismuth can be observed. These are the products of the monotectic transformation taking place at the temperature of 416 °C under the conditions of thermodynamic equilibrium. Their different sizes and inhomogeneous distribution are a consequence of a coagulation effect during slow crystallization [24]. The larger surface tension of zinc, then bismuth at the same temperature and the almost lack of mutual solubility of both elements generated the observed shapes of Bi particles.

The opposite effect took place for zinc precipitates into the (Bi) matrix. Zinc has higher melting point than bismuth, thus the embryo of (Zn) solid phase can grow as a primary precipitate. The growth orientation was limited by heat transfer directions. Moreover, the considerably larger size of the zinc precipitates was caused by a relatively long time of diffusion.

At the border of both zones, from the side of the zone rich in zinc, one can observe a strip devoid of bismuth precipitates. This effect is analogous to the halo effect observed around weakly coherent precipitates of slowly crystallized alloys. The same zone also exists for an alloy crystallized in supergravity conditions.

At the interface, from the bismuth-rich side, there are few irregular zinc precipitates with oval shapes.

Figure 5 shows a longitudinal section of the Zn-25 wt% Bi alloy sample crystallized under 600 *g* supergravity. Strong segregation has occurred and as a result, most of the bismuth was located in the left part of the sample (the bottom edge of the drawing is parallel to the centrifugal force direction). The segregation boundary between (Zn) and (Bi) zones is flat. This is a specific area with different chemical composition, changed along 1 mm distance. Both sides of this line concern alloy compositions from the opposite sides of the phase diagram. Non-equilibrium crystallization generates into this zone all possible morphologies for the Zn–Bi system.

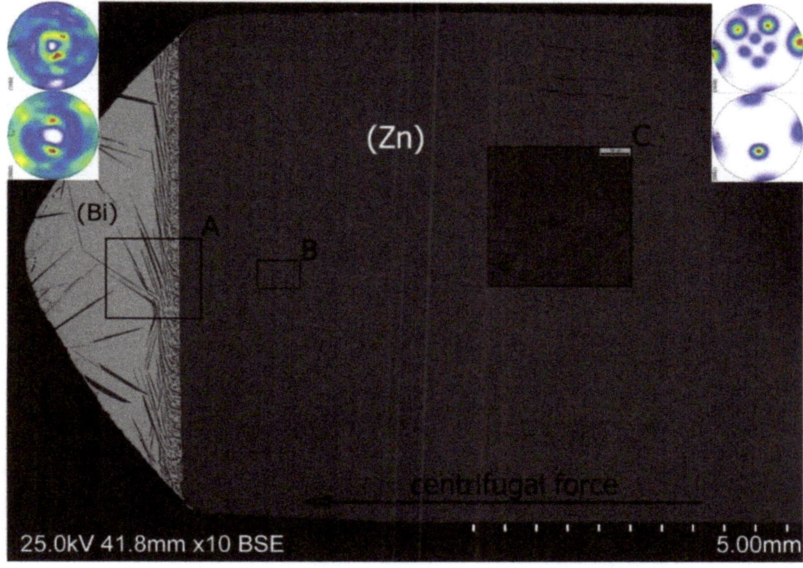


**Figure 5.** Microstructure of the Zn-25 wt% Bi alloy on longitudinal section along centrifugal force direction (600 *g*). Marked areas (A, B, C) in the figure are presented in a bigger magnification in the next figures. Macro-textures from: (Zn) phase are included in the right upper corner and (Bi) phase are included in the left upper corner.

To examine the mutual distribution of Zn and Bi, an EDS linear analysis of the chemical composition of the obtained structure was carried out along the centrifugal force direction as shown in Figure 6.

The microhardness of regions rich in zinc and bismuth was measured. In the bismuth-rich zone, the average microhardness HV0.01 was equal to 17.6 with standard deviation of 4.9, while in the zinc-rich zone, the microhardness was 58.5 with standard deviation of 5.1.

The analysis of the macrostructure showed the existence of a few characteristic zones, marked A, B, C in Figure 5.

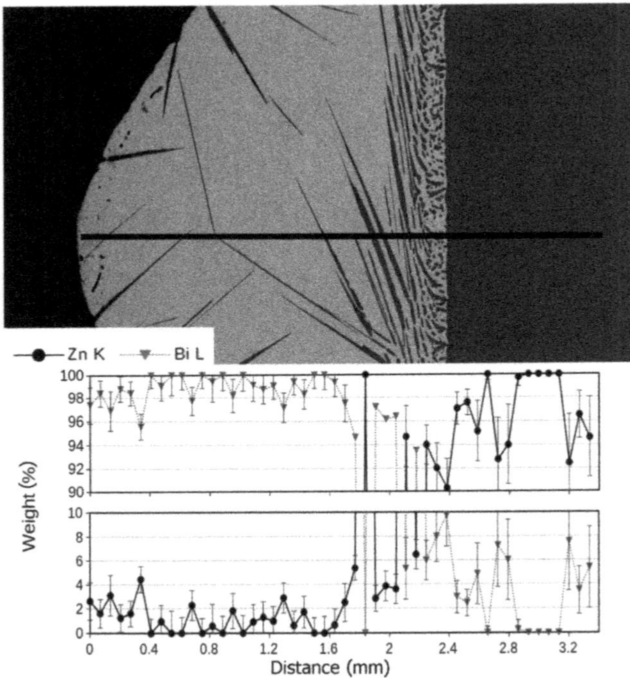


**Figure 6.** Linear EDS analysis of Bi and Zn content at the tip of the sample. Supergravity crystallization introduced a strong segregation effect.

Moving from the left tip of the sample in Figure 5, one can see large needle-shaped precipitates formed as a result of the eutectic reaction; their eutectic point is located at 2.7 wt% of Zn and at temperature of 254.5 °C (Figure 1). The nucleation sites of the needles of a hypoeutectic primary phase located on the outer edges of the sample are very interesting. Their presence indicates heterogeneous nucleation on the walls of the crucible, while large dimensions of the particles suggest their growth from the liquid as a primary phase. Moving right towards the interface, an increase in the amount of zinc is visible (Figure 6).

Close to the sedimentation area, smaller and more densely distributed precipitates appeared, characteristic of structures with a chemical composition lying near the eutectic point (Figure 7, near separation zone). At the same time, small needle-shaped precipitates fill the spaces between larger crystallites. This is shown in Figure 7a. This type of structure can be observed in a relatively narrow range at the interface. The reason is to be searched in the specific character of the Zn–Bi system. In this system, the solidus line runs practically horizontally and only from the side of high bismuth contents does its temperature exceed 254.5 °C. This allows the liquid phase to maintain its presence up to this temperature, regardless of the chemical composition which, in turn, promotes the diffusion process and maintains the crystallization kinetics at a constant level throughout the entire length of the sample.

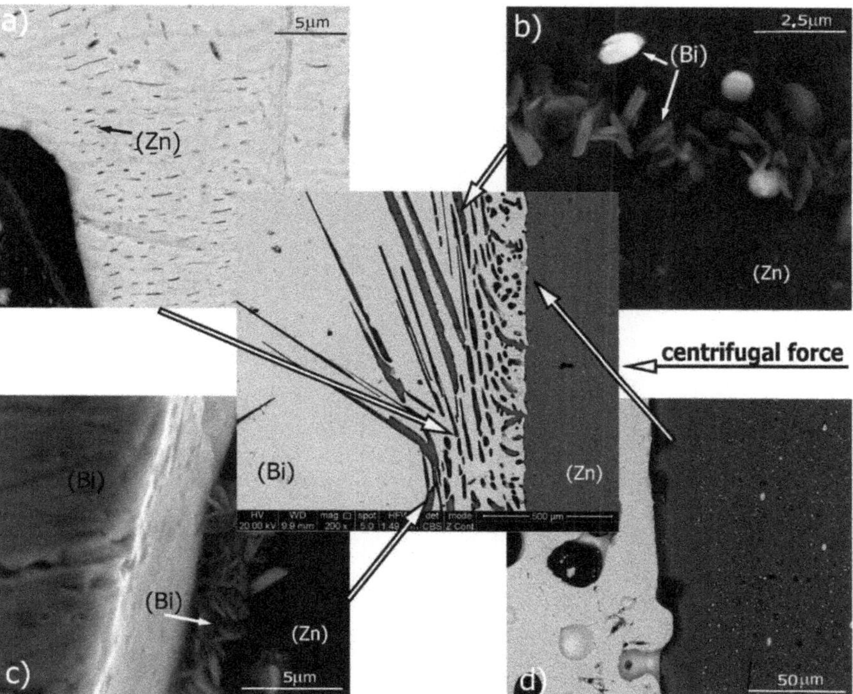


**Figure 7.** Interface between Bi-rich zone and Zn-rich zone (A-area in Figure 5). A few characteristic features can be observed: (**a**) very narrow zinc needles in Bi-rich zone, ordered parallel to G-force vector; (**b**,**c**) tiny lamellas of bismuth eutectic precipitates; and (**d**) a bismuth-free area on the Bi/Zn boundary.

At the bismuth/zinc interface, small zinc-rich plate-shaped precipitates were observed (Figure 7b,c). The EDS analysis of chemical composition showed a zinc content of more than 70 wt%, but the small size of the particles and their geometry did not allow for more accurate determination. This indicates that a second reaction occurred at the monotectic point of 98.1 wt% Zn/416 °C. Columnar Zn crystals separated by higher Bi concentration regions could be imprints of the monotectic reaction.

The chemical composition of the characteristic points in the Zn–Bi phase diagram was given on the basis of data from the equilibrium crystallizing system. In fact, the position of this point can be significantly shifted by the force of gravity. Moreover, along G-force direction, the density of alloy changed. It is a gravity segregation effect. The density change along the sample length occurred in a non-linear way, with a narrow zone on the border of the zinc and bismuth-rich areas, where the change has gradient character. Literature reports show that significant shifts of transition points on systems crystallizing under conditions of high pressure, for example, are possible [25–28].

The next detected area is shown in Figure 7d. This zone contains pure zinc and is practically devoid of bismuth. The width of the zone ranges from 10 to 20 microns. Its presence is due to the precipitation of primary phases from the solution taking place at the interface. The result is an effect similar to the "halo" phenomenon observed around the precipitates of the primary phases. It involves total removal of one of the alloy components from the area surrounding the precipitate [29].

Moving to the right along the sample axis reveals the presence of a large amount of fine spheroidal precipitates of pure bismuth (zone B marked in Figure 5 and the left side in Figure 7d). The magnification are shown in Figure 8, which is a WDS elements distribution map.

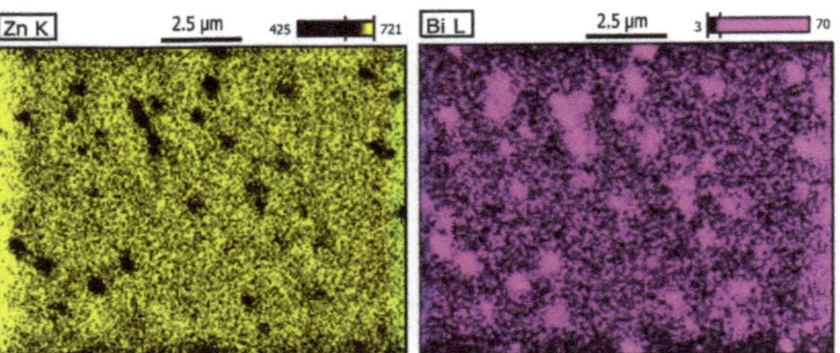


**Figure 8.** Wavelength dispersion spectroscopy (WDS) analysis in the Zn-rich area. In the zinc matrix, many dispersed (Bi) particles can be observed.

The presence of these precipitates indicates the solubility of a small amount of bismuth in solid phase of zinc. Closer analysis of the Bi–Zn phase equilibrium system indicates such a possibility and additionally, the examined crystallization process took place under conditions far from the state of equilibrium. After crystallization, while the alloy was cooling, the secondary precipitation occurred, but since it occurred in the solid phase, the centrifugal force was not able to displace the precipitated particles, hence their even distribution in this part of the sample.

### 3.2. Crystallographic and Texture Analysis

The crystallographic structures of zinc and bismuth are shown in Figure 9. Bismuth crystallizes in the rhombohedral structure with R$\overline{3}$m (166) space group. The rhombohedral lattice parameter $\alpha_{rho}$ and the rhombohedral angle $\alpha$ are shown in Figure 9a.

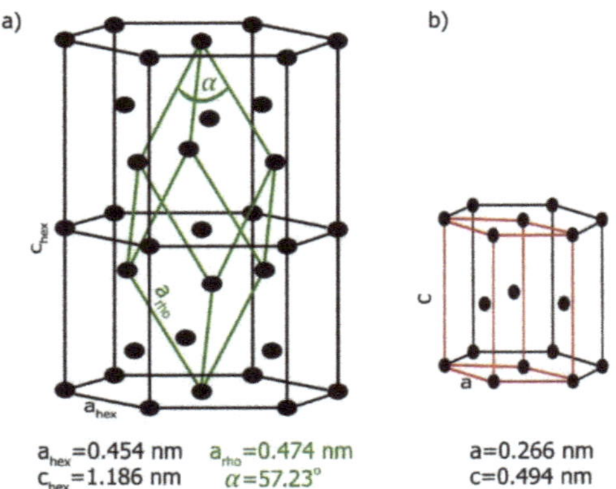


**Figure 9.** (**a**) Crystal structure of bismuth at 20 °C (rhombohedral (green) unit cell is marked into the black hexagonal lattice) and (**b**) zinc hexagonal unit cell at 20 °C (the elemental cell is marked by the red line) [30].

For the undistorted reference lattice, parameters used from PDF-2 were: $\alpha_{rho} = 0.475$ nm and $\alpha = 57.230°$. Equivalently, the structure can be described as hexagonal with in-plane and out-of-plane lattice constants $\alpha_{hex} = 0.454$ nm and $c_{hex} = 1.1862$ nm, where $c/a = 2.6093$. These lattice parameters were measured at 20 °C. Zinc crystallizes in hexagonal close pack structure with P63/mmc (194) space group.

Additionally, there was a significant difference in lattice constants. According to Hume-Rothery rules, this excludes the possibility of the formation of solid solutions, as reflected in the appearance of the Bi–Zn phase equilibrium diagram, where solid solubility practically did not exist.

For a thorough characteristics of the received structure, the XRD diffraction of regions rich in zinc and bismuth was carried out. The results are presented in Figure 10. In both regions of the studied specimen, only peaks characteristic for zinc and bismuth were registered. The high, dominating intensities of (002) and (101) peaks for zinc and (012) and (014) for bismuth indicate high texture of the material and appearance of dominating crystallographic orientations in both zones. The level of background in both measurements (for both parts of the sample) was homogeneous and there were no premises to discuss formation of an amorphous phase.

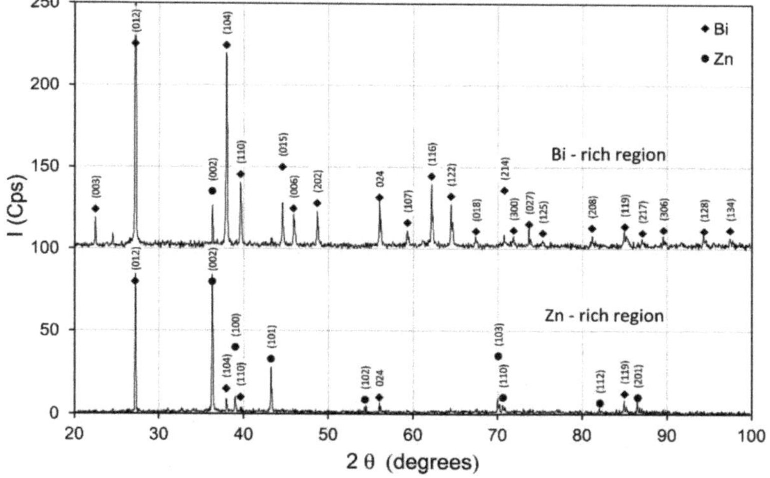


**Figure 10.** X-ray diffractograms of the Bi- and Zn-rich regions under 600 *g*.

Regardless of the homogeneity of phases and lack of complex phases in the studied regions, their microstructure is not uniform.

Regular needle-shaped structures can be observed in areas B and C of Figure 5. That region in a higher magnification is presented in Figure 11 These needles are a result of twin deformations from stress relaxation during cooling and a simultaneous decrease in centrifugal force. Zinc as a metal with the hexagonal structure has a strong tendency of mechanical twinning [31,32]. This is indirectly caused by a small number of easy slip—only three directions $\langle 11\bar{2}0 \rangle$ are lying in the (0001) base plane. Twinning structures were observed only in the right part of the sample in a rich zinc area. They were not found in an area rich in bismuth. One preferred orientation observed in Figure 11 suggests the existence of a highly ordered structure orientation.

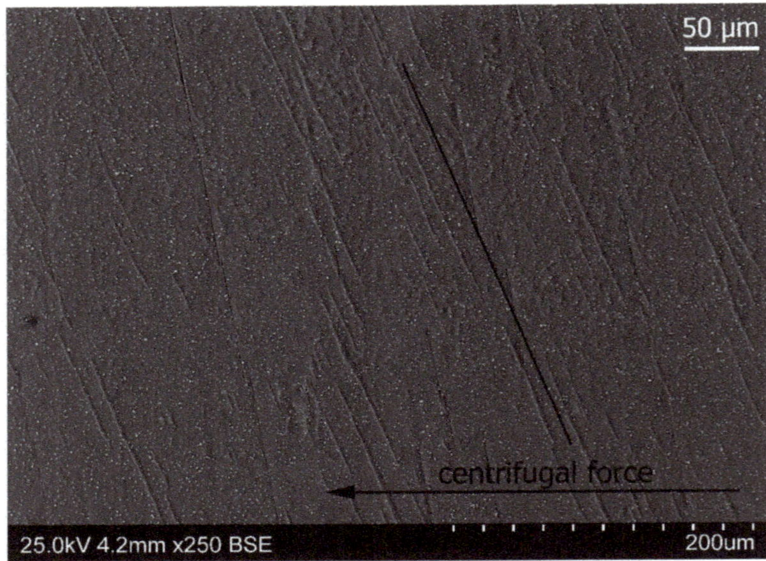


**Figure 11.** The area of needle-shaped structures in the Zn-rich zone of the sample, probably accommodation twins occurring in the central part of the sample (zones B and C, Figure 5). One preferred needle orientation suggests the existence of a highly oriented structure orientation.

To confirm this concept, an EBSD analysis was performed for area C in Figure 5. The result is shown in Figure 12 in contrasting colors by overlapping of inverse pole figure (IPF) and band contrast (BC) images. A misorientation analysis (Figure 12) was carried out, which showed the existence of areas with a mutual angle of disorientation of about 86 degrees. This angle is characteristic for twinning in zinc [33] for identification of the ordered coniferous structures observed in the upper left corner of Figure 12. Their macromorphology is similar to those observed in Figure 11. Closer crystallographic analysis of these structures requires the use of high-resolution microscopy, which is planned in the next stages of research.

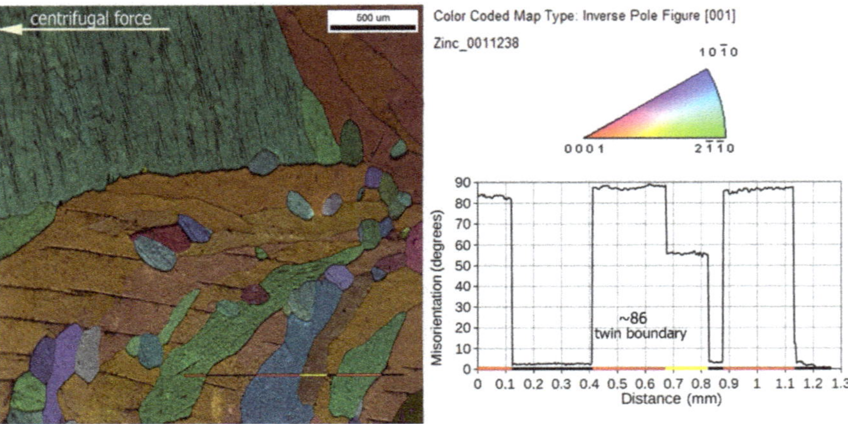


**Figure 12.** EBSD and misorientation profile obtained in the center of the 600 *g* gravity sample, marked in Figure 5 as zone C (Zn-rich zone). EBSD colored by overlapping of inverse pole figure (IPF) and band contrast (BC) images.

The studied area was characterized by a coarse-grained structure with an orientation close to ⟨0001⟩ or $\langle 10\bar{1}0 \rangle$ regions, depicted in IPF representation by orange or blue, respectively (Figure 12). It is worth noting that $\langle 10\bar{1}0 \rangle$ is parallel to the direction of centrifugal force.

For a more complete picture of crystallographic dependencies, the macrotexture of the Zn phase was measured by XRD. Two pole figures were measured respectively: (0002) and $(10\bar{1}0)$. Then, the orientation distribution function (ODF) was calculated in a Matlab-MTEX environment [34]. The results are presented in Figures 13 and 14 in the form of full pole figures on the (0002) and $(10\bar{1}0)$ planes.

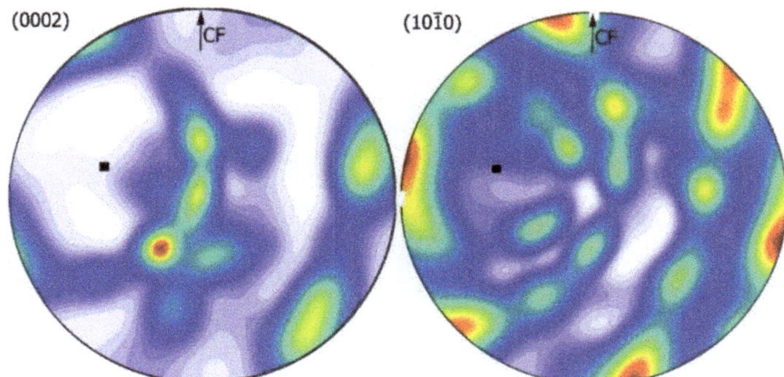


**Figure 13.** Texture analysis Zn-rich zone for 1 *g* gravity variant: full pole figures on the (0002) and $(10\bar{1}0)$ planes, recalculated in ODF. The measurement was performed with a longitudinal section of the sample.

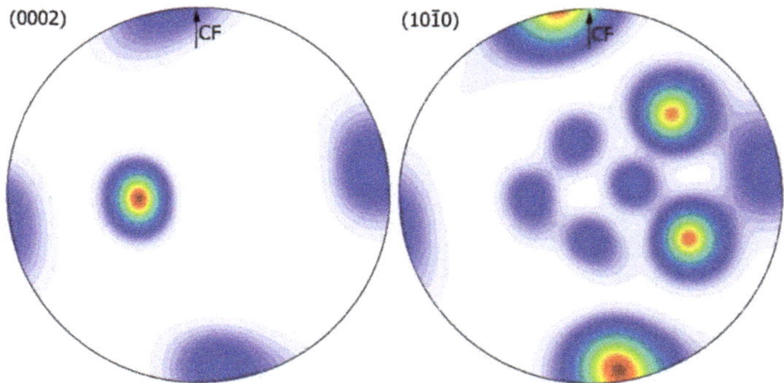


**Figure 14.** Texture analysis Zn-rich zone for 600 *g* gravity variant: full pole figures on the (0002) and $(10\bar{1}0)$ planes, recalculated in ODF. The measurement was performed with a longitudinal section of the sample.

The texture obtained for the sample at 1 *g* (normal gravity) has a character of a fibre texture $\langle 10\bar{1}0 \rangle$ type, where the poles of planes (0002) and $(10\bar{1}0)$ are arranged along small circles (Figure 13). The axis of this texture (shown by the black square) is tilted approximately 76° from the CF direction.

The texture obtained for the sample at 600 *g* (supergravity variant) (Figure 14) has a strong character with two clearly outlined components with mutual twin correlation typical of hexagonal structures [33,35,36]. The main component with the ⟨0002⟩ axis tilted by ~10 degrees from CF and the growth direction parallel to $\langle 10\bar{1}0 \rangle$ and a weaker component with mutual twin orientation of four peaks along the circumference on figure

(0002). This observation confirms the EBSD microtexture analyses carried out for the macro area.

In the pole figure (0002) Bi (Figure 15), two strong peaks can be seen in a plane normal to the direction of the CF force at a distance of about 40° from the center of the pole figure. On the other hand, in the pole figure $(10\bar{1}0)$, an increase in intensity around the normal of the figure is observed in a distributed small circle with a radius of 25° (Figure 15).

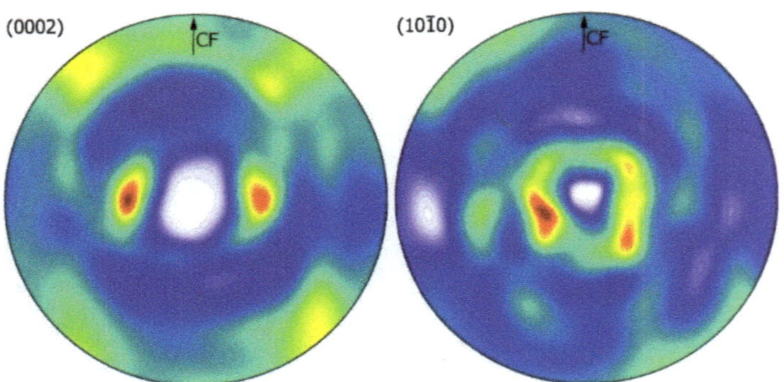


**Figure 15.** Texture analysis Bi-rich zone for 600 *g* gravity variant: full pole figures on the (0002) and $(10\bar{1}0)$ planes, recalculated in ODF. The measurement was performed with a longitudinal section of the sample.

The analysis of the texture of both sample zones shows that the interface was formed by two structures in a specific crystallographic relationship. The interfacial boundary has a complex character due to the strong gradient of the chemical composition in its area. This resulted in the occurrence of side-by-side microstructures belonging to opposite ends of the Zn–Bi equilibrium system.

## 4. Conclusions

The obtained results show the existence of the influence of gravity on the directed growth of the structure in Zn–Bi alloys. Analogous results obtained for other hexagonal alloys as Zn–Ti–Cu [13,14] subjected to the Bridgman zone crystallization method indicate a similar influence of the force of gravity and the direction of heat dissipation on the structure growth process. In the case of crystallization in supergravity conditions, a structure with two dominant orientations was obtained, the disorientation angle of which corresponds to the geometry of the twinning process in zinc. This effect was obtained under volumetric cooling conditions, without the forced temperature gradient as it was in the Bridgman process. The combination of zone crystallization and the force of gravity can give an interesting effect of organizing the structure. This process can be used in the production of complex materials using massive joints with controlled geometry.

Generally, studies of crystallization under 600 *g* supergravity conditions carried out on a monotectic Zn-25 wt% Bi alloy have shown the following:

- Gravity force strongly affects the nucleation and growth kinetics of eutectics;
- The boundary between the zinc-rich (Zn) and bismuth-rich (Bi) zones obtained for crystallization under supergravity is flat, while in the case of equilibrium crystallization, it is paraboloidal according to the curvature of the crystallization front;
- Crystallization under supergravity conditions results in high crystallographic order of the structure at both micro- and macro-scale;
- The (Zn)/(Bi) interface is formed by two structures in a specific crystallographic relationship;

- The strong gradient of the chemical composition near the separation boundary was observed;
- The microhardness of the zinc-rich area was about 58.5 HV and in the bismuth-rich area, it was decreased to 17.6 HV;
- The zinc-rich zone contained minor precipitates of parent bismuth, resulting from the mutual Bi–Zn solubility in a non-equilibrium state;
- The twins relation dominates in (Zn) crystallized under supergravity.

**Author Contributions:** Conceptualization, G.B., P.P. and P.K.; methodology, G.B., P.P., S.B. and G.M.-N.; validation, G.B.; formal analysis, G.B. and P.P.; investigation, G.B., P.P., P.K., S.B. and G.M.-N.; resources, G.B., P.P. and S.B.; data curation, G.B., P.P. and S.B.; writing—original draft preparation, G.B., P.P. and S.B.; writing—review and editing, G.B., P.P. and S.B.; visualization, G.B. and P.P.; supervision, G.B.; funding acquisition, G.B. All authors have read and agreed to the published version of the manuscript.

**Funding:** This research was funded by the AGH University of Science and Technology, grant number 16.16.180.006.

**Institutional Review Board Statement:** Not applicable.

**Informed Consent Statement:** Not applicable.

**Data Availability Statement:** The data presented in this study are available on request from the corresponding author.

**Conflicts of Interest:** The authors declare no conflict of interest.

## References

1. Rogers, J.R.; Davis, R.H. Modeling of collision and coalescence of droplets during microgravity processing of Zn-Bi immiscible alloys. *Metall. Trans. A* **1990**, *21*, 59–68. [CrossRef]
2. Wang, J.; Zhong, Y.B.; Fautrelle, Y.; Zheng, T.X.; Li, F.; Ren, Z.M.; Debray, F. Influence of the static high magnetic field on the liquid–liquid phase separation during solidifying the hyper-monotectic alloys. *Appl. Phys. A* **2013**, *112*, 1027–1031. [CrossRef]
3. Löffler, J.F.; Johnson, W.L. Crystallization of Mg–Al and Al-based metallic liquids under ultra-high gravity. *Intermetallic* **2002**, *10*, 1167–1175. [CrossRef]
4. Löffler, J.F.; Bossuyt, S.; Peker, A.; Johnson, W.L. Eutectic isolation in Mg-Al-Cu-Li(-Y) alloys by centrifugal processing. *Philos. Mag.* **2003**, *83*, 2797–2813. [CrossRef]
5. Yang, Y.; Song, B.; Song, G.; Jia, S. Removing Impurity Element of Copper from Pb-3% Cu Melt by Super Gravity. In *TMS 2015 144th Annual Meeting & Exhibition*; Springer: Cham, Switzerland, 2015. [CrossRef]
6. Zhao, L.; Guo, Z.; Wang, Z.; Wang, M. Removal of low-content impurities from Al by super-gravity. *Metall. Mater. Trans. B* **2010**, *41*, 505–508. [CrossRef]
7. Zhao, L.; Guo, Z.; Wang, Z.; Wang, M. Influences of super-gravity field on aluminum grain refining. *Metall. Mater. Trans. A* **2010**, *41*, 670–675. [CrossRef]
8. Wierzba, B.; Nowak, W.J. Phase separation during the sedimentation process in Cu–Sn alloy. *Phys. A Stat. Mech. Appl.* **2019**, *523*, 602–610. [CrossRef]
9. Chirita, G.; Soares, D.; Silva, F.S. Advantages of the centrifugal casting technique for the production of structural components with Al–Si alloys. *Mater. Des.* **2008**, *29*, 20–27. [CrossRef]
10. Zhao, J.Z.; Li, H.L.; Li, H.Q.; Xing, C.Y.; Zhang, X.F.; Wang, Q.L.; He, J. Microstructure formation in centrifugally cast Al–Bi alloys. *Comput. Mater. Sci.* **2010**, *49*, 121–125. [CrossRef]
11. Löffler, J.F.; Peker, A.; Bossuyt, S.; Johnson, W.L. Processing of metallic glass forming liquids under ultra-high gravity. *Mater. Sci. Eng. A* **2004**, *375*, 341–345. [CrossRef]
12. Hong, J. Crystal growth in high gravity. *J. Cryst. Growth.* **1997**, *181*, 459–460. [CrossRef]
13. Boczkal, G.; Mikułowski, B.; Wołczyński, W. Oscillatory structure of the Zn-Cu-Ti single crystals. *Mater. Sci. Forum* **2010**, *649*, 113–118. [CrossRef]
14. Boczkal, G.; Mikulowski, B.; Oertel, C.G.; Skrotzki, W. Work-hardening characteristics of Zn-Ti alloy single crystals. *Cryst. Res. Technol. J. Exp. Ind. Crystallogr.* **2010**, *45*, 111–114. [CrossRef]
15. Carlberg, T.; Fredriksson, H. The influence of microgravity on the solidification of Zn-Bi immiscible alloys. *Metall. Trans. A* **1980**, *11*, 1665–1676. [CrossRef]
16. Savas, M.A.; Altintas, S.; Erturan, H. Effects of squeeze casting on the properties of Zn-Bi monotectic alloy. *Metall. Mater. Trans. A* **1997**, *28*, 1509–1515. [CrossRef]

17. Cao, C.; Chen, L.; Xu, J.; Zhao, J.; Pozuelo, M.; Li, X. Phase control in immiscible Zn-Bi alloy by tungsten nanoparticles. *Mater. Lett.* **2016**, *174*, 213–216. [CrossRef]
18. Spisak, W.; Chlebicki, A.; Kaszczyszyn, M.; Szar, M.; Kozak, J.; Olma, A. Threeelectrode galvanic microcells as a new antimicrobial tool. *Sci. Rep. (Nat. Publ. Group)* **2020**, *10*, 7341. [CrossRef]
19. Okamoto, H.; Schlesinger, M.E.; Mueller, E.M. *ASM Handbook. Volume 3, Alloy Phase Diagrams*; ASM International Handbook Committee: Materials Park, OH, USA, 1992.
20. Vizdal, J.; Braga, M.H.; Kroupa, A.; Richter, K.W.; Soares, D.; Malheiros, L.F.; Ferreira, J. Thermodynamic assessment of the Bi–Sn–Zn system. *Calphad* **2007**, *31*, 438–448. [CrossRef]
21. Majumdar, B.; Chattopadhyay, K. Aligned monotectic growth in unidirectionally solidified Zn-Bi alloys. *Metall. Mater. Trans. A* **2000**, *31*, 1833–1842. [CrossRef]
22. Nogi, K.; Ogino, K.; McLean, A.; Miller, W.A. The temperature coefficient of the surface tension of pure liquid metals. *Metall. Trans. B* **1986**, *17*, 163–170. [CrossRef]
23. Aqra, F.; Ayyad, A. Surface tension of pure liquid bismuth and its temperature dependence: Theoretical calculations. *Mater. Lett.* **2011**, *65*, 760–762. [CrossRef]
24. Wołczyński, W.; Lipnicki, Z.; Bydałek, A.W.; Ivanova, A.A. Structural Zones in Large Static Ingot. Forecasts for Continuously Cast Brass Ingot. *Arch. Foundry Eng.* **2016**, *16*, 141–146. [CrossRef]
25. Schulte, O.; Holzapfel, W.B. Effect of pressure on the atomic volume of Zn, Cd, and Hg up to 75 GPa. *Phys. Rev. B* **1996**, *53*, 569. [CrossRef]
26. Shu, Y.; Hu, W.; Zhao, Z.; Wang, L.; Liu, Z.; Tian, Y.; Yu, D. Anomalous melting behavior of polycrystalline bismuth quenched at high temperature and high pressure. *Mater. Lett.* **2016**, *168*, 36–39. [CrossRef]
27. Chen, H.Y.; Xiang, S.K.; Yan, X.Z.; Zheng, L.R.; Zhang, Y.; Liu, S.G.; Bi, Y. Phase transition of solid bismuth under high pressure. *Chin. Phys. B* **2016**, *25*, 108103. [CrossRef]
28. Tonkov, E.Y.; Ponyatovsky, E.G. *Phase Transformations of Elements under High Pressure*; CRC Press: Boca Raton, FL, USA, 2018.
29. Li, K.W.; Wang, X.B.; Li, S.M.; Wang, W.X.; Chen, S.P.; Gong, D.Q.; Cui, J.L. Halo formation in binary Fe-Nb off-eutectic alloys. *High Temp. Mater. Process.* **2015**, *34*, 479–485. [CrossRef]
30. Aguilera, I.; Friedrich, C.H.; Blgel, S. Electronic phase transitions of bismuth under strain from relativistic self-consistent GW calculations. *Phys. Rev. B* **2015**, *91*, 125129. [CrossRef]
31. Wang, Y.N.; Huang, J.C. The role of twinning and untwinning in yielding behavior in hot-extruded Mg–Al–Zn alloy. *Acta Mater.* **2007**, *55*, 897–905. [CrossRef]
32. Antonopoulos, J.G.; Karakostas, T.; Komninou, P.; Delavignette, P. Dislocation movements and deformation twinning in zinc. *Acta Metall.* **1988**, *36*, 2493–2502. [CrossRef]
33. Yoo, M.H.; Lee, J.K. Deformation twinning in hcp metals and alloys. *Philos. Mag. A* **1991**, *63*, 987–1000. [CrossRef]
34. Hielscher, R.; Schaeben, H. A novel pole figure inversion method: Specification of the MTEX algorithm. *J. Appl. Cryst.* **2008**, *41*, 1024–1037. [CrossRef]
35. Sułkowski, B. Analysis of crystallographic orientation changes during deformation of magnesium single crystals. *Acta Phys. Pol. A* **2014**, *126*, 768. [CrossRef]
36. Sułkowski, B.; Chulist, R. Twin-induced stability and mechanical properties of pure magnesium. *Mater. Sci. Eng. A* **2019**, *749*, 89–95. [CrossRef]

*Article*

# Characteristics of Al-Si Alloys with High Melting Point Elements for High Pressure Die Casting

**Tomasz Szymczak [1,*], Grzegorz Gumienny [1,*], Leszek Klimek [2], Marcin Goły [3], Jan Szymszal [4] and Tadeusz Pacyniak [1]**

1. Department of Materials Engineering and Production Systems, Lodz University of Technology, 90-924 Lodz, Poland; tadeusz.pacyniak@p.lodz.pl
2. Institute of Materials Science and Engineering, Lodz University of Technology, 90-924 Lodz, Poland; leszek.klimek@p.lodz.pl
3. Department of Physical & Powder Metallurgy, AGH University of Science and Technology, 30-059 Krakow, Poland; marcing@agh.edu.pl
4. Department of Technical Sciences and Management, University of Occupational Safety Management in Katowice, 40-007 Katowice, Poland; jszymszal@wszop.edu.pl

* Correspondence: tomasz.szymczak@p.lodz.pl (T.S.); grzegorz.gumienny@p.lodz.pl (G.G.); Tel.: +48-426312276 (T.S.); +48-426312264 (G.G.)

Received: 9 October 2020; Accepted: 29 October 2020; Published: 29 October 2020

**Abstract:** This paper is devoted to the possibility of increasing the mechanical properties (tensile strength, yield strength, elongation and hardness) of high pressure die casting (HPDC) hypoeutectic Al-Si alloys by high melting point elements: chromium, molybdenum, vanadium and tungsten. EN AC-46000 alloy was used as a base alloy. The paper presents the effect of Cr, Mo, V and W on the crystallization process and the microstructure of HPDC aluminum alloy as well as an alloy from the shell mold. Thermal and derivative analysis was used to study the crystallization process. The possibility of increasing the mechanical properties of HPDC hypoeutectic alloy by addition of high-melting point elements has been demonstrated.

**Keywords:** crystallization; microstructure; mechanical properties; thermal and derivative analysis

---

## 1. Introduction

High melting point elements, such as Cr, Mo, V and W are rarely used as additions in Al-Si alloys. Research papers described two main reasons for using these additions in Al-Si alloys. The first reason is the enhancement of the precipitation hardening effect [1–4] while the second one is a reduction of the detrimental effect of iron on mechanical properties [4–9]. However, the use of Cr, Mo, V and W in Al-Si alloys is very limited considering the nature of their interaction with the main alloy constituents, i.e., Al and Si. The phase diagrams Al-Cr [10,11], Al-Mo [12], Al-V [11] and Al-W [11] show the lack of solubility or very limited solubility of high-melting point elements in aluminum. According to [10], Cr is not soluble in Al, while according to [11], the solubility of Cr in Al is negligible and amounts to 0.71 wt % at 661.5 °C. Tungsten is not soluble in Al [11]. Similar behavior shows molybdenum [11,12], while the solubility of vanadium in aluminum is maximum 0.6 wt % (0.3 at %) [11]. Very limited solubility of the above-mentioned elements in aluminum causes the precipitation of a number of intermetallic phases. Also, Si-V [13], Cr-Si [11], Mo-Si [14] and W-Si [15] phase diagrams show that the tested elements practically do not dissolve in silicon. In accordance with these phase diagrams, Cr, Mo, V and W also tend to form numerous intermetallic phases with silicon. For example, four intermetallic phases can occur in the Si-V diagram, i.e., $SiV_3$, $Si_3V_5$, $Si_5V_6$ and $Si_2V$ [13]. On the other hand, the analyzed elements offer excellent mutual solid-state solubility. Cr-Mo, Cr-V, Mo-V, Mo-W and V-W phase diagrams presented in [11,16–19] show that they form unlimited solutions. In [20], Cr-W phase diagram has been described. It shows that Cr and W form a solid solution (αCr, W) with unlimited mutual solubility of both elements. At 1677 °C, it decomposes into two solid solutions (α1)-rich in chromium and (α2)-rich in tungsten.

The presented data show that in Al-Si alloys containing Cr, Mo, V and W, intermetallic phases with Al and/or Si can crystallize. In multicomponent Al-Si alloys, the possibility of the formation of more complex phases containing other constituents, like Cu, Mg or Ni should be taken into account. Intermetallic phases can significantly increase the brittleness of Al alloys and reduce their strength and elongation. The risk of the precipitation of intermetallic phases in aluminum alloys containing Cr, Mo, V and W increases with the decreasing rate of heat transfer from the casting. Therefore, these additives are not applicable to Al-Si alloys produced in sand and ceramic molds. Additionally, in the case of hypoeutectic alloys, the formation of intermetallic phases in a higher temperature than the temperature of α(Al) phase crystallization changes the additives concentration ahead of the dendritic crystallization front. As a consequence of the high heat transfer during solidification in metal molds, the α(Al) phase can be supersaturated with high-melting point elements. High pressure die casting (HPDC) is a technology widely used in industry and characterized by intensive heat transfer from the casting. Therefore, in relation to hypoeutectic aluminum alloys containing Cr, Mo, V and W, precisely this technology should be seen as an opportunity to effectively increase the mechanical properties of castings made from the above mentioned alloys. The possibility of increasing the properties of HPDC Al-Si alloy by supersaturation of α(Al) phase with high melting point elements is the most innovative aspect of this paper.

The aim of the paper is to confront the phenomena occurring during the crystallization of Al-Si alloy with different content of high-melting point elements with their mechanical properties in the context of developing the probable mechanism of changes in the properties of the alloy. The crystallization process was investigated by thermal and derivative analysis. Optical and scanning microscopy were used to study the alloy microstructure. A point analysis of the chemical composition of selected phases in the microstructure was also performed with the use of an EDS (Energy Dispersive Spectroscopy) detector. The static tensile test and the Brinell hardness test were used to determine the mechanical properties

## 2. Materials and Methods

A typical hypoeutectic Al-Si alloy for high pressure die casting, i.e., EN-AC 46000 (EN-AC AlSi9Cu3 (Fe)), was used for the tests. It is included in the PN-EN 1706 standard. The chemical composition of the base alloy is shown in Table 1. The range of the chemical composition is derived from 15 specimens of the base alloy. The reason for such a number of specimens is the testing of 15 different combinations of high melting point elements. The number of Cr, Mo, V and W combinations means all statistically possible connections that can be used for the four analyzed elements. Therefore, each time new base alloy was smelted from ingots of EN AC-46000 alloy. The individual connections Cr, Mo, V and W have been shown in the following description of a melting technology.

**Table 1.** Chemical composition of the base EN AC-46000 alloy.

| Chemical Composition, wt % | | | | | | | | | |
|---|---|---|---|---|---|---|---|---|---|
| Si | Cu | Zn | Fe | Mg | Mn | Ni | Ti | Cr | Al |
| 8.69–9.35 | 2.09–2.43 | 0.90–1.07 | 0.82–0.97 | 0.21–0.32 | 0.18–0.25 | 0.05–0.13 | 0.042–0.049 | 0.023–0.031 | rest |

The base alloy was melted in a gas-heated shaft furnace with a maximum charge capacity of 1.5 tonnes (StrikoWestofen, Gummersbach, Germany). After smelting, it was refined inside the shaft furnace. An Ecosal Al 113S solid refiner was used for this treatment. After tapping the alloy from the furnace into a ladle, it was deslagged with an Ecremal N44 deslagging agent. After deslagging, the liquid alloy was transferred to a holding furnace set up next to an Idra 700S horizontal cold chamber pressure die casting machine (Idra, Travagliato, Italy). In the holding furnace, AlCr15, AlMo8, AlV10 and AlW8 master alloys were added into the alloy. The temperature in the holding furnace was 750 °C. After dissolution of the master alloys, the temperature of the liquid metal was reduced to a value appropriate for the test casting process. High melting point elements, i.e., chromium, molybdenum, vanadium and tungsten, were added as single elements, in double combinations (CrMo, CrV, CrW, MoV, MoW and VW), in triple combinations (CrMoV, CrMoW, CrVW and MoVW), or all the additives were added simultaneously. For individually added Cr, Mo, V and W, their content was kept in the range of 0.0–0.5 wt %, and it was increased in 0.1 wt % steps. If more than one high melting point element was added, all the elements were used in equal amount. For double combinations, the content of additives was in the range of 0.0–0.4 wt %, and in subsequent melts it was increased in 0.1 wt % steps. In the case of triple and quadruple combinations, the content of elements was in the range of 0.00–0.25 wt %, and in subsequent melts it was increased in 0.05 wt % steps. From the resultant alloys, covers of roller shutter housings with a predominantly 2 mm wall thickness were made by the high pressure die casting process.

For each chemical composition tested, three specimens were taken from one high pressure die casting to carry out the tensile test. The specimens had a flat shape and $2 \times 10$ mm$^2$ rectangular cross-section. This cross-section is recommended by PN-EN 1706 standard for testing the strength of high pressure die casting. An Instron 3382 machine (Instron, Norwood, MA, USA) was used to perform the tensile test at a speed of 1 mm/min. The test enabled the determination of the tensile strength $R_m$, the yield strength $R_{p0.2}$ and the elongation A. Alloy hardness was measured by the Brinell method using an HPO-2400 hardness tester (WPM LEIPZIG, Leipzig, Germany) according to PN EN ISO 6506. The diameter of the ball was 2.5 mm, the load was 613 N, and the static load holding time was 30 s.

Thermal and derivative analysis was used as a tool to study the solidification process. Thermal and derivative curves were recorded with a PtRh10-Pt thermocouple enclosed by a quartz tube placed in the thermal center of a probe made of a resin coated sand. Its dimensions are shown in Figure 1. The alloy was superheated to 1000 °C before the probe was filled with liquid alloy.

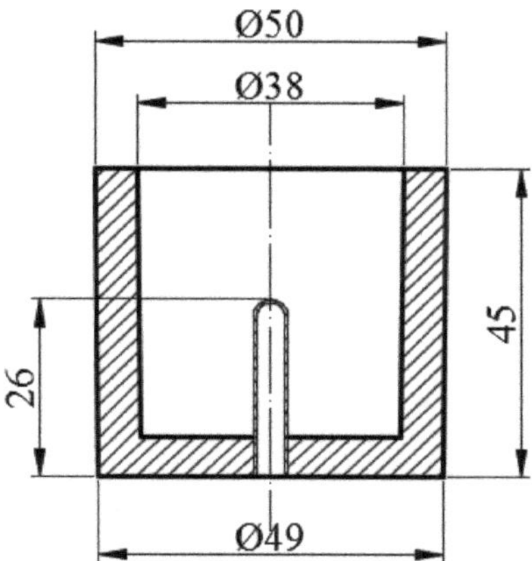


**Figure 1.** Dimensions of the resin coated sand probe (units: mm).

Metallographic specimens for microstructure examinations were prepared on specimens taken from castings made in the resin coated sand probe and by high pressure die casting. The surface of the specimens was etched with a 2% HF acid solution. The microstructure was examined using Nikon Eclipse MA200 optical microscope (Nikon, Tokyo, Japan) with ×100 and ×1000 magnification for castings made in the resin coated sand probe and by high pressure die casting process, respectively. The applied magnification ensures correct visibility of all alloy phases in the examined area.

Point microanalysis of the element concentration was performed using the Pioneer EDS detector cooperating with the HITACHI S-3000N scanning electron microscope (Hitachi, Tokyo, Japan) and the VENTAGE software from NORAN. The specimens for scanning under the scanning microscope were cut from the test castings using a hand hacksaw and water cooling. The shell mold casting specimens had the shape of a cube with a side of 10 mm. Specimens taken from high pressure die castings had the shape of a cuboid with sides 2/10/20 mm, and after cutting they were mounted in a conductive phenolic resin. After mounting, the specimen was cylindrical with a diameter of 36 mm and a height of 12 mm. The specimens were then ground using sandpaper and polished using diamond suspensions with a gradation of 9 to 1 μm.

The area of the spot examination of the elements concentration is shown in the figures of the microstructure made with the use of a scanning microscope. These images show the intermetallic phases within which the analyzes were performed. For the purpose of this study, phase precipitations of relatively large sizes were selected in order to avoid the analytical beam going beyond the area of the tested phase.

## 3. Results and Discussion

Figure 2 compares the thermal and derivative analysis curves of the EN AC-46000 base alloy and alloy containing 0.5 wt % Mo and Cr, Mo, V and W added in an amount of 0.25 wt % each.

Studies have revealed a three-stage solidification process of the base EN EN-46000 alloy from the resin coated sand probe. The first stage taking place at the highest temperature is the crystallization of α(Al) dendrites solid solution. This process is marked by the thermal effect $C_5AB$. The primary crystallization of α(Al) dendrites directly from the liquid results from Al-Si phase diagram shown in Figure 3.

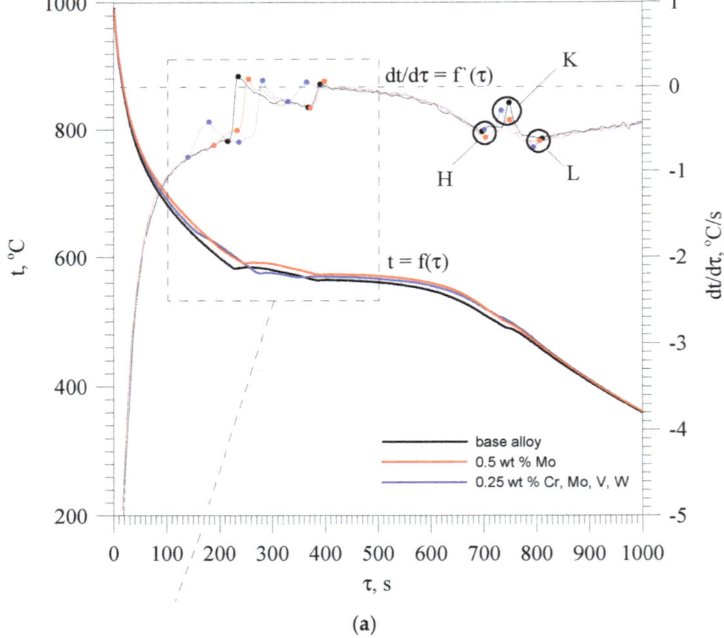


(a)

**Figure 2.** *Cont.*

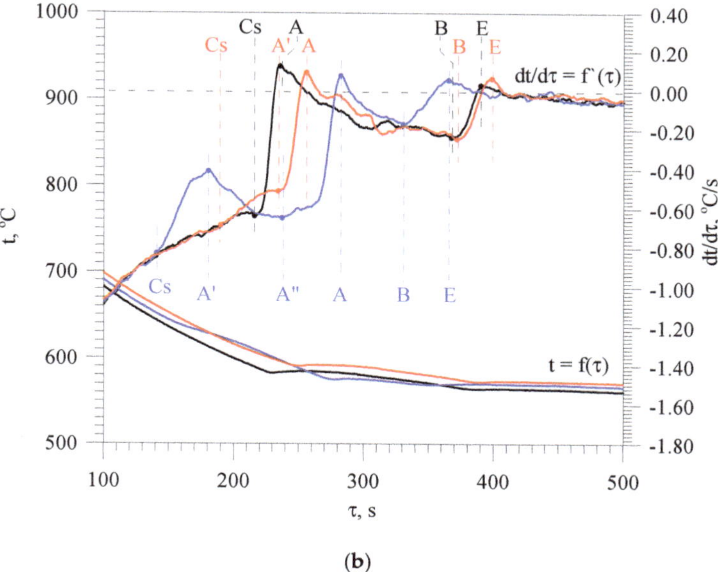


**(b)**

**Figure 2.** Thermal and derivative curves of the base alloy (black), alloy containing 0.5 wt % Mo (red) and alloy containing Cr, Mo, V and W in an amount of 0.25 wt % each (blue). (**a**) and their enlarged fragment (**b**).

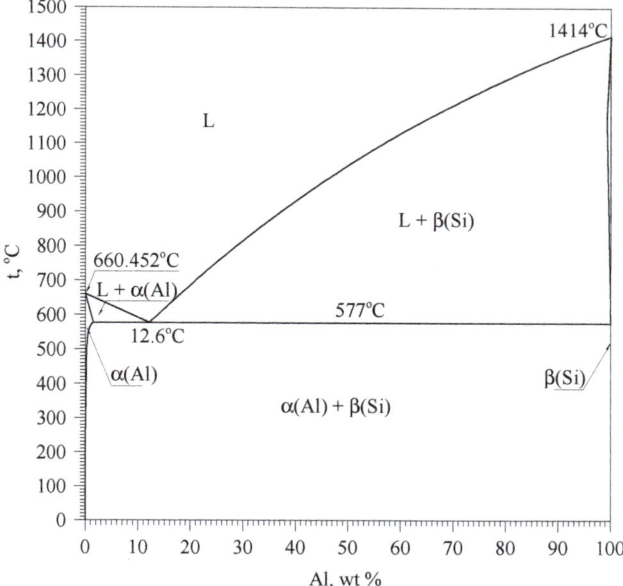


**Figure 3.** Al-Si phase diagram (data from [11,21,22]).

The presented Al-Si phase diagram shows that after the crystallization of α(Al) dendrites, the crystallization of α(Al) + β(Si) eutectic mixture takes place. The tested AlSi9Cu3(Fe) alloy is multicomponent, therefore more complex eutectic mixtures crystallize in it. In these mixtures, apart from α(Al) and β(Si) phases, there are also intermetallic phases. The crystallization of multicomponent Al-Si alloys with this type of complex eutectic mixtures was described in detail in [22–24]. In accordance with the knowledge contained therein, the authors present the further crystallization process of the analyzed alloy.

When dendrite crystallization is complete, the complex α(Al) + $Al_{15}(Fe,Mn)_3Si_2$ + β(Si) (ternary) and α(Al) + $Al_2Cu$ + AlSiCuFeMnMgNi + β(Si) (quaternary) eutectic mixtures crystallize. Thermal effects caused by the crystallization of the above mentioned eutectic mixtures have been marked with the symbols BEH and HKL, respectively. The designation of AlSiCuFeMnMgNi phase should be treated as conventional, because it does not mean an intermetallic phase containing all the elements included, but a number of phases that may contain these elements. The chemical composition of this phase depends on the diversified (as a result of microsegregation) content of elements in different areas of the residual liquid. It can be $Mg_2Si$ phase or the phases from Al-Cu-Ni, Al-Fe-Mn-Si and Al-Cu-Mg-Si systems. The microstructure of the base alloy and containing Cr, Mo, V and W is presented in detail in [25]. This paper presents only those aspects of microstructure formation that are related to the mechanism of its strengthening presented by the authors.

When the additives in an amount not exceeding 0.1 wt.% Cr, 0.4 wt % Mo or W, and 0.5 wt % V are added as single elements into the base alloy, there are no new thermal effects on the thermal and derivative curves and new phases are not formed in the microstructure. In this range of the content, the above mentioned elements tend to join the phases already existing in the base alloy. These are mainly iron-rich intermetallic phases occurring in both eutectic mixtures, designated as $Al_{15}(Fe,Mn,M)_3Si_2$, where M is any high melting point element or any combination of them. A single addition of chromium in an amount of 0.2 wt % and molybdenum or tungsten in an amount of 0.5% leads to primary crystallization of the $Al_{15}(Fe,Mn,M)_3Si_2$ phase. Vanadium in the tested content range did not cause this phase to crystallize. As a result of the crystallization of the primary $Al_{15}(Fe,Mn,M)_3Si_2$ phase on the derivative curve a thermal effect is created, which does not have a distinct local maximum. An example of such an effect is visible on the derivative curve of an alloy containing 0.5 wt % Mo (Figure 2–red color) and is described as $C_sA'$. Increasing the amount of high melting point elements produces a thermal effect with a clearly visible local maximum. This thermal effect is shown in Figure 2 (blue color), for an alloy containing 0.25 wt % Cr, Mo, V and W. It is described as $C_sA'A''$. The amount and size of this phase increases with the increase in the content of high melting point elements in the alloy. Figure 4 shows the microstructure of the base alloy and containing 0.25 wt % Mo, V and W each with marked component phases.

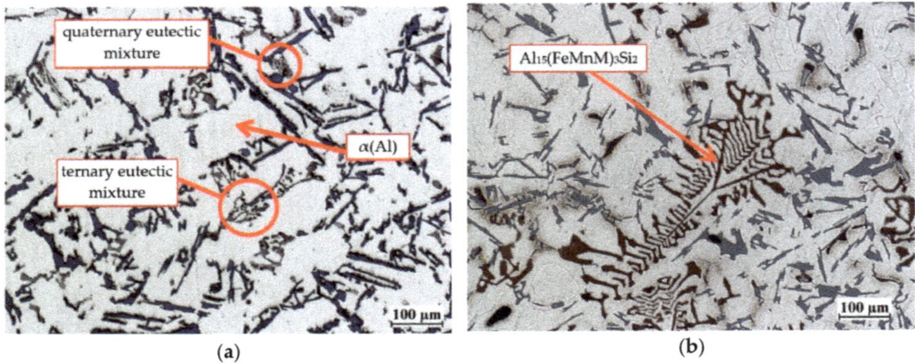


**Figure 4.** The microstructure of: (**a**) EN AC-AlSi9Cu3(Fe) base alloy, (**b**) alloy containing Mo, V and W in an amount of 0.25 wt. % each.

In Figure 4b $Al_{15}(Fe,Mn,M)_3Si_2$ phase is characterized by a skeletal morphology. For example, Figure 5 shows the chemical composition of the analogous phase in the alloy containing 0.4 wt % Cr.

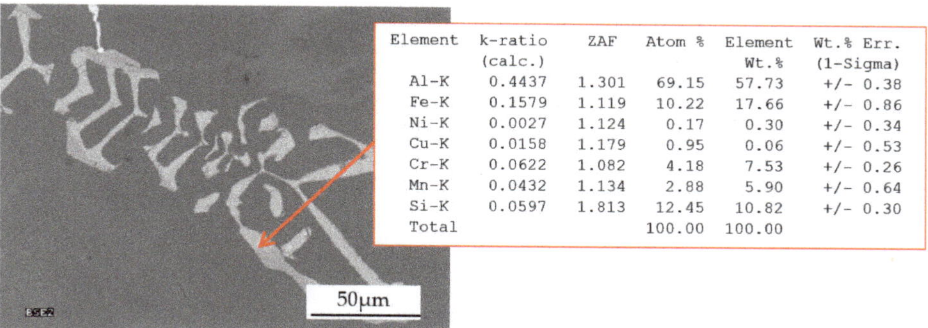

| Element | k-ratio (calc.) | ZAF | Atom % | Element Wt.% | Wt.% Err. (1-Sigma) |
|---|---|---|---|---|---|
| Al-K | 0.4437 | 1.301 | 69.15 | 57.73 | +/- 0.38 |
| Fe-K | 0.1579 | 1.119 | 10.22 | 17.66 | +/- 0.86 |
| Ni-K | 0.0027 | 1.124 | 0.17 | 0.30 | +/- 0.34 |
| Cu-K | 0.0158 | 1.179 | 0.95 | 0.06 | +/- 0.53 |
| Cr-K | 0.0622 | 1.082 | 4.18 | 7.53 | +/- 0.26 |
| Mn-K | 0.0432 | 1.134 | 2.88 | 5.90 | +/- 0.64 |
| Si-K | 0.0597 | 1.813 | 12.45 | 10.82 | +/- 0.30 |
| Total | | | 100.00 | 100.00 | |



**Figure 5.** An example of the area with $Al_{15}(Fe,Mn,M)_3Si_2$ phase in the alloy containing 0.4 wt % Cr from the shell mold and the results of the point measurement of the chemical composition within it.

There is a high concentration of Al (57.7 wt %) and an increased concentration of Fe (17.7 wt %) in the phase; Si (10.8 wt %); Cr (7.53 wt%) and Mn (5.9 wt %). Its phase chemical composition indicates that it is $Al_{15}(Fe,Mn,Cr)_3Si_2$ phase. The occurrence of this phase in alloys containing Cr has been confirmed in [26–28]. The chemical analysis of $Al_{15}(Fe,Mn,M)_3Si_2$ phase detected in alloys containing additionally Mo, V and W showed the presence of all these elements. Thus, the crystallization of these intermetallics causes the depletion of the remaining liquid in high melting point elements. A relatively large amount of the primary $Al_{15}(Fe,Mn,M)_3Si_2$ phase causes the depletion of the liquid in high melting point elements leading to crystallization the classic plate $\alpha$(Al) + $\beta$(Si) eutectic mixture instead of the ternary one. Thus, high melting point elements first increase their concentration ahead of the crystallization front of $\alpha$(Al) phase, but then reduces it as a result of the primary crystallization of $Al_{15}(Fe,Mn,M)_3Si_2$ phase. The possibility of supersaturation of $\alpha$(Al) dendrites is related to the concentration of the dissolved substance ahead of the dendrite crystallization front [29] and the rate of heat transfer during the solidification process. In [30,31], general relationships (1) and (2) were developed to link the crystal nucleation rate and growth rate, respectively, with the supersaturation of the crystal with the substance on the crystallization front:

$$B = k_b \times \Omega^b, \quad (1)$$

$$V = k_v \times \Omega^v, \quad (2)$$

where:

B-crystal nucleation rate,
V-crystal growth rate,
$\Omega$-supersaturation,
$k_b$ and $k_v$-original constant rate of crystal nucleation and growth, respectively,
and v-supersaturation exponent for crystal nucleation and growth, respectively.

From the above mentioned relationships it follows that an increase in the crystal nucleation rate and growth rate increases the degree of supersaturation. It is also generally known that that the higher rate of heat transfer leads to the higher rate of the nucleation and growth of the crystal. Therefore, two factors can accelerate supersaturation–a high concentration of high melting point elements on the dendritic crystallization front and the high rate of heat transfer during crystallization. Based on 84 thermal and derivative tests for the base alloy as well as all variants of the high melting point elements used, the average heat transfer rate during the crystallization of α(Al) phase was calculated. The value obtained was 0.13 °C/s. This, relatively low, heat transfer rate in castings from shell molds creates rather modest possibilities of α(Al) supersaturation with high melting point elements. However, the extensive thermal and derivative tests showed an increase in the crystallization start temperature $C_s$ as a result of addition high melting point elements and increasing their content in the alloy. This was the case for all analyzed combinations of these elements. Figure 6 shows the effect of high-melting elements on the crystallization start temperature ΔtCs, for individually and simultaneously added Cr, Mo, V and W.

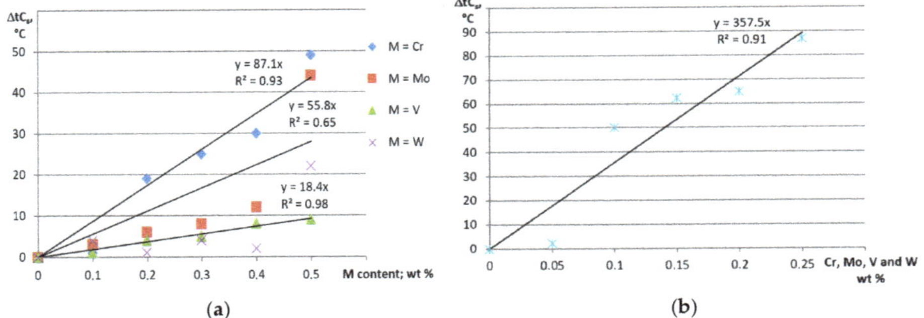


**Figure 6.** Changes in the crystallization temperature $\Delta tC_s$ caused by the variable content of Cr, Mo, V or W added: (**a**) as single elements and (**b**) jointly; M–any high melting point element tested, i.e., Cr, Mo, V or W.

Figure 6a shows that all individually added high melting point elements increase the liquidus temperature. This is mainly due to the pre-dendritic crystallization of $Al_{15}(Fe,Mn,M)_3Si_2$ phase, which occurs when 0.2 wt % Cr and 0 5 wt.% Mo and W are added (thermal effects $C_sA'$ or $C_sA'A''$). Crystallization of this phase is responsible for a rapid increase in the temperature $\Delta tC_s$. However, for each high melting point element analyzed, the increase in $C_s$ temperature was also obtained in the range of their concentration that did not cause pre-dendritic crystallization of $Al_{15}(Fe,Mn,M)_3Si_2$ phase. This tendency is clearly visible when V, Mo and W are added. Vanadium did not cause pre-dendritic crystallization of $Al_{15}(Fe,Mn,M)_3Si_2$ phase, while Mo and W did not cause this phase to crystallize to 0.4 wt %. In the above-mentioned range of high melting point elements, the increase of $\Delta tC_s$ is relatively mild. When Cr, Mo, V and W are simultaneously added (Figure 6b), up to 0.05 wt % each, the increase in ΔtCs is also initially mild, but starting with the 0.10 wt %, a rapid jump and faster growth occur. This rapid increase in $\Delta tC_s$ is due to the pre-dendritic crystallization of the $Al_{15}(Fe,Mn,M)_3Si_2$ phase. The increase in the liquidus temperature at the content of high melting point elements insufficient to start the primary crystallization of $Al_{15}(Fe,Mn,M)_3Si_2$ phase indicates transfer of a certain amount of these elements to α(Al) dendrites. Much better conditions for supersaturation of the α(Al) solid solution with high melting point elements are provided by the high pressure die casting (HPDC) technology, where the rate of heat transfer from the casting is much higher. The rate of heat transfer during crystallization of the α(Al) dendrites in HPDC alloy was determined from relationship (3) presented in [32]:

$$SDAS=39.4 \times V\hat{}(-0.317), \tag{3}$$

where:

SDAS-secondary dendrite arms spacing in α(Al) phase, μm;
V–the rate of heat transfer during the crystallization of α(Al) dendrites, °C/s.

The measured average SDAS in the examined HPDC alloy was 10.65 μm. The rate of heat transfer calculated for SDAS from relationship (3) amounts to ~55 °C/s. This value is over 400 times higher than the value obtained for alloy from shell mold. The high rate of heat transfer from the HPDC alloy produces more refined microstructure compared with the microstructure obtained in shell mold. Figure 7 shows the microstructure of HPDC EN AC-46000 alloy containing Cr, V and W, each in an amount of 0.25 wt %. The constituent phases are also marked.

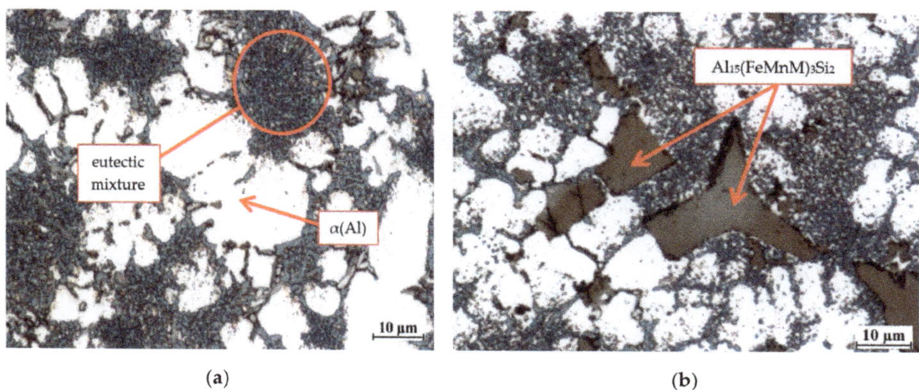


**Figure 7.** The microstructure of: (**a**) base alloy and (**b**) alloy containing Cr, V and W in an amount of 0.25 wt % each.

The microstructure of the die cast base alloy consists of α(Al) phase dendrites and eutectic mixture. The eutectic mixture consists of α(Al) and β(Si) solid solutions and relatively fine intermetallic phases. Diffraction tests have shown the presence of intermetallic phases: $Al_2Cu$, $Al_2CuMg$ and other phases that crystallize in Al-Fe-Si, Al-Cu-Fe and Mn-Ni-Si phase systems. The addition of a certain amount of high melting point elements causes the pre-dendritic crystallization of the $Al_{15}(Fe,Mn,M)_3Si_2$ phase. With the increasing content of above mentioned elements, the amount of this phase also increases. Figure 7b shows this phase assume the form of polygons or stars. The size of the precipitates, depending on the content of high melting point elements, is from a few to ~50 µm. Analysis of the chemical composition of the pre-dendritic $Al_{15}(Fe,Mn,M)_3Si_2$ phase has shown that it can "assimilate" all high melting point elements tested. Figure 8 presents the results of EDS point analysis carried out within the area of the pre-dendritic phase in the alloy containing Cr, Mo, V and W in an amount of 0.25 wt.% each.

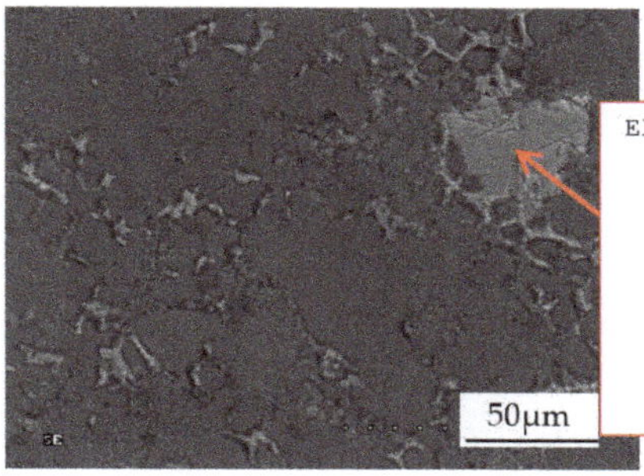

| Element | k-ratio (calc.) | ZAF | Atom % | Element Wt.% | Wt.% Err. (1-Sigma) |
|---|---|---|---|---|---|
| Al-K | 0.4655 | 1.222 | 69.88 | 56.86 | +/- 0.36 |
| Si-K | 0.0557 | 1.633 | 10.74 | 9.10 | +/- 0.39 |
| Mo-L | 0.0248 | 1.499 | 1.28 | 3.72 | +/- 0.24 |
| V -K | 0.0267 | 1.108 | 1.93 | 2.96 | +/- 0.33 |
| Cr-K | 0.0597 | 1.077 | 4.11 | 6.44 | +/- 0.46 |
| Mn-K | 0.0369 | 1.128 | 2.51 | 4.17 | +/- 0.64 |
| Fe-K | 0.1419 | 1.112 | 9.37 | 15.78 | +/- 0.81 |
| W -M | 0.0053 | 1.874 | 0.18 | 0.99 | +/- 1.05 |
| Total | | | 100.00 | 100.00 | |

50µm

**Figure 8.** An example of $Al_{15}(FeMnM)_3Si_2$ phase in HPDC alloy containing Cr, Mo, V and W in an amount of 0.25 wt % each and the results of a point analysis of the chemical composition within it.

The EDS analysis shows that the tested phase is most effective in assimilating Cr (6.44 wt %), slightly less effective in Mo (3.72 wt %) and V (2.96 wt %), and the least effective in W (0.99 wt %). The total concentration of Cr, Mo, V and W in this phase is ~14 wt %. The "assimilation" of chromium, molybdenum, vanadium and tungsten by the $Al_{15}(Fe,Mn,M)_3Si_2$ phase causes the depletion of these elements in the liquid. Since the $Al_{15}(Fe,Mn,M)_3Si_2$ phase crystallizes directly in front of the $\alpha$(Al) dendrites, in the liquid ahead of the dendritic crystallization front, similar changes occur in the concentration of Cr, Mo, V and W as in the alloy from the shell mold. Chromium, molybdenum, vanadium and tungsten in the range of concentration that did not cause the pre-dendritic crystallization of $Al_{15}(Fe,Mn,M)_3Si_2$ phase leads to an increase in the concentration of these elements ahead of the crystallization front of $\alpha$(Al) dendrites. Further increase in the content of above mentioned elements to a level that can trigger the crystallization of the pre-dendritic $Al_{15}(Fe,Mn,M)_3Si_2$ phase reduces the Cr, Mo, V and W concentration ahead of the crystallization front of $\alpha$(Al) dendrites. The described changes in the concentration of Cr, Mo, V and W ahead of the crystallization front of $\alpha$(Al) dendrites can significantly affect the mechanical properties of HPDC alloy. The relatively high cooling rate during the crystallization of $\alpha$(Al) dendrites, (~55 °C/s), leads to supersaturation of the $\alpha$(Al) phase with chromium, molybdenum, vanadium and tungsten, and may increase the mechanical properties of HPDC alloy. This is confirmed by the results of the statistical analysis of the effect of Cr, Mo, V and W on the tensile strength $R_m$; yield strength $R_{p0.2}$; elongation A and Brinell hardness of HPDC alloy presented in [33–35]. Statistical analysis was performed at the significance level p ($\alpha$) = 0.05. The analysis of variance (ANOVA) test for the main effects was used to assess the influence of high melting point elements on the mechanical properties. The values of strength properties in [33–35] were presented in the standardized and dimensionless form. The results of the statistical analysis showed that each high melting point element can cause an increase in $R_m$, $R_{p0.2}$; A and HB. The data presented in these papers shows an unambiguously analogous nature of the impact of the tested high melting point elements on the properties of this alloy. For the purposes of this paper, standardized quantities were decoded to the form of real values, which can be expressed in the units of measurement proper for the tested mechanical properties. Figure 9 presents an example of the effect of high melting point elements on the $R_m$, $R_{p0.2}$; A and HB. The error bars shown in the graphs represent the values of the Standard Error of the Mean (SEM). The small SEM values shown in (Figure 9) prove the correctness of the results, which indicate the effect of high melting point elements on the tested properties.

Figure 9a shows that Cr at the level of 0.05 wt % allows to obtain the tensile strength $R_m$ = 274.4 MPa. It is a value by ~13% higher in relation to the base alloy. Other high melting point elements make it possible to obtain a high $R_m$ value with a content 0.05–0.10 wt %. 0.05 wt % vanadium has the greatest effect on the increase in $R_m$. This elements results in $R_m$ = 277 MPa, which means an increase by ~14% relative to the base alloy.

The highest yield strength is obtained with the addition of high melting point element in an amount of 0.05 wt %. The element most effective in increasing the value of $R_{p0.2}$ is vanadium (Figure 9b). Its presence produces $R_{p0.2}$ = 126.8 MPa, which means an increase by ~12% relative to the base alloy. The highest elongation is usually obtained when the amount of high melting point element is 0.15 wt %. The greatest impact on the increase in elongation has 0.15 wt % Mo. The elongation is then 5.30%, which gives a 40% increase relative to the base alloy. The data presented in Figure 9c shows that chromium in an amount of 0.15 wt % is also very effective in increasing the elongation to A = 5.21%. Relative to the base alloy, it is a 36% increase.

To obtain high values of the Brinell hardness, the optimal content of high melting point elements is 0.05 wt % for chromium, molybdenum and vanadium, and 0.5 wt % for tungsten. The impact of V and W on HB levels is shown in Figure 9d. The greatest impact on HB increase has 0.5 wt.% W (117 HB). The relative increase is rather small and amounts to 5%.

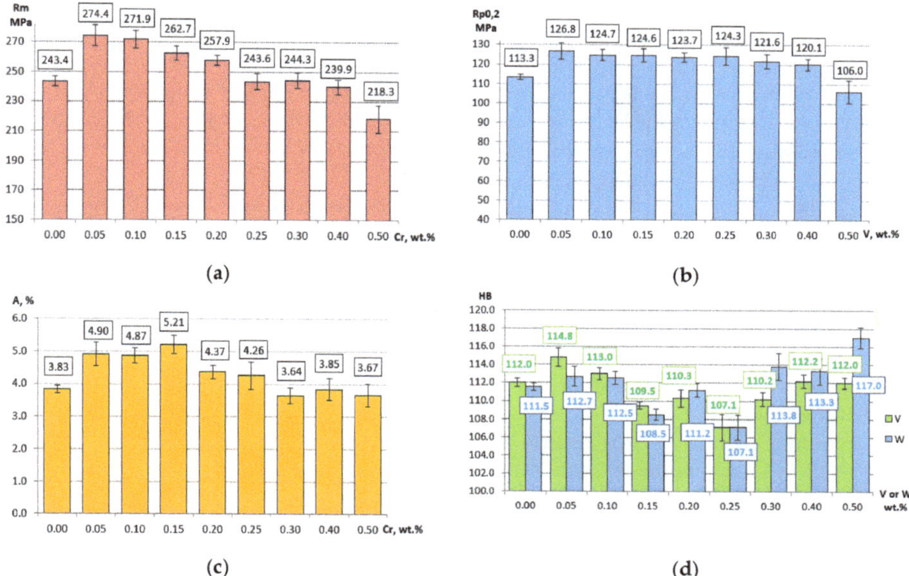


**Figure 9.** Examples of the effect of selected high melting point elements on the mechanical properties: (**a**) the effect of Cr on $R_m$ level; (**b**) the effect of V on $R_{p0.2}$ level; (**c**) the effect of Cr on A level, (**d**) the effect of V and W on HB level.

The highest values of $R_m$ and $R_{p0.2}$ are the result of a relatively high supersaturation of α(Al) solid solution with these elements. The effect of this phenomenon on the strength properties is most beneficial, when the high melting point elements are added in an amount insufficient to start the primary crystallization of $Al_{15}(Fe,Mn,M)_3Si_2$ phase. The supersaturation of α(Al) dendrites is also beneficial for the hardness of alloy containing 0.05 wt % Cr, Mo or V. The highest hardness obtained at 0.5 wt % W was due to the presence of relatively large precipitates of $Al_{15}(Fe,Mn,W)_3Si_2$ phase. The largest increase in elongation accompanying the addition of high melting point elements in an amount of 0.15 wt % was the result of primary crystallization of relatively small and compact precipitates of $Al_{15}(Fe,Mn,M)_3Si_2$ phase, which reduced the concentration of high melting point elements as well as Fe and Mn ahead of the crystallization front of α(Al) dendrites. Crystallization of $Al_{15}(Fe,Mn,M)_3Si_2$ phase in alloy with high melting point elements can reduce the supersaturation of α(Al) dendrites compared with the alloy without them. This leads to an increase in the alloy plasticity.

Statistical analysis in [33–35] gives no evidence supporting the occurrence of a synergistic effect between simultaneously introduced high melting point elements and alloy properties. However, the results of mechanical tests carried out on various combinations of the co-introduced high melting point elements indicate the possibility of obtaining much higher values of $R_m$, $R_{p0,2}$ and A than the values obtained in a statistical analysis carried out for the same additives but introduced as single elements. The largest increase in $R_m$ was obtained with the addition of vanadium and tungsten added in an amount of approximately 0.2 wt % each. The value of $R_m$ was then 299 MPa, which means that it was higher by ~50% relative to the base alloy. The same addition of V and W also caused the largest (over twofold) increase in an elongation. The maximum increase in $R_{p0.2}$, i.e., by ~21%, was due to the use of Cr and W added in an amount of 0.1 wt.% each.

## 4. Conclusions

From the data presented in this paper, the following conclusions emerge:

1. Proper amount of high melting point elements added into HPDC hypoeutectic alloy as well as alloy from shell molds results in pre-dendritic crystallization of the $Al_{15}(Fe,Mn,M)_3Si_2$ phase.
2. High intensity of the heat transfer during HPDC process enables supersaturation of α(Al) dendrites with high melting point elements.
3. The addition of Cr, Mo, V and W as well as pre-dendritic crystallization of the $Al_{15}(Fe,Mn,M)_3Si_2$ phase change concentration of the above mentioned elements ahead of the crystallization front of α(Al) dendrites, resulting in its various supersaturation during HPDC process.
4. A significant increase in mechanical properties of the tested hypoeutectic alloys was obtained by supersaturation of the α(Al) phase with high melting point elements during HPDC process.
5. Changes in mechanical properties of the tested alloys were largely affected by supersaturation of the α(Al) phase, and also by the size of precipitates and content of the pre-dendritic $Al_{15}(Fe,Mn,M)_3Si_2$ phase.
6. In the future, to prove the correctness of the presented hypothetical mechanism of strengthening of HPDC alloy by supersaturation of α(Al) phase with high melting point elements, analyzes should be carried out to show the presence of these elements in α(Al) phase with their different contents.

**Author Contributions:** Conceptualization, T.S., G.G. and T.P.; methodology, T.S., G.G. and T.P.; software, L.K. and J.S.; validation, T.S. and G.G.; formal analysis, T.S., G.G. and J.S.; investigation, T.S. G.G., L.K. and M.G.; resources, T.S.; data curation, T.S.; writing—original draft preparation, T.S. and G.G.; writing—review and editing, T.S. and G.G.; visualization, T.S.; supervision, T.S. and G.G.; project administration, T.P.; funding acquisition, T.P. All authors have read and agreed to the published version of the manuscript.

**Funding:** This research received no external funding.

**Conflicts of Interest:** The authors declare no conflict of interest.

## References

1. Li, Y.; Yang, Y.; Wu, Y.; Wei, Z.; Liu, X. Supportive strengthening role of Cr-rich phase on Al–Si multicomponent piston alloy at elevated temperature. *Mater. Sci. Eng. A* **2011**, *528*, 4427–4430. [CrossRef]
2. Kim, H.Y.; Han, S.W.; Lee, H.M. The influence of Mn and Cr on the tensile properties of A356–0.20Fe alloy. *Mater. Lett.* **2006**, *60*, 1880–1883. [CrossRef]
3. Farkoosh, A.R.; Chen, X.G.; Pekguleryuz, M. Dispersoid strengthening of a high temperature Al–Si–Cu–Mg alloy via Mo addition. *Mater. Sci. Eng. A* **2015**, *620*, 181–189. [CrossRef]
4. Elhadari, H.A.; Patel, H.A.; Chen, D.L.; Kasprzak, W. Tensile and fatigue properties of a cast aluminum alloy with Ti, Zr and V additions. *Mater. Sci. Eng. A* **2011**, *528*, 8128–8138. [CrossRef]
5. Mahta, M.; Ememy, M.; Daman, A.; Keyvani, A.; Campbell, J. Precipitation of Fe rich intermetallics in Cr- and Co-modified A413 alloy. *Int. J. Cast. Met. Res.* **2005**, *18*, 73–79. [CrossRef]
6. Timelli, G.; Bonollo, F. The influence of Cr content on the microstructure and mechanical properties of $AlSi_9Cu_3$(Fe) die-casting alloys. *Mater. Sci. Eng. A* **2010**, *528*, 273–282. [CrossRef]
7. Mbuya, T.O.; Odera, B.O.; Ng'ang'a, P.S. Influence of iron on castability and properties of aluminium silicon alloys: Literature review. *Int. J. Cast Met. Res.* **2003**, *16*, 451–465. [CrossRef]
8. Lin, B.; Li, H.; Xu, R.; Xiao, H.; Zhang, W.; Li, S. Effects of Vanadium on Modification of Iron-Rich Intermetallics and Mechanical Properties in A356 Cast Alloys with 1.5 wt.% Fe. *J. Mater. Eng. Perform.* **2019**, *28*, 475–484. [CrossRef]
9. Mrówka-Nowotnik, G. Intermetallic Phases Examination in Cast $AlSi_5Cu1Mg$ and $AlCu_4Ni_2Mg_2$ Aluminium Alloys in As-Cast and T6 Condition. In *Recent Trends in Processing and Degradation of Aluminium Alloys*; Ahmad, Z., Ed.; IntechOpen: Rijeka, Croatia, 2011; pp. 19–40.
10. Okamoto, H. Al-Cr (Aluminum-Chromium). *J. Phase Equilibria Diffus.* **2008**, *29*, 111–112. [CrossRef]
11. Alloy Phase Diagrams. In *ASM Handbook*; ASM Intertional: Novelty, MO, USA, 1992; Volume 3.
12. Okamoto, H. Al-Mo (Aluminum-Molybdenum). *J. Phase Equilibria Diffus.* **2010**, *31*, 492–493. [CrossRef]
13. Okamoto, H. Si-V (Silicon-Vanadium). *J. Phase Equilibria Diffus.* **2010**, *31*, 409–410. [CrossRef]

14. Okamoto, H. Mo-Si (Molybdenum-Silicon). *J. Phase Equilibria Diffus.* **2011**, *32*, 176. [CrossRef]
15. Guo, Z.; Yuan, W.; Sun, Y.; Cai, Z.; Qiao, Z. Thermodynamic Assessment of the Si-Ta and Si-W Systems. *J. Phase Equilibria Diffus.* **2009**, *30*, 564–570. [CrossRef]
16. Venkataraman, M.; Neumann, J.P. The Cr-Mo (Chromium-Molybdenum) System. *Bull. Alloy Phase Diagr.* **1987**, *8*, 216–220. [CrossRef]
17. Smith, J.F.; Bailey, D.M.; Carlson, O.N. The Cr-V (Chromium-Vanadium) System. *Bull. Alloy Phase Diagr.* **1982**, *2*, 469–473. [CrossRef]
18. Zheng, F.; Argent, B.B.; Smith, J.F. Thermodynamic Computation of the Mo-V Binary Phase Diagram. *J. Phase Equilibria* **1999**, *20*, 370–372. [CrossRef]
19. Okamoto, H. V-W (Vanadium-Tungsten). *J. Phase Equilibria Diffus.* **2010**, *31*, 324. [CrossRef]
20. Nagender Naidu, S.V.; Sriramamurthy, A.M.; Rama Rao, P. The Cr-W (Chromium-Tungsten) System. *Bull. Alloy Phase Diagr.* **1984**, *5*, 289–292. [CrossRef]
21. Murray, J.L.; McAlister, A.J. The Al-Si (Aluminum-Silicon) System. *Bull. Alloy Phase Diagr.* **1984**, *5*, 74–84. [CrossRef]
22. Pietrowski, S. *Al-Si Alloys*; Lodz University of Technology Publishing House: Lodz, Poland, 2001.
23. Belov, N.A.; Eskin, D.G.; Aksenov, A.A. *Multicomponent Phase Diagrams: Applications for Commercial Aluminum Alloys*; Elsevier: London, UK, 2005.
24. Glazoff, M.V.; Khvan, A.V.; Zolotorevsky, V.S.; Belov, N.A.; Dinsdale, A.T. *Casting Aluminum Alloys*; Elsevier: Oxford, UK, 2019.
25. Szymczak, T.; Gumienny, G.; Klimek, L.; Goły, M.; Pacyniak, T. Microstructural characteristics of AlSi9Cu3(Fe) alloy with high melting elements. *Metals* **2020**, *10*, 1278. [CrossRef]
26. Robles Hernández, F.C.; Sokolowski, J.H. Thermal analysis and microscopical characterization of Al–Si hypereutectic alloys. *J. Alloy. Compd.* **2006**, *419*, 180–190. [CrossRef]
27. Matejka, M.; Bolibruchova, D. Influence of Remelting AlSi9Cu3 Alloy with Higher Iron Content on Mechanical Properties. *Arch. Foundry Eng.* **2018**, *18*, 25–30.
28. Shabestari, S.G. The effect of iron and manganese on the formation of intermetallic compounds in aluminum–silicon alloys. *Mater. Sci. Eng. A* **2004**, *383*, 289–298. [CrossRef]
29. Chai, G.; Bäckerud, L.; Rolland, T.; Arnberg, L. Dendrite coherency during equiaxed solidification in binary aluminum alloys. *Metall. Mater. Trans. A* **1995**, *26A*, 965–970. [CrossRef]
30. Nývlt, J. Kinetics of nucleation in solutions. *J. Cryst. Growth* **1968**, *3*, 377–383.
31. Nemdili, L.; Koutchoukali, O.; Mameri, F.; Gouaou, I.; Koutchoukali, M.S.; Ulrich, J. Crystallization study of potassium sulfate-water system, metastable zone width and induction time measurements using ultrasonic, turbidity and 3D-ORM techniques. *J. Cryst. Growth* **2018**, *500*, 44–51. [CrossRef]
32. Wu, X.; Zhang, H.; Ma, Z.; Jia, L.; Zhang, H. Effect of Holding Pressure on Microstructure and Mechanical Properties of A356 Aluminum Alloy. *J. Mater. Eng. Perform.* **2018**, *27*, 483–491. [CrossRef]
33. Szymczak, T.; Szymszal, J.; Gumienny, G. Statistical methods used in the assessment of the influence of the Al-Si alloy's chemical composition on its properties. *Arch. Foundry Eng.* **2018**, *18*, 203–211.
34. Szymczak, T.; Szymszal, J.; Gumienny, G. Evaluation of the effect of the Cr, Mo, V and W content in an Al-Si alloy used for pressure casting on its proof stress. *Arch. Foundry Eng.* **2018**, *18*, 105–111.
35. Szymczak, T.; Szymszal, J.; Gumienny, G. Evaluation of the Effect of Cr, Mo, V and W on the Selected Properties of Silumins. *Arch. Foundry Eng.* **2018**, *18*, 77–82.


**Publisher's Note:** MDPI stays neutral with regard to jurisdictional claims in published maps and institutional affiliations.

*Article*

# Study of Natural and Artificial Aging on AlSi9Cu3 Alloy at Different Ratios of Returnable Material in the Batch

**Dana Bolibruchová [1], Marek Matejka [1,*], Alena Michalcová [2] and Justyna Kasińska [3]**

1 Faculty of Mechanical Engineering, Department of Technological Engineering, University of Zilina, Univerzitná 8215/1, 010 26 Žilina, Slovakia; danka.bolibruchova@fstroj.uniza.sk
2 Department of Metals and Corrosion Engineering, University of Chemistry and Technology in Prague, Technická5, 166 28 Prague 6, Czech Republic; michalca@vscht.cz
3 Department of Metal Science and Materials Technology, Kielce University of Technology, Al. Tysiąclecia Państwa Polskiego 7, 25 314 Kielce, Poland; kasinska@tu.kielce.pl
* Correspondence: marek.matejka@fstroj.uniza.sk; Tel.: +421-41-513-2772

Received: 4 September 2020; Accepted: 9 October 2020; Published: 13 October 2020

**Abstract:** Aluminum alloys currently play an important role in the production of castings in various industries, where important requirements include low component weight, reduction of the environmental impact and, above all, reduction of production costs of castings. One way to achieve these goals is to use recycled aluminum alloys. The effect of natural and artificial aging of AlSi9Cu3 alloy with different ratios of returnable material in the batch was evaluated by a combination of optical, scanning, transmission microscope and mechanical tests. An increase in the returnable material in the batch above 70% resulted in failure to achieve the minimum value required by the standard for tensile strength and ductility. The application of artificial aging had a positive effect on the microstructure and thus on the mechanical properties of experimental alloys. By analyzing the results from TEM, it can be stated that in the given cases there is a reduced efficiency of $\theta'$-$Al_2Cu$ precipitate formation with an increase of the returnable material in the batch and in comparison with artificial aging, which is manifested by low mechanical properties.

**Keywords:** Al-Si-Cu secondary aluminum alloy; returnable material; natural and artificial aging; Cu precipitate; transmission electron microscopy; mechanical properties

## 1. Introduction

At present, more than half of the aluminum castings produced come from recycled materials. A substantial part of the use (approximately 70%) of recycled aluminum is in the production of aluminum alloys cast mainly for the automotive industry. Secondary aluminum metallurgy as well as many other recycling processes are all important issues from both an economic and environmental point of view. The recycling process enables the economy of raw materials and energy savings [1–3]. Another positive aspect of aluminum recycling is the impact on the environment. The production of secondary aluminum alloys releases only 5% of greenhouse gases compared to the production of primary aluminum, as the processes associated with mining, ore refining and smelting are eliminated. For this reason, today foundries use in the batch an increasing amount of remelted material, which can make up several tens of percent of the total batch [4,5].

Despite the knowledge gained so far about the effect of multiple remelted material, there is no clear opinion on its effect. At present, determining the correct amount of remelted material in the batch is a major problem for foundries. The use of remelted return material carries with itself a high risk of a decrease in the overall quality of the casting, even though the recycled material used meets all the requirements for the input material. For this reason, it is very important to understand the remelting process and the subsequent use of the remelted material in the batch. The main goal of the paper is a more detailed description of the method and the remelted material influence in the batch on the structure and mechanical properties after natural and artificial aging. Determining the correct ratio with respect to the amount of remelted material in the batch is currently one of the key factors in increasing the competitiveness of foundries. Despite the fact that this is an important problem, it is not given sufficient attention from a scientific point of view and there are very few professional publications on the issue. The results of the experiments present how the quality of castings decreases with increasing amount of remelted material in the batch.

Al-Si-Cu alloys are the most widely used type of aluminum alloys. They make up about half of the total production of aluminum castings and are mainly used in applications for the automotive and aerospace industries. These are most often sub-eutectic (exceptionally eutectic) alloys with a content of 6 to 13 wt.% Si and 1 to 5 wt.% Cu [6]. Thanks to copper, the alloy has good mechanical properties and excellent machinability, but copper reduces the resistance to corrosion. However, this is sufficient for use in the automotive industry. Copper also allows heat treatment by hardening to form the intermetallic $Al_2Cu$ compound. The alloy has a low tendency to crack and shrink. To achieve these properties, it is advantageous to keep the silicon content in the upper tolerance range for the alloy [7].

Due to the increasing use of recycled aluminum alloys for demanding castings, their quality is considered a key factor and it is therefore necessary to find the right compromise between price and final quality. The quality of an aluminum alloy made from recycled materials can be affected mainly by a change in the morphology of the individual structural components. The change in morphology is most often caused by a change in the chemical composition or by the effect of multiple remelting of the alloy [8–10].

Different methods are used to neutralize the negative effects of remelting or an increase in the returnable material in the batch. One of the possible methods for Al-Si-Cu-based alloys is the application of precipitation hardening—aging. This is a diffusion process that takes place in the structure of the casting (often after solution annealing), in which the solid solution is depleted by additive elements. The supersaturated solid solution contains a larger amount of an additive dissolved than is thermodynamically advantageous for a given alloy at a given temperature. Such a solid solution has a natural tendency to reach the state with the lowest free enthalpy, with the following disintegration. This decay most often takes place by a heterogeneous phase transformation, i.e., precipitation [11–14].

The mechanical and physical properties of the casting change during precipitation from the supersaturated solid solution. Sensitive structural properties such as hardness, ductility, yield strength or corrosion resistance are highly dependent on the distribution of the individual phases in the structure. The greatest hardening is caused by the initial stages of precipitation. The values of yield strength, tensile strength and hardness are reduced by reaching the critical size of the precipitates in the precipitation-strengthened alloys. Precipitation can also affect the properties of the alloy in an undesirable way, such as the formation of structural instability or tempering brittleness.

The precipitation process begins with the diffusion of the additive element into microscopic regions richer in the given element, and a new phase is nucleated in them. The growth of these nuclei produces coherent precipitates referred to as Guinier–Preston zones (GPZ). Coherence means that these regions are part of the solid crystal lattice, thereby deforming the lattice and causing internal stresses in it. The induced stresses are the cause of the increasing strength and hardness of the alloy. The second stage is the formation of particles of non-equilibrium (GP II) coherent precipitates $\theta''$. The deformation fields present in the lattice surrounding these coherent particles inhibit and prevent the movement of dislocations, thus leading to the most significant increase in strength. In the further process, the phase begins to form separate units with a gradual loss of coherence. The formations cease to be crystallographically connected to the original supersaturated solid solution, and generally partially coherent acicular-shaped particles with their own crystal lattice are formed. Such a transition precipitate—phase $\theta'$, gradually changes size and morphology into acicular shapes.

The final stage of the aging process is the gradual loss of coherence and the formation of an incoherent particle of plate-like or lenticular shape. The incoherent hardening phase no longer has a crystalline bond to the α (Al) solid solution. At this stage, the strength and hardness decrease and the alloy becomes over-aged. A schematic representation of the precipitation process is shown in Figure 1. Maximum strength characteristics after hardening are obtained in the region of coherent precipitates θ″ and at the beginning of the precipitation of the partially coherent phase θ′ [15–20].

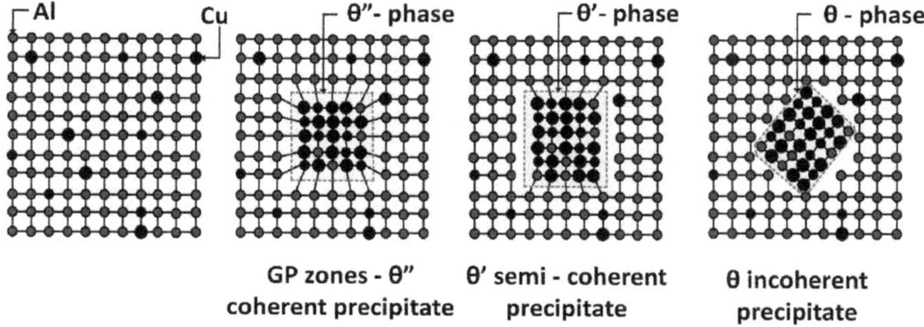


**Figure 1.** Schematic representation of the precipitation process [7].

## 2. Materials and Methods

### 2.1. Experimental Material

A sub-eutectic alloy AlSi9Cu3 (EN AC-46,000, A226) was chosen to evaluate the effect of increasing the ratio of returnable material in the batch. The primary AlSi9Cu3 alloy is characterized by medium mechanical characteristics, good strength at elevated temperatures, good foundry properties and corrosion resistance. Due to its properties, it is mainly used in automotive industry products, such as in cylinder heads and engine blocks, crankshaft housings and other components [6–8].

Two types of AlSi9Cu3 alloy were used in the experiment. The first type was a commercial purity AlSi9Cu3 secondary alloy ingots purchased from the company Dor, Považská Bystrica, Slovakia. The other type of AlSi9Cu3 alloy was prepared by remelting a typical foundry returnable material, such as ingot remnants, gating and riser systems. The foundry returnable material was remelted and cast into the shape of ingots. The remelting of the returnable material with a total batch weight of 95 kg took place in an electric resistance furnace with a volume of 100 kg in a steel crucible with applied protective graphite coating. The chemical composition of the alloy according to the standard (EN 1706) of the secondary alloy (commercial purity) and the alloy prepared by remelting the foundry returnable material is given in Table 1.

**Table 1.** Chemical composition of primary AlSi9Cu3 by standard, secondary alloy and returnable materials of AlSi9Cu3 alloy (wt.%).

| Elements | Si | Fe | Cu | Mn | Mg | Ni | Zn | Ti | Cr |
|---|---|---|---|---|---|---|---|---|---|
| Primary AlSi9Cu3 (EN 1706) | 8.0–11.0 | 0.6–1.1 | 2.0–4.0 | 0.55 | 0.15–0.55 | 0.55 | 1.20 | 0.20 | 0.15 |
| Commercial purity AlSi9Cu3 | 9.563 | 1.081 | 2.206 | 0.184 | 0.426 | 0.092 | 1.160 | 0.038 | 0.027 |
| Returnable AlSi9Cu3 | 9. 294 | 1.674 | 2.074 | 0.184 | 0.348 | 0.129 | 1.016 | 0.034 | 0.113 |

### 2.2. Experimental Methods

In the next part of the experiment, five alloys were cast in succession with the designation 20–80; 50–50; 70–30; 80–20; and 90–10, where the first number indicates the percentage of returnable material and the second number the proportion of secondary alloy in the batch. Each batch weighed 12.5 kg.

Melting was performed in an electric resistance furnace (LAC, Židlochovice, Czech Republic) with a T15 type regulator with a capacity of 15 kg in a graphite crucible, which was treated with a protective coating. The casting took place at a temperature of 750 ± 5 °C. The samples were cast from each melt for structural analysis and mechanical property tests (10 samples) under the same conditions. Casting was performed into a metal mold with a temperature of 150 ± 5 °C. We did not refine, inoculate or modify the alloys in any way in the process, and only the oxide membranes were mechanically removed before casting. The chemical composition of the newly formed alloys is given in Table 2. Structural analysis and mechanical characteristics of the samples were evaluated after natural aging (NA—approximately 160 h at 20 °C) and after artificial aging T5 (AA—at 200 ± 5 °C for 4 h and cooling in water with a temperature of 60 ± 5 °C). Chemical composition was measured using arc spark spectroscopy (Bunker-Q2 ION, Kalkar, Germany).

**Table 2.** Chemical composition of experimental alloys AlSi9Cu3 alloy (wt.%).

| Elements | Si | Fe | Cu | Mn | Mg | Ni | Zn | Ti | Cr |
|---|---|---|---|---|---|---|---|---|---|
| 20–80 | 9.507 | 1.294 | 2.197 | 0.231 | 0.391 | 0.122 | 1.044 | 0.035 | 0.049 |
| 50–50 | 9.418 | 1.419 | 2.173 | 0.223 | 0.361 | 0.134 | 1.041 | 0.033 | 0.072 |
| 70–30 | 9.245 | 1.569 | 2.02 | 0.209 | 0.344 | 0.108 | 0.961 | 0.031 | 0.112 |
| 80–20 | 9.415 | 1.617 | 2.08 | 0.206 | 0.358 | 0.156 | 1.07 | 0.032 | 0.101 |
| 90–10 | 9.291 | 1.643 | 2.143 | 0.199 | 0.357 | 0.127 | 1.046 | 0.032 | 0.106 |

Samples (10 mm × 10 mm) for metallographic evaluation were prepared by standard metallographic procedures (coarse and fine wet grinding, polishing on an automatic instrument using a diamond emulsion, and etching). For light microscopy purposes, samples were etched with 20 mL $H_2SO_4$ + 100 mL distilled water to enhance the ferrous intermetallic phases.

For SEM (scanning electron microscope) and EDX (energy dispersive X-ray) analysis, samples were etched with 0.5% HF solution. For deep etchings, an etchant consisting of 36 mL HCl + 100 mL $H_2O$ was used, etching the surface of the samples for approximately 30 s, to reveal the three-dimensional morphology of eutectic silicon and intermetallic phases. ($\alpha$-phase) lattice residues were removed by vigorous rinsing with alcohol-based liquid. The microstructure of the experimental material was evaluated using a NEOPHOT 32 optical microscope and SEM observations with EDX analysis using a VEGA LMU II scanning electron microscope (Tescan, Brno, Czech Republic) connected to energy dispersive X-ray spectroscopy (BruckerQuantax EDX analyzer, Bunker, Kalkar, Germany).

Observation of the structure by TEM (transmission electron microscopy) and by the experimental technique of SAED (Selected Area Electron Diffraction) was performed on a Jeol 2200 FS device (JEOL Ltd., Tokyo, Japan). Samples were prepared by re-polishing using a Gatan PIPs (precision ion polishing system, Gatan, Pleasanton, CA, USA). The principle consists in de-dusting the sample with a stream of ionized argon. The samples were circular in shape with a diameter of 3 mm and a thickness of 70 micrometers. At the beginning of re-polishing, an angle of 8° was used, and when an orifice appeared in the sample, the angle was gradually reduced to 4° and 2° on both sides of the sample. This ensured that the area around the orifice was thin enough to allow the material to be observed. For TEM, the sample is transparent to a thickness of 100 nm. Such a procedure makes it possible to maintain more massive edges and to have the center of the sample transparent.

The evaluation of microhardness of structural components was performed on a Hanemann device type Mod 32 at a load of 10 p. The resulting value presents the average value from 20 measurements.

The tensile strength test was performed according to EN 42 0310 using a WDW 20 device (Jnkason, Jinan, China). The maximum load of the device was 20 kN at a constant crosshead feed rate of 2 mm/min. The evaluation of ductility $A_{50}$ was performed on the MC electronic manual extensometer device (Jnkason, Jinan, China). The cast samples were mechanically machined into test bars with a diameter of 10 mm in five pieces for each condition. The Brinell method on an INNOVATEST NEXUS 3002XLM-INV1 hardness tester (Innovatest, Borgharenweg, Maastricht, The Netherlands) was used to measure hardness. A 5 mm diameter bead with a load of 125 kP (1226 N) was pressed into the prepared sample for 15 s. A minimum of five measurements were made for each sample.

Twelve castings were cast in each melt. Finished castings were removed from the mold after approximately 60 s in an effort to maintain the same solidification conditions for each casting.

## 3. Results

### 3.1. Mechanical Properties

The influence of applying natural and artificial aging on AlSi9C3 alloy with different amounts of returnable material in the batch on mechanical properties (Ultimate tensile strength—UTS, Yield strength—YS, Ductility—$A_{50}$ and Brinell hardness—HBW) is shown in Figures 2 and 3. The numerical value represents the average value of the five measurements for both alloy states.

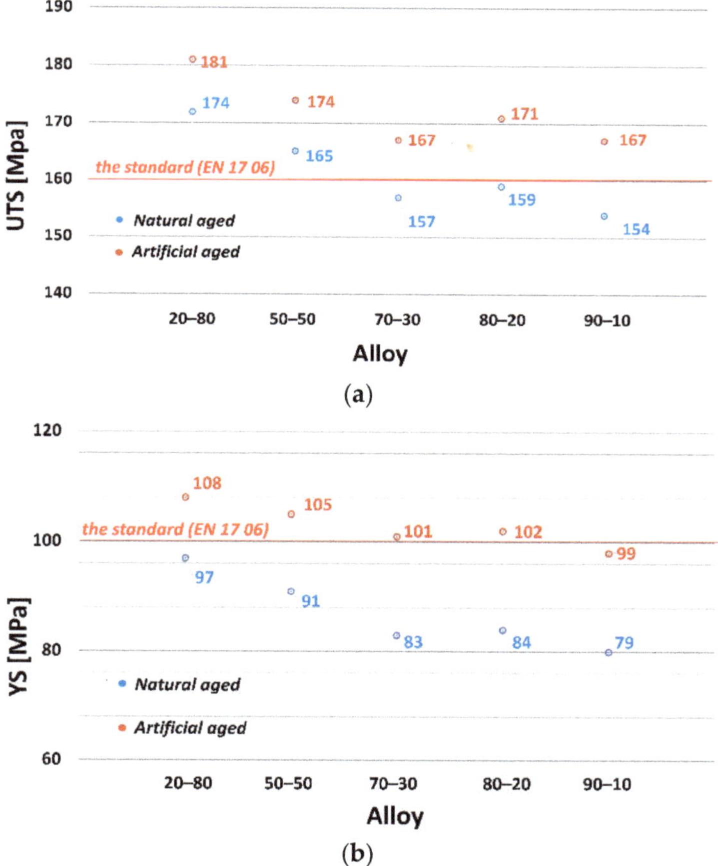


**Figure 2.** Relationship between: (**a**) Ultimate strength tensile; (**b**) Yield strength and experimental alloy and the state the alloy.

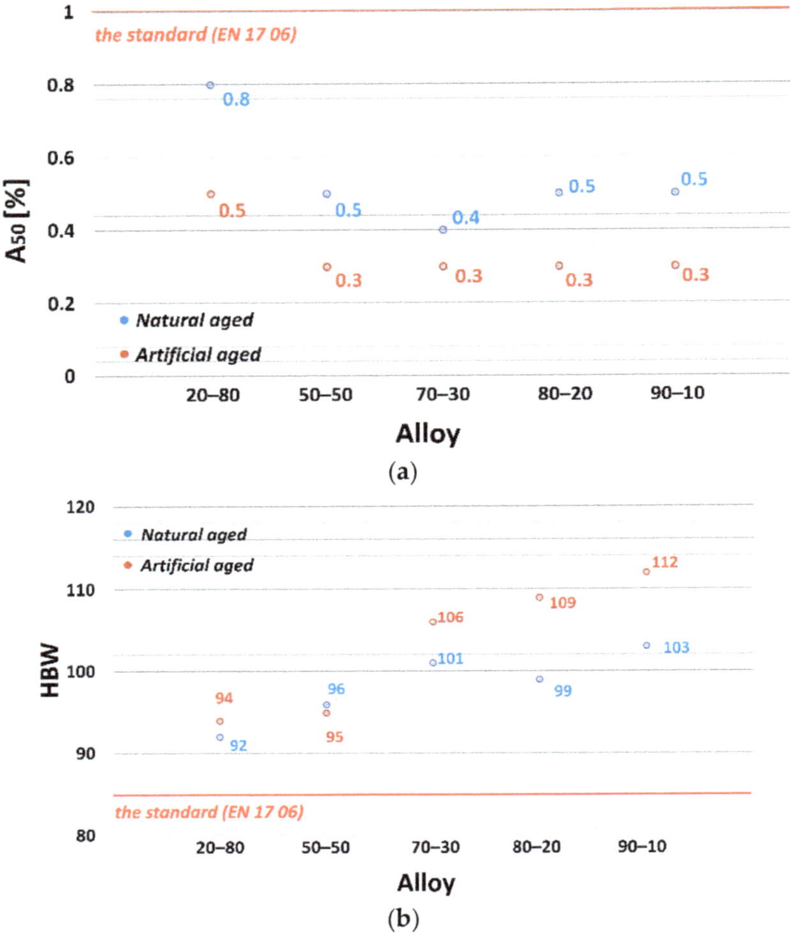


**Figure 3.** Relationship between: (**a**) Elongation; (**b**) Brinell hardness and experimental alloy and the state the alloy.

The best values were obtained for the alloy with a 20% returnable material content, UTS = 174 MPa (Figure 2a) and YS = 97 MPa (Figure 2b). By increasing the proportion of returnable material to 70%, there was a gradual decrease in the observed characteristics. From the alloys with the ratios 70–30 to 90–10 there was a stabilization and the values ranged at approximately the same levels. The lowest values of UTS = 154 MPa and YS = 79 MPa were measured for the alloy with the highest amount of returnable material—the 90–10 alloy. By applying artificial aging (T5) (red dots on the graph), an increase was achieved in tensile strength and agreed yield strength in all alloys. The resulting values of UTS and YS show a slight increase in artificial aging (200 °C/4 h). For the 20–80 alloy, the increase in UTS was approximately by 4% due to artificial aging, and for alloys with 70% and higher ratios, the increase in UTS was approximately by 8% compared to natural aging. The red line in the graphs indicates the minimum values required by the standard (EN 17 06) for gravity-cast AlSi9Cu3 alloy.

A decreasing trend was also observed in the evaluation of ductility (Figure 3a). Due to the increasing proportion of returnable material in the batch, there was a significant decrease in ductility values. From the maximum measured value of $A_{50}$ = 0.8% (20–80 alloy) to the value of $A_{50}$ = 0.4%

(70–30 alloy). After artificial aging, there was a decrease in ductility in all cases. The minimum ductility $A_{50}$ = 0.3% was measured for alloys with 50% and higher proportion of returnable material in the batch. The resulting HBW hardness values (Figure 3b) had an increasing trend due to the increase in the returnable material in the batch. The increase in the HBW hardness can be attributed to the increasing weight % of Fe in the alloys, which predetermines the increased number of iron phases in the structure and thus the increase in hardness [10,21]. Artificial aging, as expected, increased HBW hardness values. Maximum HBW hardness values were measured for the 90–10 alloy, namely, HBW = 103 (natural aging) and HBW = 112 (artificial aging). The red line in the graphs indicates the minimum values required by the standard (EN 17 06) for gravity cast AlSi9Cu3 alloy.

### 3.2. Microstructure

#### 3.2.1. Natural Aging

The structure of the subeutectic AlSi9Cu3 alloy consists of α-phase (pale gray), eutectic (dark gray) and intermetallic phases with different chemical compositions, most often based on Cu and Fe. In the microstructure of the 20–80 experimental alloy, eutectic grains are present in the so-called unmodified shape as hexagonal plate crystals with unoriented distribution. Iron-based intermetallic phases crystallized in interdendritic regions, most frequently as $Al_5FeSi$ phases in the plate morphology (aciculars in metallographic cutting) (Figures 4a and 5a). Copper-rich intermetallic phases were primarily excreted near eutectic silicon grains, especially aciculars of iron phases, mainly as $Al_2Cu$ in the form of Al-$Al_2$Cu-Si containing about 24 wt.% Cu. The increased proportion of returnable material resulted in a local thickening of the eutectic silicon grains, an increase in the lengths of the iron phase aciculars, and in their local thickening (Figures 4b,c and 5b). Cu-based intermetallic phases were also observed in alloys with 70% and higher returnable material content with increased Fe content as $Al_7FeCu_2$ (Figure 5c).

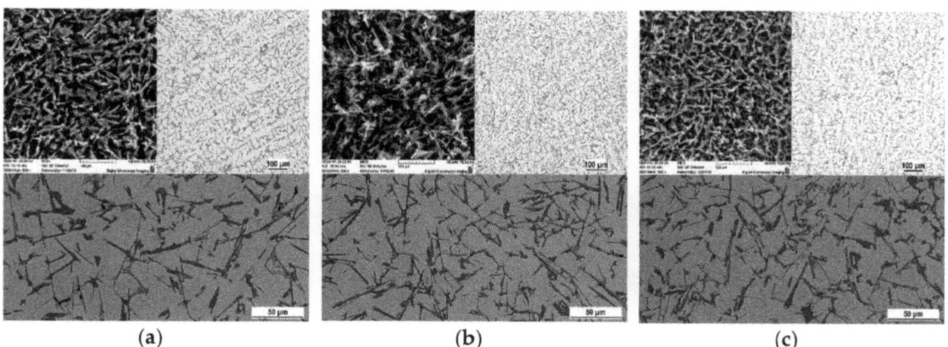


(a) (b) (c)

**Figure 4.** Microstructure of AlSi9Cu3 alloy depend on the ratio of returnable material in the batch after natural aging, SEM; (**a**) 20–80 alloy, (**b**) 50–50 alloy, (**c**) 90–10 alloy.

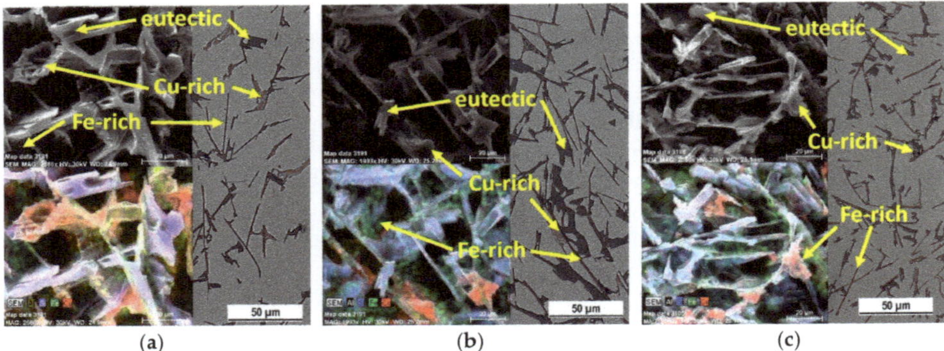


(a) (b) (c)

**Figure 5.** Changes of eutectic Si, Cu-rich and Fe-rich phases of AlSi9Cu3 alloy depend on the ratio of returnable material in the batch after natural aging, SEM; (**a**) 20–80 alloy, (**b**) 50–50 alloy, (**c**) 90–10 alloy.

### 3.2.2. Artificial Aging

The application of artificial aging did not significantly change the morphology of eutectic silicon. For the 20–80 alloy, a hexagonal plate formation with well-recognizable twinning could be observed (Figure 6a). Artificial aging caused only local growth of the arched edges without subsequent thinning and fragmentation. Similarly, without heat treatment, eutectic silicon in the 50–50, 70–30 and 90–10 alloys is present in an undirected distribution of plates, or as unmodified eutectic (Figure 6b,c).

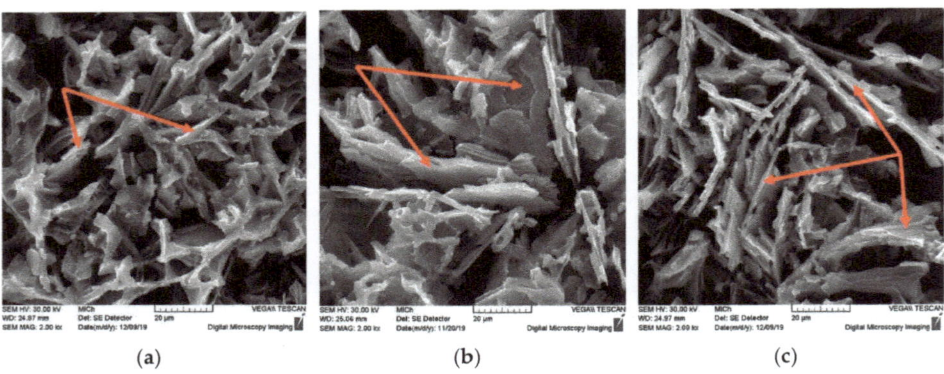

(a) (b) (c)

**Figure 6.** Changes of eutectic Si of AlSi9Cu3 alloy depend on the ratio of returnable material in the batch after artificial aging, SEM; (**a**) 20–80 alloy, (**b**) 50–50 alloy, (**c**) 90–10 alloy.

Artificial aging had no significant effect on the morphology of the copper- and iron-rich phases (Figure 7). Cu-based intermetallic phases are present as $Al_2Cu$ with tetragonal crystal lattice and $Al_7FeCu_2$ [22,23]. In all experimental alloys, artificial aging resulted in a local shortening of Fe-based intermetallic phase lengths compared to natural aging.

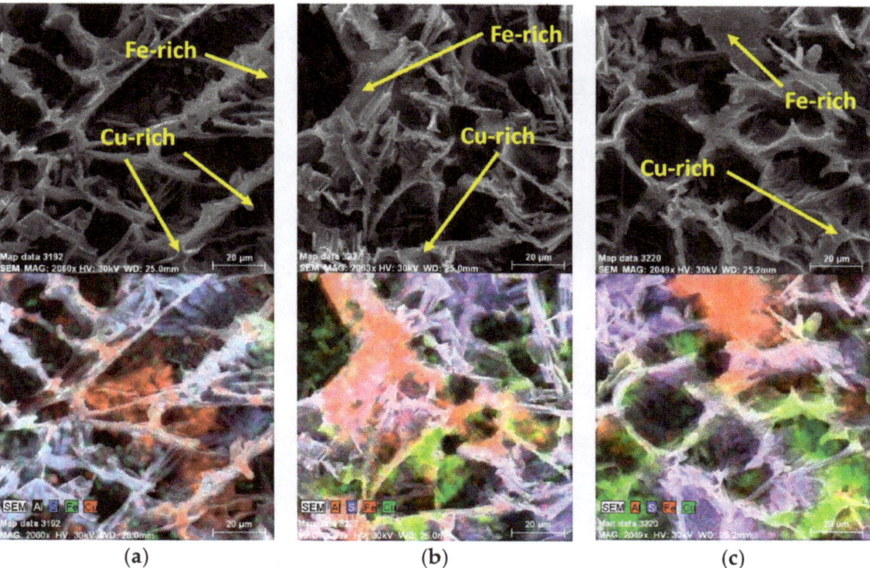


**Figure 7.** Changes of Cu-rich and Fe-rich phases of AlSi9Cu3 alloy depend on the ratio of returnable material in the batch after artificial aging, SEM; (**a**) 20–80 alloy, (**b**) 50–50 alloy, (**c**) 90–10 alloy.

### 3.3. Substructure

Two types of alloys were selected for transmission electron microscopy (TEM), with the lowest (20–80) and highest (90–10) ratios of returnable material in the batch after natural and artificial aging. Experimental alloys 20–80 and 90–10 represent alloys with the best or the worst results of mechanical tests. The use of TEM is the most effective way to observe the occurrence of the metastable phase $\theta'$-$Al_2Cu$, which is responsible for precipitation hardening in the alloy and fundamentally affects the mechanical properties. The elevated temperature precipitation follows the same mechanism for crystalline [24] and also for amorphous alloys [25]. At first, the non-stoichiometric clusters are formed due to diffusion. This local density variation serves as nuclei for precipitates growth. The precipitates follow during their growth crystallography of matrix grain [24] or even crystallography of the substrate in case of amorphous matrix [26]. The formation of precipitates can be predominantly proven by SAED because of its sufficient spatial resolution.

#### 3.3.1. 20–80 Alloy

In the alloy with the lowest amount of returnable material after natural aging, precipitates are present in the morphology of thicker rods with an approximate length of 800 nm (Figure 8). The presence of a structural defect is indicated at the bottom of the image (dark spots). These are mainly clusters of dislocations, or small-angular grain boundaries, but the shown defects could also be introduced by the sample preparation process (mechanical grinding, ion polishing).

The application of EDX mapping showed that the rod particles are primarily formed by Si (Figure 9). In more detail, and using the HAADF (high annular angular dark field) mode, the presence of the $\theta'$-$Al_2Cu$ phase was shown in the acicular morphology located in close proximity to the Si particles (Figure 9b). In addition to the acicular morphology, the $\theta'$-$Al_2Cu$ phase is also in the morphology of the sharp-edged plate formation (Figure 9a). The sharp-edged $\theta'$-$Al_2Cu$ phase was also observed in the subsequent evaluation, especially near the Si- and Fe-based phases. Spot EDX analysis showed an increased presence of Cu in phases designated 1, 2 and 3 (Figure 10).

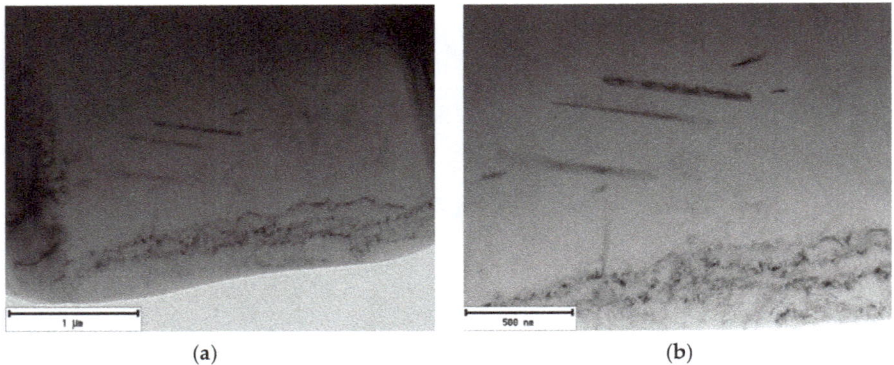


**Figure 8.** Substructure of 20–80 alloy after natural aging, TEM; (**a**) Overview (**b**) Detailed view.

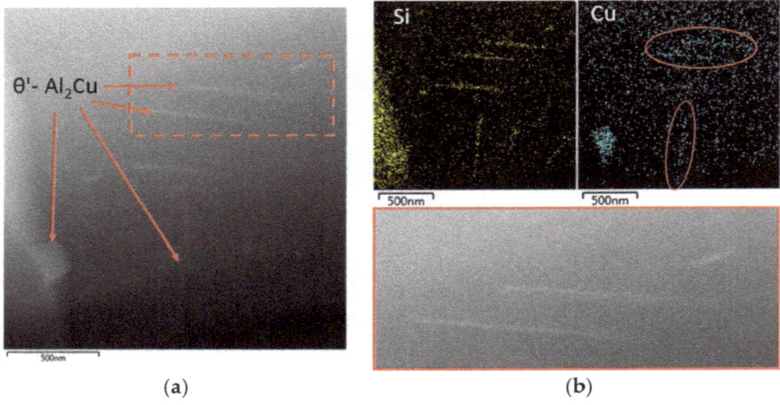


**Figure 9.** Substructure of 20–80 alloy after natural aging focus on θ'-$Al_2Cu$ phases, TEM, STEM/HAADF; (**a**) 20–80 alloy, (**b**) Morphology of the θ'-$Al_2Cu$ phase and distribution of Si and Cu elements and detail of needle morphology of θ'-$Al_2Cu$ phase.

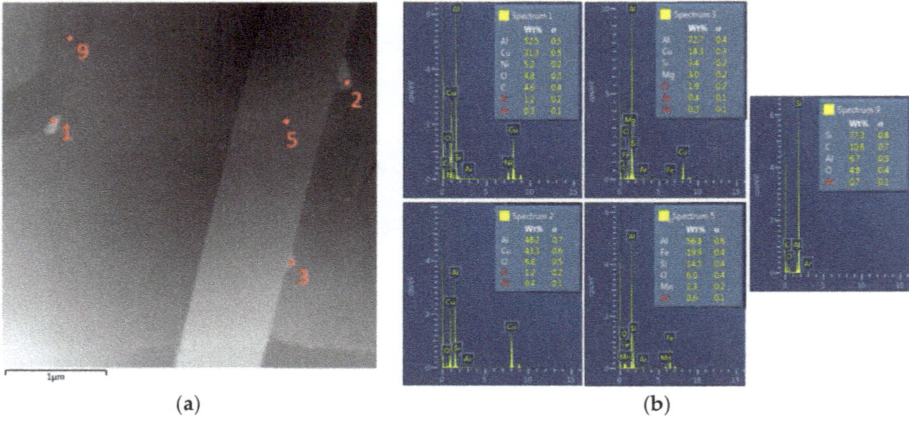


**Figure 10.** Point EDX analysis of 20–80 alloy substructure after natural aging; (**a**) Places of point EDX analysis, (**b**) Results of chemical composition of structural components.

The substructure of the 20–80 alloy after artificial aging, which exhibited "ideal" properties, is characterized by a large number of perpendicularly oriented acicular particles. It is a rapid increase in these particles compared to the alloy after natural. Structural disorders (especially clusters of dislocations) are present to approximately the same extent as in natural aging (Figure 11). Similar to the 20–80 alloy, after natural aging, perpendicularly oriented acicular particles are exhibited by θ′-$Al_2Cu$ phases and Si-rich phases. During the application of mapping, phases were observed again in acicular and also plate morphology (Figure 12).

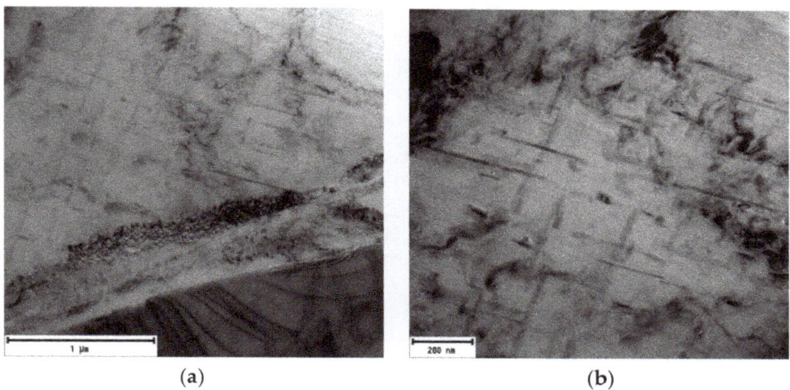


(**a**) (**b**)

**Figure 11.** Substructure of 20–80 alloy after artificial aging, TEM; (**a**) Overview (**b**) Detailed view.

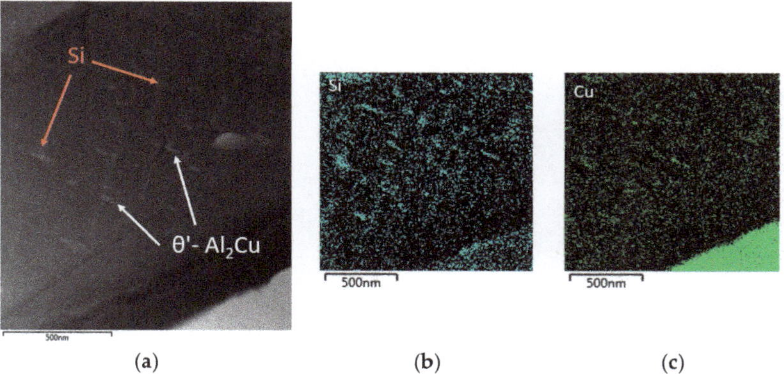


(**a**) (**b**) (**c**)

**Figure 12.** Substructure of 20–80 alloy after artificial aging TEM; (**a**) θ′-$Al_2Cu$ and Si phases, (**b**) mapping of Si (**c**) mapping of Cu.

The experimental technique of SAED (selected area electron diffraction) forming a part of the transmission electron microscope was used to evaluate the substructure of the 20–80 alloy after artificial aging. SAED provides information on crystallography and orientation of crystalline materials (Figure 13). The regularity of the maxima (dashes) of precipitates θ′ can be observed in the SAED image, which indicates their regular orientation towards the matrix and is exactly crystallographically given.

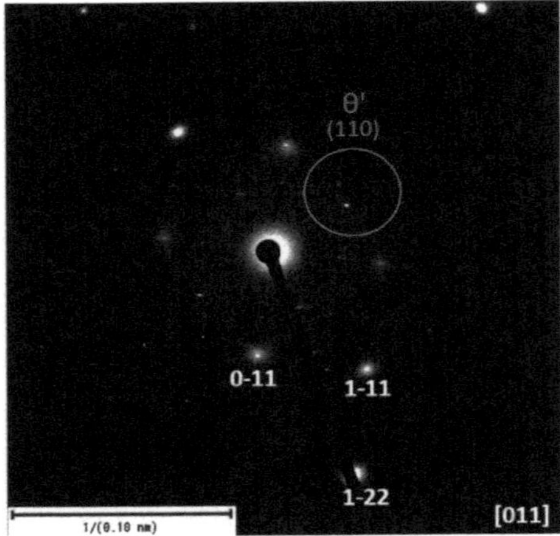


**Figure 13.** SAED of θ′-$Al_2Cu$ phases of 20–80 alloy after artificial aging.

### 3.3.2. 90–10 Alloy

In the substructure of the alloy with the highest amount of returnable material after natural aging, similarly to the 80–20 alloy, the presence of a smaller number of rod formations with a length of 500 to 1000 nm and lattice defects of the dark cluster can be seen (Figure 14). Application of artificial aging in the 90–10 alloy allows us to observe increased occurrence of particles in acicular morphology, but also in plate-like one with rounded edges (Figure 15).

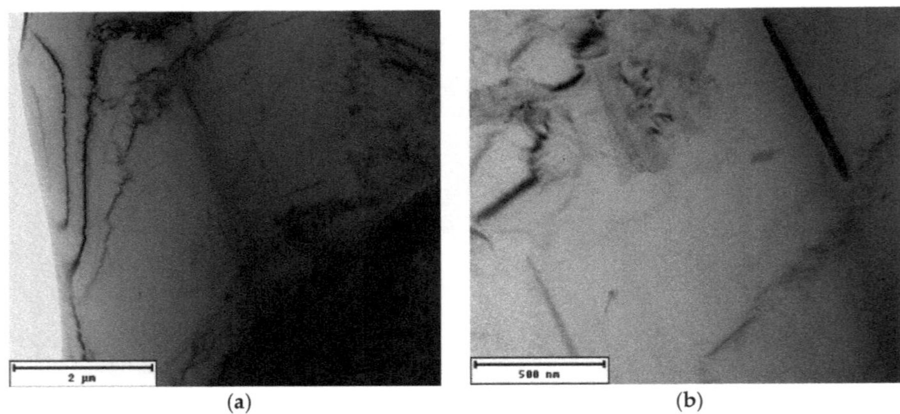


**Figure 14.** Substructure of 90–10 alloy after natural aging, TEM; (**a**) Overview (**b**) Detailed view.

Element mapping confirmed the presence of θ′-$Al_2Cu$ precipitates in both alloys (both after natural and artificial aging) (Figure 16). In the 90–10 alloy after artificial aging, Si-based particles were also present in the substructure along with the precipitates. (Figure 16b).

The microhardness evaluation of the primary α-phase confirmed the increased presence of the θ′-$Al_2Cu$ phase after the application of artificial aging that is responsible for precipitation hardening in AlSi9Cu3 alloys. The trend of the microhardness values was decreasing, with increasing ratio of returnable material in the batch (with the exception of the 80–20 alloy) (Figure 17). The application of artificial aging resulted in increased values of all alloys, which confirms the increased ability to form precipitates in the substructure of experimental alloys.

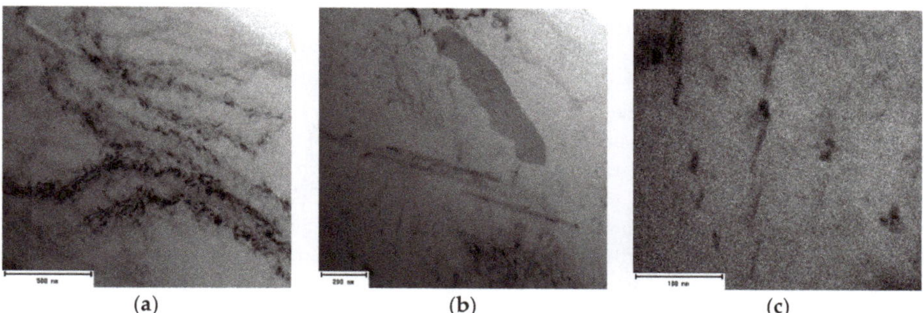


**Figure 15.** Substructure of 90–10 alloy after artificial aging, TEM; (**a**) Overview (**b**) Detailed view (**c**) High resolution.

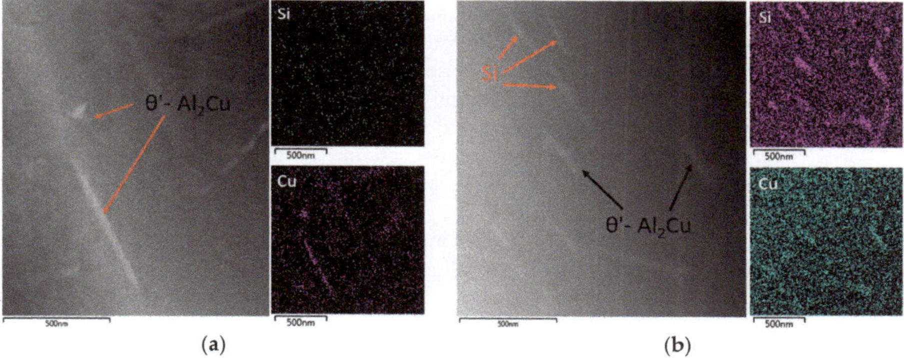


**Figure 16.** Substructure of 90–10 alloy focus on θ′-$Al_2Cu$ phases, TEM; (**a**) natural aging, (**b**) artificial aging.

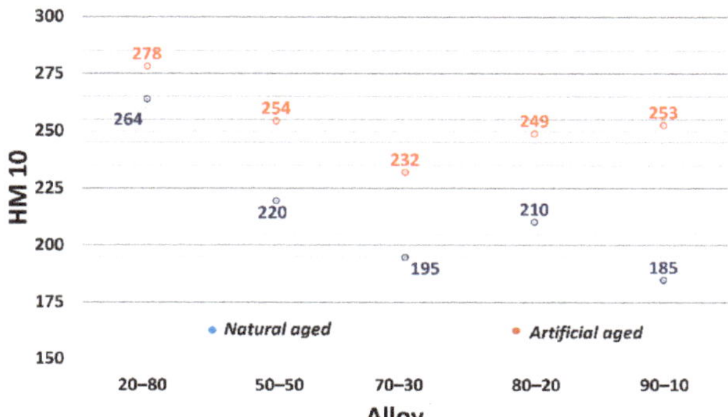


**Figure 17.** Dependence of the microhardness of the primary α-phase on AlSi9Cu3 alloys from the different ratio of remelted returnable material in the batch.

## 4. Discussion

The evaluation of the obtained results confirmed the expected decrease in the overall quality of the AlSi9Cu3 alloy due to the increasing proportion of recycled material in the batch from 20% to 90%. The negative effect of the recycled material increase in the batch was already manifested at the amount of 50% (alloy 50–50). The microstructure of alloy 50–50 is characterized by an increased presence of harmful iron phases and by the degraded grains of Si compared to an alloy with a lower amount of recycled material in the batch (alloy 20–80). By further increasing the amount of recycled material in the batch to the level of 70% to 90%, a large amount of harmful iron phases was formed, which had a major impact on the gradual decrease of the investigated mechanical characteristics except for the hardness of these alloys (70–30, 80–20 and 90–10) [27–29]. Compared to the EN 17 06 standard, only the alloys with a high proportion of returnable material (70–30, 80–20, 90–10) did not reach the required minimum tensile strength, and the minimum agreed yield strength was not reached by any investigated alloy after natural aging.

The application of artificial aging had a positive effect on the length of the acicular iron phases and local refinement of eutectic silicon, and thus on the resulting mechanical properties of the investigated alloys, which in all cases exceeded the minimum value required by the standard when assessing tensile strength and agreed yield strength. On the contrary, it had a negative effect only when evaluating the ductility of alloys with different amounts of returnable material in the batch.

The use of transmission electron microscopy confirmed the decreasing ability of the AlSi9Cu3 alloy to age naturally with an increase in the returnable material in the batch. The substructure of the alloy with 90% returnable material (the 90–10 alloy) is characterized by only a very small number of coherent and semi-coherent $\theta'$-$Al_2Cu$ phases compared to the substructures of the 20–80 alloy. The reduced presence of curable $Al_2Cu$ phases was also confirmed in the microhardness measurement, by decreasing the values of the primary $\alpha$-phase with a gradual increase in the returnable material content in the batch.

By applying artificial aging (200 °C/4 h), a rapid increase of $\theta'$-$Al_2Cu$ phases in acicular morphology was observed in the substructure of the 20–80 alloy. The increased ability to form coherent and semi-coherent $\theta'$-$Al_2Cu$ phases using artificial aging was also shown in the alloy with the highest amount of returnable material (the 90–10 alloy).

## 5. Conclusions

The aim of this study is to describe the change in the effectiveness of natural and artificial aging on the AlSi9Cu3 alloy with an increasing proportion of returnable material in the batch. The results of the study confirm that the morphology of the eutectic silicon degrades and the lengths of the iron phases increase due to the increase in the returnable material in the batch. The results also show a decrease in the ability to form $\theta'$-$Al_2Cu$ precipitates in the structure of alloys with 50% and higher proportions of returnable material in the batch. This suppressed the ability of the self-hardening process that is characteristic of AlSi9Cu3 alloy. Degradation of morphology and suppression of the self-hardening process were accompanied by a decrease in tensile strength, agreed yield strength and ductility. The above results show that although the use of returnable material in the batch as a substitute for primary alloys has a significant economic and environmental aspect in foundry production, a negative decrease in mechanical properties of experimental alloys due to reduced efficiency of $\theta'$-$Al_2Cu$ precipitate formation was confirmed.

The application of artificial aging had a positive effect on the overall quality of experimental alloy castings, thus expanding the possibilities of using castings created mostly by recycled material. Artificial aging had a positive effect on the microstructure and substructure, which was subsequently reflected in the improvement of the mechanical properties of the castings. Increased efficiency of artificial aging compared to natural aging was manifested especially in alloys with 50% and higher proportion of returnable material in the batch (which gradually lost the ability to self-cure).

Artificial aging led to increased diffusion of the additive element into microscopic areas and thus to nucleation of a new phase, which restored the alloys' ability to form precipitates in the substructure.

**Author Contributions:** Conceptualization, D.B. and M.M.; methodology, M.M., D.B. and J.K.; software, M.M. and J.K.; validation, D.B., M.M. and A.M.; formal analysis, M.M., J.K. and D.B.; investigation, M.M., J.K. and A.M.; resources, A.M. and D.B.; data curation, M.M. and A.M.; writing—original draft preparation, D.B. and M.M.; writing—review and editing, D.B., A.M. and M.M.; visualization, M.M. and J.K.; supervision, D.B.; project administration, D.B.; funding acquisition, D.B. All authors have read and agreed to the published version of the manuscript.

**Funding:** The article was created as part of the VEGA grant agency project: 1/0494/17. The authors thank the agency for their support. The authors acknowledge the assistance provided by the Research Infrastructure NanoEnviCz, supported by the Ministry of Education, Youth and Sports of the Czech Republic under Project No. LM2018124.

**Conflicts of Interest:** The authors declare no conflict of interest.

## References

1. Das, K.S.; Green, J.A.S. Aluminum Industry and Climate Change-Assessment and Responses. *JOM* **2010**, *62*, 27–31. [CrossRef]
2. International Aluminum Institute. *Global Aluminum Recycling: A Cornerstone of Sustainable Development*; International Aluminum Institute: London, UK, 2009.
3. Ciu, J.; Roven, H.J. Recycling of automotive aluminum. *Trans. Nonferrous Met. Soc. China* **2010**, *20*, 2057–2063. [CrossRef]
4. Gaustad, G.; Olivetti, E.A.; Kirchain, R. Improving aluminum recycling: A survey of sorting and impurity removal technologies. *Resour. Conserv. Recycl.* **2012**, *58*, 79–87. [CrossRef]
5. Green, J.A.S. *Aluminum Recycling and Processing for Energy Conservation and Sustainability*; ASM International: Russell Township, OH, USA, 2007.
6. Kasińska, J.; Bolibruchová, D.; Matejka, M. The influence of remelting on the properties of AlSi9Cu3 alloy with higher iron content. *Materials* **2020**, *13*, 575. [CrossRef]
7. ASM International. *Metals Handbook Metallography and Microstructures*; ASM International: Russell Township, OH, USA, 2004; Volume 9.
8. Bolibruchová, D.; Pastirčák, R. *Foundry Metallurgy of Non-Ferrous Metals*; University of Žilina: Žilina, Slovakia, 2018. (In Slovak)
9. Medlen, D.; Bolibruchová, D. The influence of remelting on the properties of AlSi6Cu4 alloy modified by antimony. *Foundry Eng.* **2012**, *12*. [CrossRef]
10. Taylor, J.A. Iron-containing intermetallic phase in Al-Si based casting alloys. *Procedia Mater. Sci.* **2012**, *1*, 19–33. [CrossRef]
11. Kuchariková, L.; Tillová, E.; Matvija, M.; Belan, J.; Chalupová, M. Study of the precipitation hardening process in recycled Al-Si-Cu cast alloys. *Met. Mater.* **2017**, *62*. [CrossRef]
12. Yuliang, Z.; Weiwen, Z.; Chao, Y.; Datong, Z.; Zhi, W. Effect of Si on Fe-rich intermetallic formation and mechanical properties of heat-treated Al–Cu–Mn–Fe alloys. *J. Mater. Res.* **2018**, *33*, 898–911. [CrossRef]
13. Capuzzi, S.; Timelli, G.; Fabrizi, A.; Bonollo, F. Influence of Ageing Heat Treatment on Microstructure and Mechanical Properties of a Secondary Rheocast AlSi9Cu3(Fe) Alloy. *Mater. Sci. Forum* **2015**, *828–829*, 212–218. [CrossRef]
14. Labisz, K.; Mariusz, K.; Dobrzanski, L.A. Phases morphology and distribution of the Al-Si-Cu alloy. *J. Achiev. Mater. Manuf. Eng.* **2009**, *37*, 309–316.
15. Vicen, M.; Fabian, P.; Tillová, E. Self-Hardening AlZn10Si8Mg Aluminium Alloy as an Alternative Replacement for AlSi7Mg0.3 Aluminium Alloy. *Arch. Foundry Eng.* **2017**, *17*. [CrossRef]
16. Andersen, S.J.; Marioara, C.D.; Friis, J.; Wenner, S.; Holmestad, R. Precipitates in aluminium alloys. *Adv. Phys. X* **2018**, *3*, 790–813. [CrossRef]
17. Zhang, X.; Hashimoto, T.; Lindsay, J.; Zhou, X. Investigation of the De-alloying Behaviour of θ-phase ($Al_2Cu$) in AA2024-T351 Aluminium Alloy. *Corros. Sci.* **2016**, *108*, 85–93. [CrossRef]
18. He, H.; Yi, Y.; Huang, S.; Zhang, Y. Effects of deformation temperature on second-phase particles and mechanical properties of 2219 Al-Cu alloy. *Mater. Sci. Eng. A* **2017**, *712*. [CrossRef]

19. Bachagha, T.; Daly, R.; Escoda, R.; Suñol, J.J.; Khitouni, M. Amorphization of Al50(Fe2B)30Nb20 Mixture by Mechanical Alloying. *Metall. Mater. Trans. A* **2013**, *44*. [CrossRef]
20. Belov, N.A.; Aksenov, A.A.; Eskin, D.G. *Multicomponent Phase Diagrams-Applications for Commercial Aluminum Alloys*; Elsevier: Oxford, UK, 2005.
21. Belov, N.A.; Aksenov, A.A.; Eskin, D.G. *Iron in Aluminum Alloys: Impurity and Alloying Element*; CRC Press: London, UK, 2002.
22. Campbell, J. *Complete Casting Handbook: Metal Casting Processes, Metallurgy, Techniques and Design*; Butterworth-Heinemann: Oxford, UK, 2011.
23. Makhlouf, M.B.; Bachagha, T.; Suñol, J.J.; Dammak, M.; Khitouni, M. Synthesis and Characterization of Nanocrystalline Al-20 at. % Cu Powders Produced by Mechanical Alloying. *Metals* **2016**, *6*, 145. [CrossRef]
24. Fatmi, M.; Ghebouli, B.; Ghebouli, M.A.; Chihi, T.; Ouakdi, E.; Heiba, Z.A. Tudy of Precipitation Kinetics in Al-3.7 wt% Cu Alloy during Non-Isothermal and Isothermal Ageing. *Chin. J. Phys.* **2013**, *51*. [CrossRef]
25. Ast, J.; Ghidelli, M.; Durst, K.; Göken, M.; Sebastiani, M.; Korsunsky, A.M. A review of experimental approaches to fracture toughness evaluation at the micro-scale. *Mater. Des.* **2019**, *173*. [CrossRef]
26. Ghidelli, M.; Sebastiani, M.; Collet, C.; Guillemet, R. Determination of the elastic moduli and residual stresses of freestanding Au-TiW bilayer thin films by nanoindentation. *Mater. Des.* **2016**, *106*, 436–445. [CrossRef]
27. Zhang, L.; Gao, J.; Damoah, L.; Robertson, D.G. Removal of iron from aluminum: A review. *Miner. Process. Extr. Metall. Rev.* **2012**, *33*. [CrossRef]
28. Podprocka, R.; Bolibruchova, D. Iron Intermetallic Phases in the Alloy Based on Al-Si-Mg by Applying Manganese. *Arch. Foundry Eng.* **2017**, *17*, 217–221. [CrossRef]
29. Freitas, B.; Otani, L.B.; Kiminami, C.S.; Botta, W.J.; Bolfarini, C. Effect of iron on the microstructure and mechanical properties of the spray-formed and rotary-swaged 319 aluminum alloy. *Int. J. Adv. Manuf. Technol.* **2019**, *102*. [CrossRef]

*Article*

# Non-Standard T6 Heat Treatment of the Casting of the Combustion Engine Cylinder Head

**Jacek Pezda [1,*] and Jan Jezierski [2]**

[1] Department of Manufacturing Technology and Automation, University of Bielsko-Biala, Willowa 2, 43-309 Bielsko-Biała, Poland

[2] Department of Foundry Engineering, Silesian University of Technology, Ul. Towarowa 7, 44-100 Gliwice, Poland; jan.jezierski@polsl.pl

* Correspondence: jpezda@ath.bielsko.pl

Received: 23 August 2020; Accepted: 14 September 2020; Published: 16 September 2020

**Abstract:** The introduction of new design solutions of cast components to the powertrain systems of passenger cars has resulted in an increased demand for optimization of mechanical properties obtained during heat treatment, assuring—at the same time—a suitable level of production capacity and limitation of manufacturing costs. In this paper, research results concerning non-standard T6 heat treatment of a combustion engine cylinder head made of AlSi7Cu3Mg alloy are presented. It has been confirmed that the optimal process of heat treatment of this component, taking into consideration the criterion of material hardness, involves solutioning at a temperature of 500 °C for 1 h, and then aging for 2 h at 175 °C. As a result, HBS10/1000/30 hardness in the range of 105–130 was obtained, which means an increase from 35% to 60% in comparison to the as-cast, depending on the position of the measurement and spheroidization of precipitations of eutectic silicon.

**Keywords:** cylinder heads; heat treatment; Brinell hardness; automotive industry

---

## 1. Introduction

Aluminum and its alloys are used in various applications within the automotive industry. They are used for casting the majority of powertrain components of passenger cars, replacing cast iron in the applications such as cylinder blocks and cylinder heads, in order to reduce their mass.

Nearly 20% of $CO_2$ emitted by humanity comes from transport. Reduction of emission of pollutants and consumption of energy, as well as the increase of recycling volumes [1], affect the natural environment and, in this aspect, the use of aluminum alloys in the automotive industry has become reasonable [2]. Due to economic and political pressure to reduce fuel consumption and $CO_2$ emissions, engineering efforts aimed at reducing the mass of automotive structures and designs have been significantly strengthened [3,4], while in the recent decades, a specific engineering solution was developed, based on the intensive use of aluminum in the form of modified or new alloys [5–9], modern casting methods [10–14] and T6 heat treatment, involving solution heat treatment, water quenching, natural and artificial aging [15–18].

Production of molded automotive components (cylinder heads mostly) involves mainly the alloys from the 3xx.x series (Al-Si-Mg, Al-Si-Cu, Al-Si-Cu-Mg) [5,9,16,19,20]. Such alloys are characterized by perfect castability [10,14], relatively low mass (which affects the reduction of fuel consumption due to reduced total mass of a vehicle), as well as good mechanical [5,6,16,20,21] and technological properties. Nearly 100% of engine pistons, almost 75% of cylinder heads, 85% of exhaust manifolds and gearbox casings, as well as other power transmission elements (rear axles, differential housings, driveshafts etc.), are produced in the form of castings [2,5,22]. Aluminum castings are also used in chassis components, in up to 40% of wheels and mounts, components of brake systems, suspension and steering systems, as well as instrumentation dashboards [2,5,23,24].

A cylinder head made as a casting of silumin is one of the characteristic elements manufactured in the automotive industry. Mechanical and thermal loads impose special requirements on the structure of the cylinder head and, for this reason, the process of the casting of the cylinder head is very complicated. It also results from the fact that the geometry of the cylinder head is predominated by the system of working medium exchange, a system of cooling ducts, camshafts, and valves control mechanism incorporated inside. Moreover, the cylinder heads incorporate inlet and outlet valves, the layout and quantity of which (in case of 35 kW/L output, two valves are sufficient) are related to the shape of the combustion chamber and position of a spark plug in the center of the combustion chamber. Modern cylinder heads mounted in heavy-duty engines are equipped with even three, four, or five valves each, and possibly two spark plugs.

The cylinder heads, in the majority of the cases, are manufactured using the gravity casting process, poured into metal molds with their shape reflecting the external shape of the cylinder head, while inlet and outlet ports of the cooling system are modeled by sand cores positioned inside the molds [9,16,25]. Operational requirements imposed on heavy-duty cylinder heads, resulting from actual trends in engine development (downsizing), generate higher working temperatures and combustion pressures, enforcing the necessity of new technological solutions to assure strength and hardness in ambient and increased temperatures (up to 250 °C) [26,27] while maintaining high volumes of production.

The standard type of heat treatment performed according to the recommendations requires long-lasting solutioning and aging of the alloys. The ASTM B917-01 standard [28] is designated for the 319 alloys (with a chemical composition similar to the EN AC-AlSi7Cu3Mg alloy) cast into permanent molds and recommends up to 12 h of soaking at a temperature of 505 °C, hot water quenching, and then 2 to 5 h of aging at 155 °C, as recommended by the ASM Handbook [29]. On the other hand, 8 h of soaking at a temperature of 505 °C and aging for 2–5 h at temperatures between 150–155 °C is recommended by the Heat Treater's Guide [30], while Zolotorevsky [31], in the case of the AK8M3 (AlSi8Cu3) alloy, advises heating at 500 °C for 5–7 h and aging at 180 °C for 5–10 h.

There are also publications pointing at the possibility of improving mechanical properties of aluminum alloys by even 30%, such as tensile strength, yield strength, and hardness, based on the heat treatment characterized by non-standard parameters of solutioning and aging operations [32–35].

In the paper, results of the research are presented, concerning the assessment of effects of shortened T6 heat treatment of the cylinder head obtained from a process of gravity casting, defined in terms of obtained Brinell Hardness (HB) hardness of the alloy on surfaces from the cylinder bores and camshafts side, as well as HB hardness in selected cross-sections.

## 2. Materials and Methods

AlSi7Cu3Mg alloy, belonging to the Al-Si-Cu-Mg group of alloys, is used, among others, to the production of castings of cylinder heads and cylinder blocks [9,36–38], as well as in other applications within the automotive industry, mainly in process of pouring into sand molds (engine crankcases, oil pans) and metal molds.

The investigated alloy of chemical composition as presented in Table 1, supplied directly by the manufacturer of the castings, was melted in an electric resistance furnace at a temperature of 720–760 °C.

**Table 1.** Chemical composition of the investigated alloy (% wt.).

| | Si | Fe | Cu | Zn | Ti | Mn | Ni | Sn | Pb | Cr | Mg | Al |
|---|---|---|---|---|---|---|---|---|---|---|---|---|
| **AlSi7Cu3Mg** | 7.5 | 0.5 | 3.0 | 0.8 | 0.03 | 0.28 | 0.04 | 0.01 | 0.03 | 0.01 | 0.4 | balance |

Chemical analysis of the alloy was performed by using a method of optical emission spectrometry with inductively coupled plasma on the PerkinElmer optical emission spectrometer, model Optima 4300 Dv.

In the next step, permanent molds (Figure 1), purposed to the casting of the test samples used in static tensile tests, were poured with the investigated alloys. The temperature of the mold was kept at a constant level between 220–250 °C.

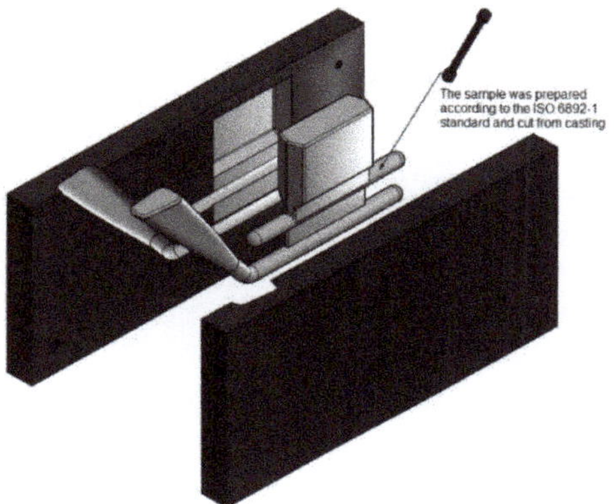


**Figure 1.** Metal mold for the pouring of the test pieces.

The course of crystallization and heating processes, as well as the melting of the investigated alloy, with marked ranges of the solutioning and aging temperatures, are presented in Figure 2.

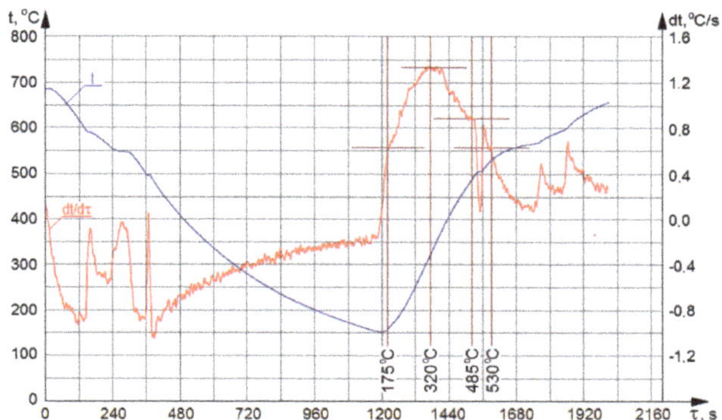


**Figure 2.** DTA curves for the investigated alloy.

The DTA (derivative and thermal analysis) method involves continuous recording of the temperature of the alloy in course of its crystallization, which enables the generation of $t = f(\tau)$ curve, depicting the course of thermal processes. Simultaneously, a curve illustrating the derivative $dt/d\tau$, highlighting less pronounced changes occurring on the thermal curve $t = f(\tau)$, is being plotted. Information obtained from the analysis of the course of the curves from the DTA method allows us to determine, with high accuracy, not only the melting temperature of a given alloy, but also melting temperatures present in an alloy of low-melting phases [39–42]. The solidification and melting process of the alloy were recorded with the use of a fully automated Crystaldimat analyzer, enabling measurement of the temperature of the test piece during solidification and melting. Standard DTA diagram is plotted during measurement for the process of solidification and melting of the alloy.

The heat treatment including solutioning operation, followed by the rapid cooling of the material in the water at a temperature of 20 °C (poured test pieces) and 20/60 °C (castings of the cylinder heads), had taken place; and then, artificial aging with cooling in the air.

The test pieces molded from AlSi7Cu3Mg alloy were soaked at temperatures of 485–530 °C and aged at temperatures of 175–320 °C. The time of the soaking operation amounted from 30 to 180 min, whereas the time of the aging operation amounted from 2 to 8 h. Basing on the assumed trivalent plan of the investigations with four variables (Table 2), heat treatment operations were performed for 27 systems (associations of temperature and time of solutioning and aging treatments).

**Table 2.** Investigations plan of the heat treatment of the EN AC-AlSi7Cu3Mg alloy.

| Number of the System in the Experimental Plan | Solutioning | | Aging | |
|---|---|---|---|---|
| | Temperature, °C | Time, h | Temperature, °C | Time, h |
| 1 | 485 | 0.5 | 175 | 2 |
| 2 | | | 250 | 8 |
| 3 | | | 320 | 5 |
| 4 | | 1.5 | 175 | 8 |
| 5 | | | 250 | 5 |
| 6 | | | 320 | 2 |
| 7 | | 3 | 175 | 5 |
| 8 | | | 250 | 2 |
| 9 | | | 320 | 8 |
| 10 | 510 | 0.5 | 175 | 8 |
| 11 | | | 250 | 5 |
| 12 | | | 320 | 2 |
| 13 | | 1.5 | 175 | 5 |
| 14 | | | 250 | 2 |
| 15 | | | 320 | 8 |
| 16 | | 3 | 175 | 2 |
| 17 | | | 235 | 8 |
| 18 | | | 320 | 5 |
| 19 | 530 | 0.5 | 175 | 5 |
| 20 | | | 250 | 2 |
| 21 | | | 320 | 8 |
| 22 | | 1.5 | 175 | 2 |
| 23 | | | 250 | 8 |
| 24 | | | 320 | 5 |
| 25 | | 3 | 175 | 8 |
| 26 | | | 250 | 5 |
| 27 | | | 320 | 2 |

Solutioning and aging operations of the test pieces and the castings were performed in a resistance furnace. Measurement of temperature was performed with the use of Ni-NiCr thermocouples of K type, with ±5 °C accuracy, directly in a chamber of the furnace, and concerned directly temperature of the test piece and the castings, temperature near heating elements (control system of the furnace), as well as temperature in the chamber of the furnace. Recordings of temperature in the chamber of the furnace and temperature of the test piece were performed continuously.

Heat treatment of the cylinder heads consisted of heating the casting in the furnace to solutioning temperature (500–515 °C), soaking at this temperature for 1 h, and then cooling in water (20 °C for cylinder heads I–III and 60 °C for cylinder head IV) and aging at temperature 175 °C for two hours.

Cylinder heads from the EN AC-AlSi7Cu3Mg alloy for the four-cylinder spark-ignition engine were produced in the technology of gravity casting into metal molds. Due to the restrictions imposed by the copyrights concerning the design of these cylinder heads, for the sake of this article, tested cylinder heads are presented in the form of drawings prepared based on 3D renders of the castings, made with the use of eviXscan Pro+ scanner, with an accuracy of 0.01–0.025 mm (according to VDI/VDE 2634 standard, part 2 [43]) and editable models from Geomagic Design X program.

Measurement of the Brinell hardness was performed in compliance with the PN-EN ISO 6506-1:2014 standard [44] with the use of Brinell hardness tester of PRL 82 type, with a steel ball of 10 mm diameter, under 9800 N load sustained for 30 s. In the case of molded test pieces, the hardness was measured on milled heads of the test pieces; while in the case of the castings of cylinder heads—surfaces destined to measure the hardness, as shown in Figures 3 and 4.

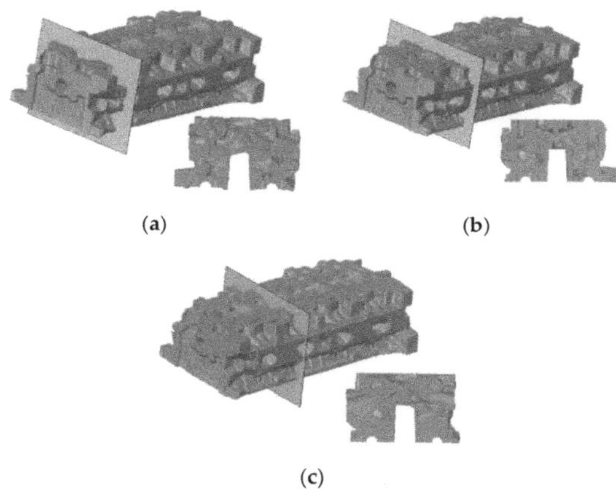

**Figure 3.** Surfaces for measurement of hardness of cylinder head casting: (**a**) cross-section 1 (55 mm from the face side), (**b**) cross-section 2 (115 mm from the face side), (**c**) cross-section 3 (160 mm from the face side).

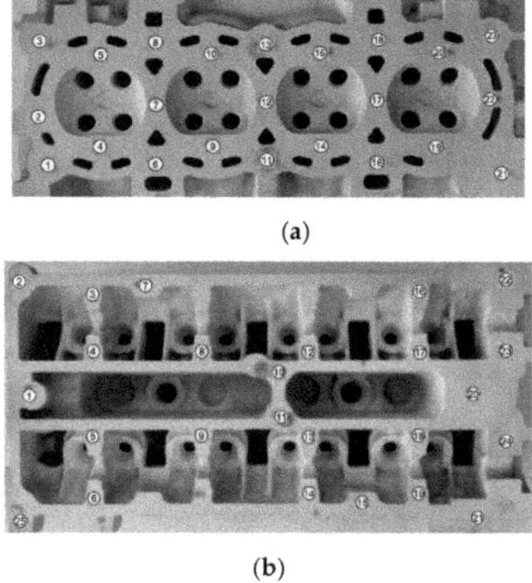

**Figure 4.** Positions of hardness measurement on milled surfaces of the cylinder head casting: (**a**) from the combustion chamber side, (**b**) from the camshafts' side.

The HBS10/1000/30 hardness of the material was also measured on milled surfaces of the casting of the cylinder head from the combustion chamber side (Figure 4a) and the side of camshafts (Figure 4b).

"Statistica" ver. 13 software package, by StatSoft (Krakow, Poland), was used to write dependencies and plot diagrams showing the effect of heat treatment parameters on resultant hardness.

## 3. Results and Discussion

### 3.1. Heat Treatment of Cast Test Pieces

The HBS10/1000/30 hardness of raw (non-heat-treated) alloy was included within 82 ± 1.8 HB limits. Following heat treatment, the HBS10/1000/30 hardness of the investigated alloy amounted from 52 to 138.

Making a comparison of the results of the raw alloy and the alloy after heat treatment (Figure 5), the highest increase of the hardness amounted to 138 HBS10/1000/30, confirmed for the system no. 7 (solutioning temperature 485 °C; solutioning time 180 min; aging temperature 175 °C; aging time 5 h), and hardness of 129 HBS10/1000/30 for the system no. 4 (solutioning temperature 485 °C; solutioning time 90 min; aging temperature 175 °C; aging time 8 h) and no. 13 (solutioning temperature 510 °C; solutioning time 90 min; aging temperature 175 °C; aging time 5 h). The test pieces from the systems identified by no. 1, 10, 19, and 25 were characterized by a slightly lower hardness (110–121 HBS10/1000/30), for which the aging temperature was equal to 175 °C. The lowest (within limits 52–62 HBS10/1000/30) hardness was obtained for the systems identified by no. 9, 15, 21, which were characterized by a high aging temperature (320 °C) for 8 h, which resulted in the reduction of obtained hardness concerning the raw casting.

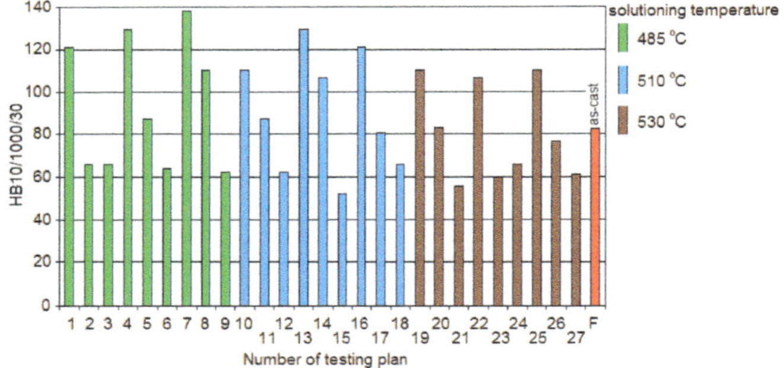


**Figure 5.** Effect of the heat treatment (of test samples) on the hardness of the AlSi7Cu3Mg alloy.

Obtained results of performed investigations allowed us to formulate a dependency (1) in the form of a second-degree polynomial, depicting the effect of the heat treatment parameters on change of the HBS10/1000/30 hardness of the investigated alloy.

$$HB = -1044.73 + 5.519t_s - 60.12 \cdot 10^{-4}t_s^2 + 107.58\tau_s - 0.43\tau_s^2 - 1.96t_a + 8.58 \cdot 10^{-4}t_a^2 + 7.73\tau_a - 0.653\tau_a^2 \\ -0.2t_s\tau_s + 24.5 \cdot 10^{-4}t_st_a + 0.002t_s\tau_a - 0.01\tau_st_a + 0.338\tau_s\tau_a - 0.02t_a\tau_a \quad (1)$$

where: $t_s$—solutioning temperature, $\tau_s$—solutioning time, $t_a$—aging temperature, $\tau_a$—aging time. Correlation coefficient $R^2 = 0.96$; correct. $R^2 = 0.9$

Figure 6 presents spatial diagrams of the influence of temperature and time of solutioning and aging treatments on the HBS10/1000/30 hardness of the investigated alloy at preset parameters of the solutioning ($t_s$—500 °C and $\tau_s$—1 h) and aging ($t_a$—175 °C and $\tau_a$—2 h).

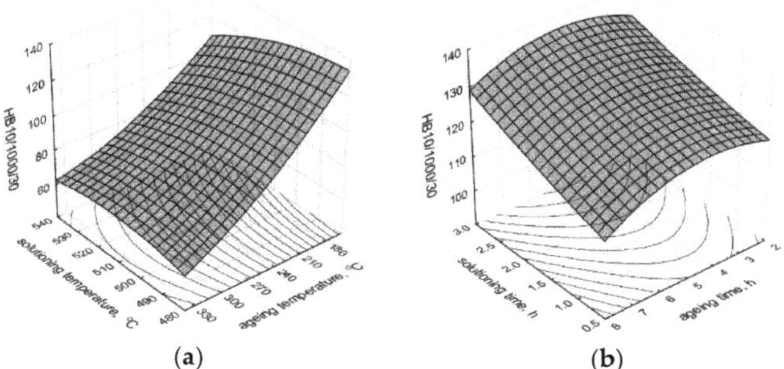


**Figure 6.** Influence of the heat treatment parameters on the hardness of the alloy. (**a**) $t_s$ and $t_a$, (**b**) $\tau_s$ and $\tau_a$.

Taking into account obtained results of the measurements of the molded test pieces, it was ascertained that to obtain a considerable increase in the hardness, the alloy should be solutioned for 1–2 h at a temperature between 500–515 °C and aged for 2–5 h at a temperature below 180 °C (which should result in obtained HBS10/1000/60 hardness at the level of 120–130 (increase at the level of 50%)). Such hardness for the alloy of 319 brands with Mg additive (with a chemical composition similar to the investigated one) was obtained by the authors of the study [32] after solutioning at temperature 495 °C for 8 h and aging at temperature 170 °C for 4–8 h.

A similar range of solutioning temperatures (500–520 °C) was recommended by the author of the publication [45], while Han [46] took the temperature of 520 °C as an initial stage of the melting of $Al_5Mg_8Cu_2Si_6$ phase, dissolution of block-type precipitations of $Al_2Cu$ and spheroidization of precipitations of Si, with a direct effect on desired mechanical properties. According to Gauthier [47], solutioning temperature above 515 °C results in partial melting of copper phase on boundaries of grains; whereas, in the case of the alloy with 0.5% additive of Mg, Samuel [45] and Ouellet [48] note the beginning of partial melting of Al5Mg8Si6Cu2 and $Al_2Cu$ phases as early as at temperature 505 °C, resulting in the worsening of mechanical properties of the alloy. Górny [49] recommends the AlSi5Cu3Mg alloy to be solutioned at 510 °C for 5 h and aged for 10 h at temperature 170 °C, while the AlSi5Cu2 alloy at 490–495 °C and aged for 10–15 h at a temperature of 150–160 °C. A completely different approach, namely two-stage solutioning at temperature 495 °C for 2 h and at 515 °C for 4 h, followed by aging at temperature 250 °C for 3 h, resulting in an optimal combination of strength and ductility (HB > 98), compared to traditional single-stage solutioning at 495 °C for 8 h, was proposed by Sokolowsky in his study [33].

In Figure 7, the microstructure of the investigated alloy before and after the proposed T6 heat treatment for selected systems from the investigations plan (Table 2) is presented.

Before the heat treatment, the microstructure of the alloy (Figure 7a) is characteristic of the eutectic Al + Si with small fibrous precipitations of Si and rounded contours of the plastic phase α(Al), which characterizes alloys after the modification [50,51]. The microstructure of the alloy after performed heat treatment, characteristic of the HBS10/1000/60 hardness at level 61 (Figure 7b) is characterized by distinctly coagulated large precipitations of Si. This is connected with a high temperature of solutioning (530 °C) and aging (320 °C), which promotes improved plasticity of the alloy with simultaneous reduction of mechanical properties, which is characteristic for the so-called over-aging of the alloy. However, the microstructure (Figure 7c) of the alloy after performed heat treatment and characteristic of high hardness (129 HBS10/1000/30) features precipitations of Si occurring in inter-dendritic areas of Al phase (on boundaries of grain), which feature rounded shapes and/or form of spheroidal precipitations.

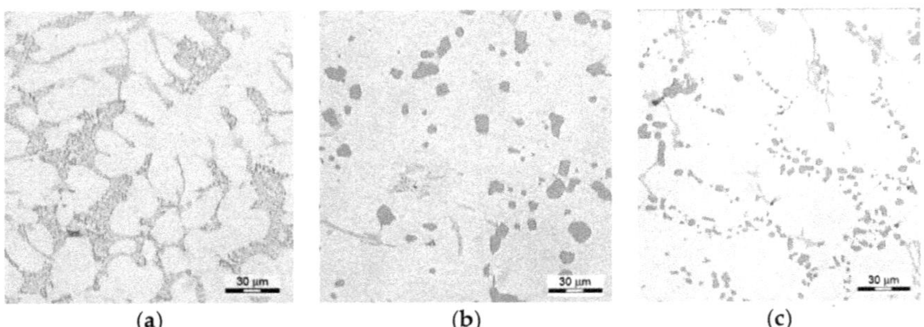


**Figure 7.** The microstructure of the alloy: (**a**) before the heat treatment—81 HBS, (**b**) for the system no. 27 (Table 2)—61 HBS, (**c**) for the system no. 13 (Table 2)—129 HBS.

## 3.2. Heat Treatment of the Cylinder Heads

In Table 3, the obtained ranges of the HB10/1000/30 hardness on the surface of the casting of the cylinder head (Figure 4) are presented.

**Table 3.** HBS10/1000/30 hardness of the casting of the cylinder head.

| Casting of the Cylinder Head | S (Raw Casting) | I | II | III | IV |
|---|---|---|---|---|---|
| Surface A * | 76–84 | 129–138 | 125–129 | 129–138 | 107–114 |
| Surface B * | 72–78 | 117–129 | 110–114 | 121–129 | 101–106 |

* surface on the cylinder head as in Figure 4.

Increase of hardness of the material of the cylinder heads marked by no. I–IV from 37 to 66%, compared to the raw casting, was obtained due to performed heat treatment operations. The difference in the hardness between analyzed surfaces results from the structure of the cylinder head's material, which, in this case, is determined by the foundry method of the cylinder head (position of pouring risers from camshaft side) and, as in the case of the cylinder head no. IV, by a higher temperature of solutioning medium, limiting the rate of the cooling [52].

In Figures 8–10, distributions of the HBS10/1000/30 hardness on selected cross-sections of the cylinder head (by Figure 3) are presented.

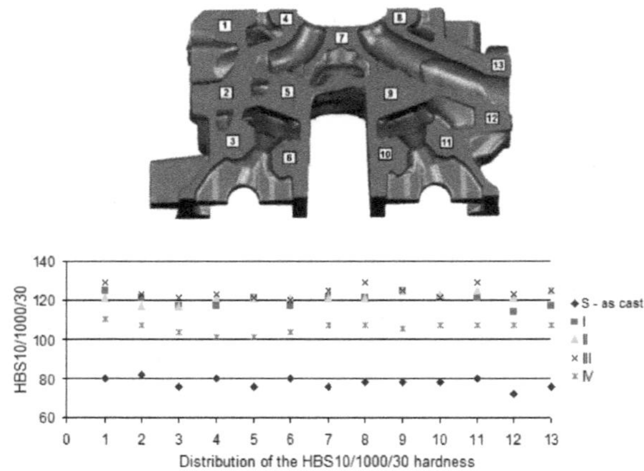


**Figure 8.** Areas of hardness measurement on the casting of the cylinder head—cross-section no. 1.

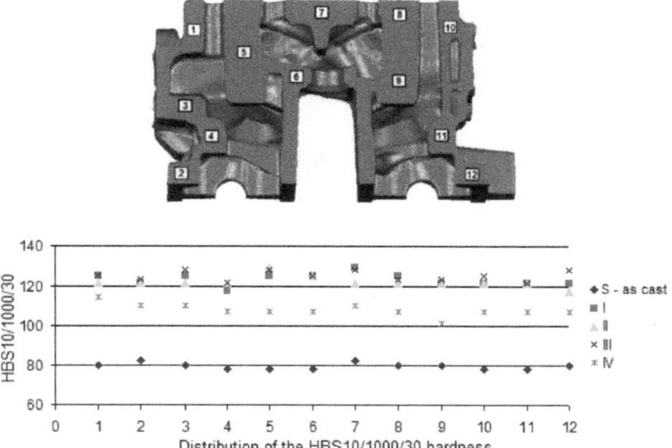


**Figure 9.** Areas of hardness measurement on the casting of the cylinder head—cross-section no. 2.

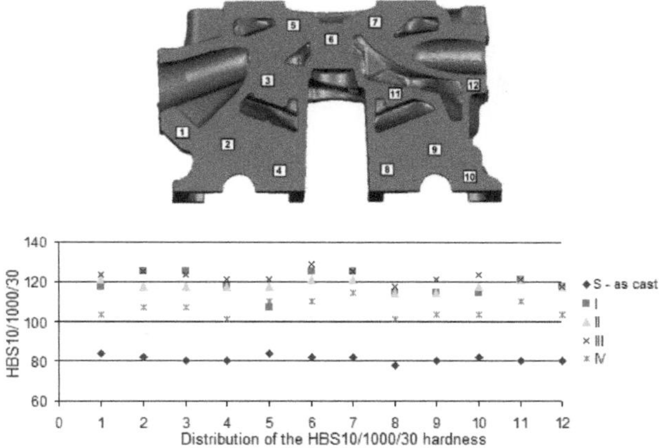


**Figure 10.** Areas of hardness measurement on the casting of the cylinder head—cross-section no. 3.

Obtained values of HBS10/1000/30 hardness on the cross-sections (Figures 8–10) for the raw casting (without the heat treatment) were included within a range from 78 to 81. After performed heat treatment, the hardness was improved (increase to 100–130 HBS). The highest values of the HBS hardness can be seen in the locations where the highest cooling rate of the alloy occurs during crystallization, which has a positive effect on the results obtained after the heat treatment, which confirms the effect of the initial structure of the alloy on a run of hardening precipitation process [53,54], as well as the process of spheroidization of eutectic silicon, even though the 319.0 alloys are more resistant to the spheroidization, which can be attributed to lower solutioning temperature of 319 alloys (495 °C) in comparison with 356 alloys (540 °C) [55]. Attention should be paid to the fact that obtaining values of the hardness of the cylinder heads made of the AlSi7Cu3Mg alloy at the level of 101–107 HB required aging at 210–230 °C for 90–240 min (after solutioning at 498 °C for 120–480 min) [37], whereas a stable level of 135 HB was possible to be obtained after 24 h of solutioning of the casting at 485 °C and 5 h of aging at 180 °C [56].

In Figure 11, microstructures in selected areas of the cross-section no. 2 of the cylinder head casting before and after the heat treatment are shown.

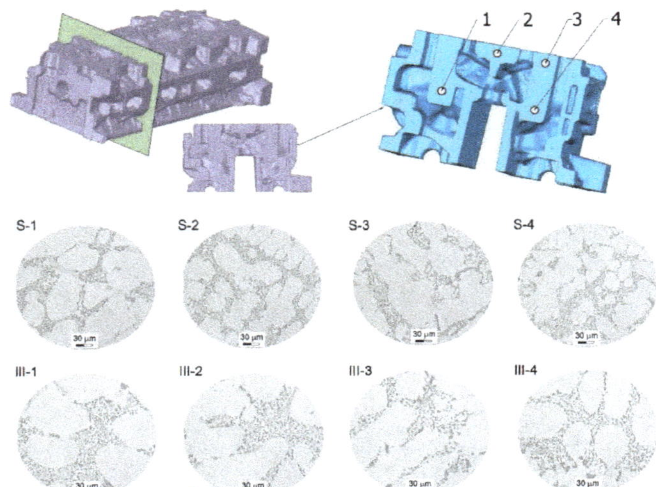


**Figure 11.** Microstructure in a selected area of the cross-sections of the cylinder head: S—the casting without heat treatment, III—the casting after the heat treatment ($t_s$ = 500 °C, $\tau_s$ = 1 h, $t_a$ = 175 °C, $\tau_a$ = 2 h).

The structure of the alloy in selected areas of the cross-section of the cylinder head (Figure 7) results from different cooling rates of the alloy during the crystallization process, hence, after the heat treatment, differences are also visible, mainly within the scope of the morphology of the precipitations of phase α(Al) and β(Si), in which precipitations are more dispersed in areas of its bigger crumbling in the eutectics α(Al) + β(Si) of the alloy before the heat treatment. Moreover, short soaking at a temperature of 500 °C, as well as aging at 175 °C has enabled obtaining a change in the form of the precipitations of silicon on the complete cross-section of the casting.

## 4. Conclusions

In terms of casting cylinder heads for combustion engines, it is possible to find such combination of the T6 heat treatment parameters (temperature, as well as the time of solutioning and aging treatments), which allows obtaining the highest value of HBS hardness with simultaneous consideration of the limited time of treatment operations.

Heat treatment performed on the test pieces poured from the alloy (from which the cylinder head of the combustion engine was produced) allows the determination of the tendency of the changes in the alloy's hardness within the assumed range of the investigations.

Solutioning of the casting at a temperature of 500 °C for 1 h and then aging for 2 h at 175 °C has allowed obtaining HBS10/1000/30 hardness at the level of 105–130, which has assured its increase within limits 35–60%, in comparison to the raw casting.

Limiting the time of heat treatment of the combustion engine's cylinder head, in case of a high volume of production of such type of casting, will allow an increase in the manufacturing capacity and decrease in the consumption of thermal energy.

**Author Contributions:** Conceptualization, J.P.; methodology, J.P.; software, J.P., and J.J.; validation, J.P., and J.J.; formal analysis, J.P., and J.J.; investigation, J.P.; resources, J.P.; data curation, J.P., and J.J.; writing—original draft preparation, J.P.; writing—review and editing, J.P. and J.J.; visualization, J.P., and J.J.; supervision, J.P.; project administration, J.P.; funding acquisition, J.J., and J.P. All authors have read and agreed to the published version of the manuscript.

**Funding:** This research received no external funding.

**Conflicts of Interest:** The authors declare no conflict of interest. The funders had no role in the design of the study; the collection, analyses, or interpretation of data; the writing of the manuscript, or in the decision to publish the results.

## References

1. Kucharikova, L.; Tillová, E.; Bokůvka, O. Recycling and properties of recycled aluminium alloys used in the transportation industry. *Transp. Probl.* **2017**, *11*, 117–122. [CrossRef]
2. Hirsch, J.; Hirsch, J. Recent development in aluminium for automotive applications. *Trans. Nonferr. Met. Soc. China* **2014**, *24*, 1995–2002. [CrossRef]
3. Pană, G.M.; Grigorie, L.D. The Issue of Car Body Manufacture in Unibody Aluminum Alloy Design. *Appl. Mech. Mater.* **2018**, *880*, 183–188. [CrossRef]
4. Henriksson, F.; Johansen, K. On Material Substitution in Automotive BIWs—From Steel to Aluminum Body Sides. *Procedia CIRP* **2016**, *50*, 683–688. [CrossRef]
5. Padmanaban, D.A.; Kurien, G. Silumins: The Automotive Alloys. *Adv. Mater. Process.* **2012**, *170*, 28–30.
6. Peta, K. Research on mechanical properties of aluminum alloys used in automotive industry. *Inż. Mater.* **2017**, *1*, 4–7. [CrossRef]
7. Pysz, S.; Maj, M.; Czekaj, E. High-Strength Aluminium Alloys and Their Use in Foundry Industry of Nickel Superalloys. *Arch. Foundry Eng.* **2014**, *14*, 71–76. [CrossRef]
8. Shabani, M.O.; Mazahery, A. Automotive copper and magnesium containing cast aluminium alloys: Report on the correlation between Yttrium modified microstructure and mechanical properties. *Russ. J. Non Ferr. Met.* **2014**, *55*, 436–442. [CrossRef]
9. Camicia, G.; Timelli, G. Grain refinement of gravity die cast secondary AlSi7Cu3Mg alloys for automotive cylinder heads. *Trans. Nonferr. Met. Soc. China* **2016**, *26*, 1211–1221. [CrossRef]
10. Kaufman, J.G.; Rooy, E.L. *Casting Properties, Processes, and Application*; ASM International: Materials Park, OH, USA, 2004; pp. 27–31.
11. Szymanek, M.; Augustyn, B.; Kapinos, D.; Żelechowski, J.; Bigaj, M. Al-Si-Re Alloys Cast by the Rapid Solidification Process/Stopy Al-Si-Re Odlewane Metodą Rapid Solidification. *Arch. Met. Mater.* **2015**, *60*, 3057–3062. [CrossRef]
12. Rosso, M. Thixocasting and rheocasting technologies, improvements going on. *J. Achiev. Mater. Manuf. Eng.* **2012**, *54*, 110–119.
13. Hong, C.P.; Kim, J. Development of an Advanced Rheocasting Process and Its Applications. *Solid State Phenom.* **2006**, 44–53. [CrossRef]
14. Merchán, M.; Egizabal, P.; De Cortazar, M.G.; Irazustabarrena, A.; Galarraga, H. Development of an Innovative Low Pressure Die Casting Process for Aluminum Powertrain and Structural Components. *Adv. Eng. Mater.* **2018**, *21*, 1–6. [CrossRef]
15. Akhtar, M.; Qamar, S.Z.; Muhammad, M.; Nadeem, A. Optimum heat treatment of aluminum alloy used in manufacturing of automotive piston components. *Mater. Manuf. Process.* **2018**, *33*, 1874–1880. [CrossRef]
16. Molina, R.; Amalberto, M.; Rosso, M. Mechanical characterization of aluminium alloys for high temperature applications Part 2: Al-Cu, Al-Mg alloys. *Metall. Sci. Technol.* **2011**, *29*, 5–13.
17. Manente, A.; Timelli, G. Optimizing the Heat Treatment Process of Cast Aluminium Alloys. In *Recent Trends in Processing and Degradation of Aluminium Alloys*; IntechOpen: London, UK, 2011; pp. 197–220.
18. Lumley, R.; O'Donnell, R.; Gunasegaram, D.; Givord, M. Heat Treatment of High-Pressure Die Castings. *Met. Mater. Trans. A* **2007**, *38*, 2564–2574. [CrossRef]
19. Fan, K.L.; He, G.; Liu, X.; Liu, B.; She, M.; Yuan, Y.; Yang, Y.; Lu, Q. Tensile and fatigue properties of gravity casting aluminum alloys for engine cylinder heads. *Mater. Sci. Eng. A* **2013**, *586*, 78–85. [CrossRef]
20. Molina, R.; Amalberto, P.; Rosso, M. Mechanical characterization of aluminium alloys for high temperature applications Part 1: Al-Si-Cu alloys. *Metal. Sci. Technol.* **2011**, *29*, 5–15.
21. Pezda, J. Influence of heat treatment parameters on the mechanical properties of hypoeutectic Al-Si-Mg alloy. *Metalurgija* **2014**, *53*, 221–224.
22. Castella, C. Self Hardening Aluminum Alloys for Automotive Applications. Ph.D. Thesis, Politecnico di Torino, Turin, Italy, 2015.

23. Das, S.; Siddiqui, R.; Bartaria, V. Evaluation of Aluminum Alloy Brake Drum for Automobile Application. *IJSTR* **2013**, *2*, 96–102.
24. Tocci, M.; Pola, A.; La Vecchia, G.; Modigell, M. Characterization of a New Aluminium Alloy for the Production of Wheels by Hybrid Aluminium Forging. *Procedia Eng.* **2015**, *109*, 303–311. [CrossRef]
25. Köhler, E.; Klimesch, C.; Bechtle, S.; Stanchev, S. Cylinder head production with gravity die casting. *MTZ Worldw.* **2010**, *71*, 38–41. [CrossRef]
26. Wang, Q.; Hess, D.; Yan, X.; Caron, F. Evaluation of a New High Temperature Cast Aluminum for Cylinder Head Applications. In Proceedings of the 122nd Metalcasting Congress, Fort Worth, TX, USA, 3–5 April 2018; Paper 18-034.
27. Jeong, C.-Y. High Temperature Mechanical Properties of Al–Si–Mg–(Cu) Alloys for Automotive Cylinder Heads. *Mater. Trans.* **2013**, *54*, 588–594. [CrossRef]
28. *ASTM B917/B917M-01. Standard Practice for Heat Treatment of Aluminum-Alloy Castings from All Processes;* ASTM International: West Conshohocken, PA, USA, 2001.
29. *ASM Handbook Volume 4: Heat Treating;* ASM International: Materials Park, OH, USA, 1991; pp. 1881–1883.
30. Chandler, H. *Heat Treater's Guide: Practices and Procedures for Nonferrous Alloys;* ASM International: Materials Park, OH, USA, 1996; p. 252.
31. Zolotorevsky, V.S.; Belov, N.A.; Glazoff, M.V. Casting Aluminum Alloys. Elsevier: Oxford, UK, 2007; pp. 500–503.
32. Tavitas-Medrano, F.J.; Mohamed, A.; Gruzleski, J.E.; Samuel, F.H.; Doty, H.W. Precipitation-hardening in cast AL–Si–Cu–Mg alloys. *J. Mater. Sci.* **2010**, *45*, 641–651. [CrossRef]
33. Sokolowski, J.H.; Djurdjevic, M.B.; A Kierkus, C.; O Northwood, D. Improvement of 319 aluminum alloy casting durability by high temperature solution treatment. *J. Mater. Process. Technol.* **2001**, *109*, 174–180. [CrossRef]
34. Zhang, B.-R.; Zhang, L.; Wang, Z.; Gao, A. Achievement of High Strength and Ductility in Al–Si–Cu–Mg Alloys by Intermediate Phase Optimization in As-Cast and Heat Treatment Conditions. *Materials* **2020**, *13*, 647. [CrossRef]
35. Pezda, J. T6 Heat Treatment of Hypo-eutectic Silumins in Aspect of Improvement of Rm Tensile Strength. *Arch. Foundry Eng.* **2012**, *12*, 49–52.
36. Javidani, M.; Larouche, D. Application of cast Al–Si alloys in internal combustion engine components. *Int. Mater. Rev.* **2014**, *59*, 132–158. [CrossRef]
37. Chaudhury, S.K.; Apelian, D.; Meyer, P.; Massinon, D.; Morichon, J. Fatigue Performance of Fluidized Bed Heat Treated 319 Alloy Diesel Cylinder Heads. *Met. Mater. Trans. A* **2015**, *46*, 3015–3027. [CrossRef]
38. Torres, R.; Esparza, J.; Velasco, E.; Garcia-Luna, S.; Colás, R. Characterisation of an aluminium engine block. *Int. J. Microstruct. Mater. Prop.* **2006**, *1*, 129. [CrossRef]
39. Szymczak, T.; Gumienny, G.; Wilk-Kolodziejczyk, D.; Pacyniak, T. Effect of chromium on the crystallization process, microstructure and properties of hypoeutectic Al-Si alloy. *Trans. Foundry Res. Inst.* **2018**, *58*, 55–72. [CrossRef]
40. Smolarczyk, P.E.; Krupiński, M. Thermal-Derivative Analysis and Precipitation Hardening of the Hypoeutectic Al-Si-Cu Alloys. *Arch. Foundry Eng.* **2019**, *19*, 41–46. [CrossRef]
41. Pietrowski, S.; Pisarek, B. Computer-aided technology of melting high-quality metal alloys. *Arch. Metall. Mater.* **2007**, *52*, 481–486.
42. Piątkowski, J.; Czerepak, M. The Crystallization of the AlSi9 Alloy Designed for the Alfin Processing of Ring Supports in Engine Pistons. *Arch. Foundry Eng.* **2020**, *2*, 65–70. [CrossRef]
43. VDI/VDE 2634 Part 2; *Optical 3D-Measuring Systems—Optical Systems Based on Area Scanning;* Verein Deutscher Ingenieure e.V.: Düsseldorf, Germany, 2012.
44. ISO, 6506-1:2014. *Metallic Materials—Brinell Hardness Test—Part 1: Test Method;* International Organization for Standardization: Geneva, Switzerland, October 2014.
45. Samuel, F. Incipient melting of Al5Mg8Si6Cu2 and Al2Cu intermetallics in unmodified and strontium-modified Al–Si–Cu–Mg (319) alloys during solution heat treatment. *J. Mater. Sci.* **1998**, *33*, 2283–2297. [CrossRef]
46. Han, Y.; Samuel, A.; Doty, H.; Valtierra, S.; Samuel, F.H. Optimizing the tensile properties of Al–Si–Cu–Mg 319-type alloys: Role of solution heat treatment. *Mater. Des.* **2014**, *58*, 426–438. [CrossRef]

47. Gauthier, J.; Louchez, P.R.; Samuel, F.H. Heat treatment of 319.2 aluminium automotive alloy Part 1, Solution heat treatment. *Cast Met.* **1995**, *8*, 91–106. [CrossRef]
48. Ouellet, P.; Samuel, F.H. Effect of Mg on the ageing behaviour of Al-Si-Cu 319 type aluminium casting alloys. *J. Mater. Sci.* **1999**, *34*, 4671–4697. [CrossRef]
49. Górny, Z. *Odlewnicze Stopy Metali Nieżelaznych*; WNT: Warszawa, Poland, 1992; pp. 182–183.
50. Dahle, A.; Nogita, K.; McDonald, S.D.; Dinnis, C.; Lu, L. Eutectic modification and microstructure development in Al–Si Alloys. *Mater. Sci. Eng. A* **2005**, *413*, 243–248. [CrossRef]
51. García-Hinojosa, J. Structure and properties of Al–7Si–Ni and Al–7Si–Cu cast alloys nonmodified and modified with Sr. *J. Mater. Process. Technol.* **2003**, *143*, 306–310. [CrossRef]
52. Emadi, D.; Whiting, L.V.; Sahoo, M.; Sokolowski, J.H.; Burke, P.; Hart, M. Optimal Heat Treatment of A356.2 Alloy. In *Light Metals-Warrendale-Proceedings*; The Minerals, Metals and Materials Society: Warrendale, PA, USA, 2003; pp. 983–989.
53. Magno, I.A.B.; Souza, F.; Barros, A.D.S.; Costa, M.O.; Nascimento, J.M.; Da Rocha, O.F. Effect of the T6 Heat Treatment on Microhardness of a Directionally Solidified Aluminum-Based 319 Alloy. *Mater. Res.* **2017**, *20*, 662–666. [CrossRef]
54. Tavitas-Medrano, F.; Gruzleski, J.; Samuel, F.H.; Valtierra, S.; Doty, H. Effect of Mg and Sr-modification on the mechanical properties of 319-type aluminum cast alloys subjected to artificial aging. *Mater. Sci. Eng. A* **2008**, *480*, 356–364. [CrossRef]
55. Tash, M.; Samuel, F.H.; Mucciardi, F.; Doty, H. Effect of metallurgical parameters on the hardness and microstructural characterization of as-cast and heat-treated 356 and 319 aluminum alloys. *Mater. Sci. Eng. A* **2007**, *443*, 185–201. [CrossRef]
56. De Mori, A.; Timelli, G.; Berto, F. High Temperature Fatigue Behaviour of Secondary AlSi7Cu3Mg Alloys. In *Mechanical Fatigue of Metals: Experimental and Simulation Perspectives*; Correia, J.A.F.O., De Jesus, A.M.P., Fernandez, A.A., Calçada, R., Eds.; Springer International Publishing: Cham, Switzerland, 2019; pp. 49–55. [CrossRef]

 *materials* 

*Article*

# Shaping the Microstructure of High-Aluminum Cast Iron in Terms of the Phenomenon of Spontaneous Decomposition Generated by the Presence of Aluminum Carbide

**Robert Gilewski [1,2], Dariusz Kopyciński [1,*], Edward Guzik [1] and Andrzej Szczęsny [1]**

1 Faculty of Foundry Engineering, AGH University of Science and Technology, Al. A. Mickiewicza 30, 30-059 Kraków, Poland; gilbert-rg@o2.pl (R.G.); guz@agh.edu.pl (E.G.); ascn@agh.edu.pl (A.S.)
2 Electrical School Complex No. 1, ul. Kamienskiego 49, 30-644 Krakow, Poland
* Correspondence: djk@agh.edu.pl

**Abstract:** A suitable aluminum additive in cast iron makes it resistant to heat in a variety of environments and increases the abrasion resistance of the cast iron. It should be noted that high-aluminum cast iron has the potential to become an important eco-material. The basic elements from which it is made—iron, aluminum and a small amount of carbon—are inexpensive components. This material can be made from contaminated aluminum scrap, which is increasingly found in metallurgical scrap. The idea is to produce iron castings with the highest possible proportion of aluminum. Such castings are heat-resistant and have good abrasive properties. The only problem to be solved is to prevent the activation of the phenomenon of spontaneous decomposition. This phenomenon is related to the $Al_4C_3$ hygroscopic aluminum carbide present in the structure of cast iron. Previous attempts to determine the causes of spontaneous disintegration by various researchers do not describe them comprehensively. In this article, the mechanism of the spontaneous disintegration of high-aluminum cast iron castings is defined. The main factor is the large relative geometric dimensions of $Al_4C_3$ carbide. In addition, methods for counteracting the phenomenon of spontaneous decay are developed, which is the main goal of the research. It is found that a reduction in the size of the $Al_4C_3$ carbide or its removal lead to the disappearance of the self-disintegration effect of high-aluminum cast iron. For this purpose, an increased cooling rate of the casting is used, as well as the addition of elements (Ti, B and Bi) to cast iron, supported in some cases by heat treatment. The tests are conducted on the cast iron with the addition of 34–36% mass aluminum. The molten metal is superheated to 1540 °C and then the cast iron samples are cast at 1420 °C. A molding sand with bentonite is used to produce casting molds.

**Keywords:** high-aluminum cast iron; $Al_4C_3$ carbide; spontaneous disintegration of the casting structure

**Citation:** Gilewski, R.; Kopyciński, D.; Guzik, E.; Szczęsny, A. Shaping the Microstructure of High-Aluminum Cast Iron in Terms of the Phenomenon of Spontaneous Decomposition Generated by the Presence of Aluminum Carbide. *Materials* **2021**, *14*, 5993. https://doi.org/10.3390/ma14205993

Academic Editor: Tomasz Wróbel

Received: 31 August 2021
Accepted: 7 October 2021
Published: 12 October 2021



## 1. Introduction

Increasing the aluminum content in cast iron can reduce its density and improve its oxidation resistance and abrasion resistance. This type of cast iron can be produced from recycled materials [1–5]. High-aluminum cast iron containing 9–19% by mass Al belongs to the group of white cast iron. In the structure of this cast iron, the ε carbide is shaped, i.e., $Fe_3AlC_x$ and the ferritic metal matrix. High-aluminum cast iron is characterized by good oxidation resistance and creep resistance. It can work in an oxidizing atmosphere up to the temperature of 920 °C, and in an atmosphere containing sulfur compounds up to the temperature of 800 °C. The structure of aluminum cast iron changes significantly in the range of aluminum content of 19–25% by mass. A barrier to the application of this material is the formation of hygroscopic $Al_4C_3$ carbide [2–8]. This carbide's properties, and its morphology in the form of needles and plates with a thickness of several micrometers, makes the reaction with water (in liquid or gas form) intense and leads to the destruction of the material, so-called self-decay, as a result of volume expansion after the reaction of aluminum carbide with water particles, according to the reaction:

$$Al_4C_3 + 12H_2O \rightarrow 4Al(OH)_3 + 3CH_4 \uparrow, \quad (1)$$

$$4Al(OH)_3 \rightarrow 2Al_2O_3 + 6H_2O, \quad (2)$$

Research on Fe–Al–C alloys has shown that the process of their decomposition in the air starts from the surface layers and progresses deeper into the material [2–5,8]. The reason for this is the formation of the $Al(OH)_3$ compound, the volume of which is approximately 2.5 times greater than the specific volume of $Al_4C_3$ carbides.

Another considered process of the decomposition of high-aluminum Fe–Al–C alloys may be the issue of weakening the cohesion of the metal matrix, which occurs as a result of the interaction of hydrogen dissolved in cast iron (the phenomenon of hydrogen embrittlement), which is transferred from the atmosphere during the metallurgical process of the Fe–Al–C alloy, according to the reaction:

$$2Al + 3H_2O = Al_2O_3 + 3H_2 \quad (3)$$

The phenomenon is also known in aluminum composites with carbon reinforcement in the form of fibers and particles, in which the $Al_4C_3$ carbide appears at the stage of component production. It should also be mentioned that similar effects occur in the case of reinforcements using SiC silicon carbide, where an excessively high temperature also affects the structure of the aluminum composite. The formation of aluminum carbide can be found in magnesium composites containing aluminum, reinforced with fibers or carbon particles [9]. The destructive effect of $Al_4C_3$ on the microstructure of materials depends on the amount and the manner of its distribution in the structure of a given material. If there is no continuity of carbide precipitation on the microscale, the adverse effects of the material properties resulting from its reaction with water are negligible and concern only the surface of the product. Thus, it is necessary to control the volume fraction and the morphology of $Al_4C_3$ in cast iron in order to produce a stable, high-aluminum cast iron with a wide range of applications. This phenomenon causes high stress and breaks the continuity of the alloy matrix.

If the phenomenon of the self-decay of high-aluminum cast iron is eliminated, a good and inexpensive material from the Fe-Al intermetallic group [10–13] could be created. To date, the method of minimizing or completely preventing the formation of the undesirable $Al_4C_3$ phase in high-aluminum cast iron composites (>19% Al) has been investigated [1,5,11]. The absence of aluminum carbide in these castings significantly improves the mechanical strength.

The hypothesis was adopted in this paper that the phenomenon of high-aluminum cast iron self-decay, caused by the presence of aluminum carbide $Al_4C_3$, can be eliminated by changing the morphology of $Al_4C_3$ carbide precipitates.

## 2. Development of the Fe–Al Alloy Phase Equilibrium System and the Fe–Al–C Triple System in the Thermo-Calc Program

Phase equilibrium diagrams of Fe-Al alloys available in the literature were developed several dozen years ago. In the research, it was decided to redevelop the phase equilibrium diagram of the Fe-Al system, using numerical analysis, based on the Thermo-Calc program. The phase equilibrium diagram of the Fe-Al alloy is shown in Figure 1. It is noteworthy that this diagram shows an additional phase transformation of the alloy within the range of 33–49% mass of Al content, at a temperature of approximately 100–120 °C. This is not visible in the graphs available in the literature related to this field. During this transformation, the crystal lattice is rebuilt—the parameters of the crystal lattice change and this may also cause additional stresses that destroy the internal structure of the alloy. Results were obtained using Thermo-CALC Software version 2019b with thermodynamic database TCFE7 and with the assumption of equilibrium calculations (Figure 1).

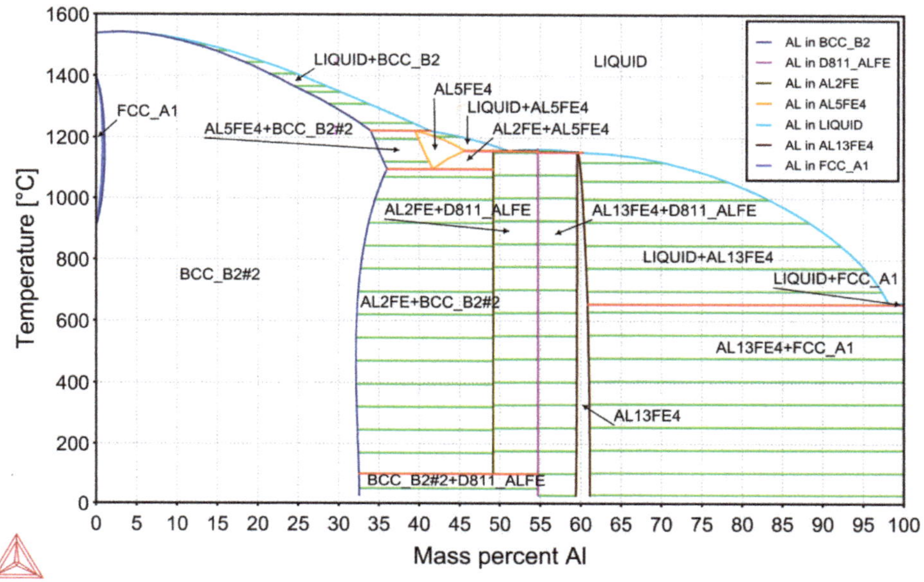


**Figure 1.** Calculated equilibrium system of Al–Fe alloys obtained in the Thermo-Calc program.

The Thermo-Calc program was also used to develop the Fe-Al-C triple system. This diagram was produced for a temperature of 1400 °C and is shown in Figure 2a. It is worth noting that, at this temperature, it is possible for graphite formation to start from the carbon content of 2.5% in the alloy. Figure 2b shows the course of the liquidus line in the Fe-Al-C alloy.

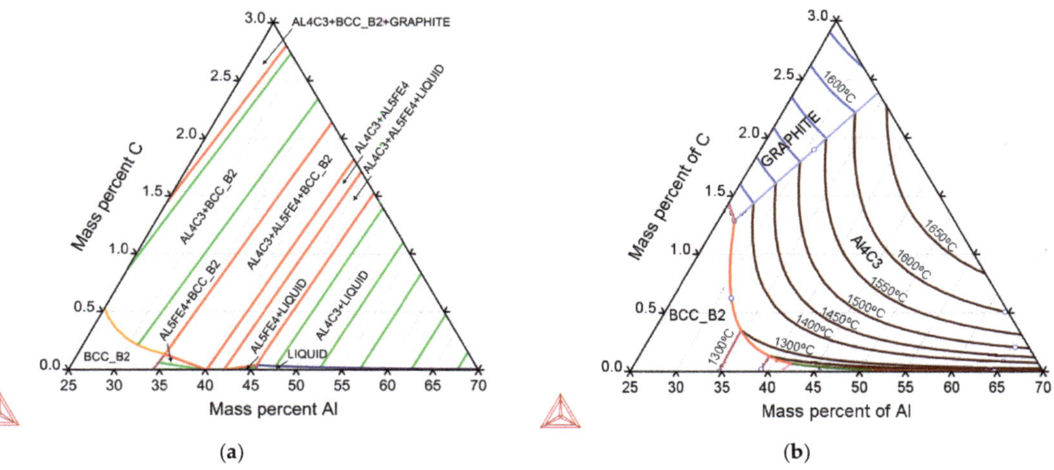


**Figure 2.** Fe–Al–C equilibrium phase diagram alloys obtained in the Thermo-Calc program at a temperature of 1400 °C (**a**) and determination of the liquidus temperature (**b**).

## 3. Methodology

A medium-frequency induction furnace with a thyristor inverter, with a capacity of 30 kg, was used to carry out the melts. The molten metal was superheated to 1540 °C and then cast iron samples were cast at 1420 °C. A molding sand with bentonite was used to make casting molds; it included quartz sand as the matrix, bentonite as a binder and water and coal dust as additives. The molds were dried at a temperature of 200 °C. One mold contained three test bars that were 30 mm in diameter and 260 mm in length.

The microscopic observations were made on a Leica optical microscope (MEF4M, Leica Microsystems, Wetzlar, Germany) using the Leica Q Win program (software version 6). The SEM microstructure and the phase composition analysis in the structure of the investigated cast iron, as well as the chemical composition analysis, were carried out on a JEOL 500LV scanning microscope (JEOL Ltd., Tokyo, Japan) with an X-ray microanalysis (EDS) attachment.

## 4. Determining the Hypothesis Concerning the Cause of the Phenomenon of Spontaneous Decay

In order to test the decomposition of high-aluminum cast iron, samples were cast with the chemical composition presented in Table 1.

**Table 1.** Chemical composition of high-aluminum cast iron.

| Element | C | Si | Mn | Al | S | Fe |
|---|---|---|---|---|---|---|
| %mass | 0.91 | 0.29 | 0.25 | 34.70 | 0.01 | remaining |

One of the samples was mounted in a specifically made bentonite mass stand and pictures were taken at intervals of 1 h. Time-lapse photos were taken with a GoPro digital camera, model YHDC5170 (San Mateo, CA, USA). Figure 3 shows the photos taken between days 1 and 21 of the study. Over a period of 21 days, the sample disintegrated.

When analyzing the photos, it could be noticed that the tested sample swelled and increased in volume over time. Therefore, it can be concluded that the disintegration of the material did not start with the surface layers, moving deeper into the material. These results indicate another cause of the spontaneous disintegration of the casting.

In order to thoroughly investigate the possible causes of decay, the decay powder and the microstructure of high-aluminum cast iron were tested. The SIEMENS X-ray diffractometer, model Kristalloflex 4H (Munich, Germany), was used. A CuK$\alpha$ = 1.54 Å lamp was used. CuK$\alpha$ radiation, vertical goniometer and counter registration of the angle and position of the reflex were used. The measurement range of the 2-theta angle was 5–135°, step: 0.05°. The time of counting a single impulse was assumed to be 1 s. The voltage was 20 kV and the current 30 mA. The experimental data were processed in order to remove the background calculation of the position, integral intensity and peak height. For this purpose, the XRayan computer program was used (software version 421). Phase identification consisted of comparing the obtained signals with reference signals contained in the database (International Centre for Diffraction Data PDF-2, 2018). The results are shown in Figure 4a. In the presented image of the diffraction pattern of the decomposition powder, no undesirable metallic phases could be found. Figure 4b shows that the powder obtained during the self-destruction of the casting consisted of aluminum and iron oxides. The SEM microstructures of the resulting powder are shown in Figure 5. The samples were previously carbon-vapor-deposited to avoid static electricity. The exact chemical composition of measuring point 1 indicated in Figure 5 contains, by mass, 3.37% C, 3.18% $O_2$, 37.50% Al, 1.85% Si, 0.01% S, 0.05% K, 0.45% Mn, 53.598% Fe. The EDS microanalysis is shown in Figure 4.

(a) (b) (c) (d)

(e) (f) (g)

**Figure 3.** High-aluminum cast iron samples—photos taken after 1 day (**a**), 9 days (**b**), 12 days (**c**), 18 days (**d**), 20 days (**e**), 21 days (**f**), 31 days (**g**).

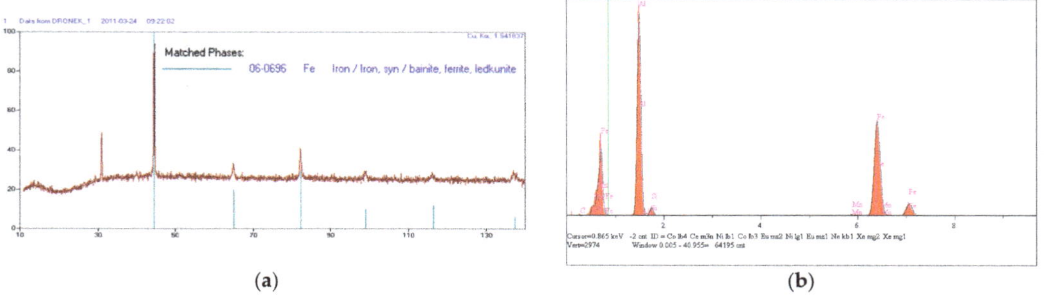


(a) (b)

**Figure 4.** Decay powder diffractogram (**a**) and EDS X-ray microanalysis (**b**).

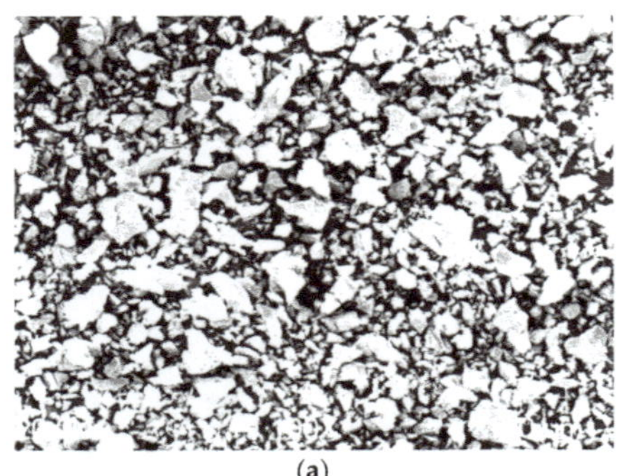

(a)

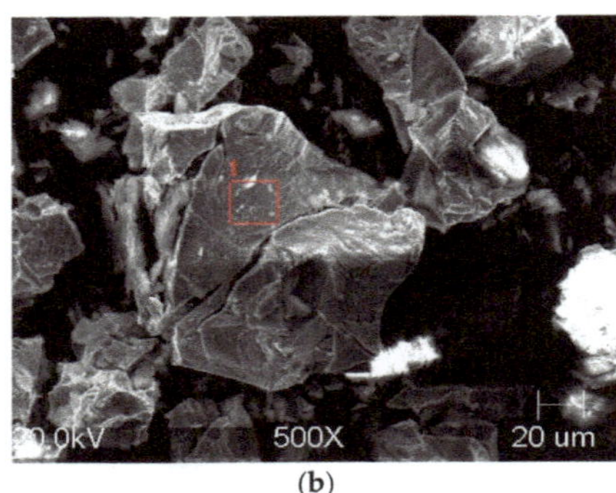


(b)

**Figure 5.** Microstructure of post-disintegration powder: different SEM microscope magnifications (**a**,**b**) and the site of EDS X-ray microanalysis (**b**).

The metallographic analysis of the fracture of the cast samples showed that the precipitation of the $Al_4C_3$ carbide was characterized by a lamellar morphology. These precipitations, with longitudinal and transverse dimensions in excess of 100 µm, had a thickness of 3–8 µm. The SEM microstructure of the fractures of the samples without and with a visible fracture is shown in Figures 6 and 7.

(a) (b)

**Figure 6.** Sample fracture made of high-aluminum cast iron; SEM microstructure—magnification 1000× (**a**), 4000× (**b**).

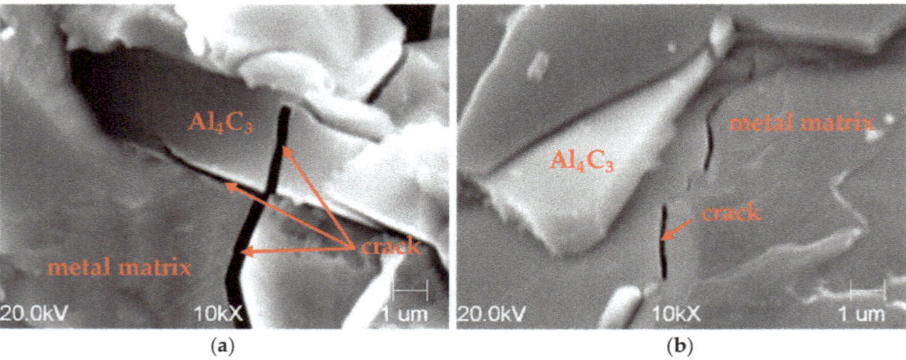


(a) (b)

**Figure 7.** Sample fracture made of high-aluminum cast iron: visible cracks in carbide and metal matrix (**a**) and metal matrix (**b**) SEM microstructure.

In Figure 7, the fracture of the metal matrix and the $Al_4C_3$ carbide can be seen.

Figure 8 shows the precipitation of the $Al_4C_3$ carbide in the form of plates with a length exceeding 100 μm and a thickness of approximately 5–8 μm. These plates sometimes interpenetrated each other (Figure 8). Carbide dimensions of this size suggest the presence of high tensile stresses in the matrix during its contraction while cooling. These forces were large enough to destroy the $Al_4C_3$ carbides.

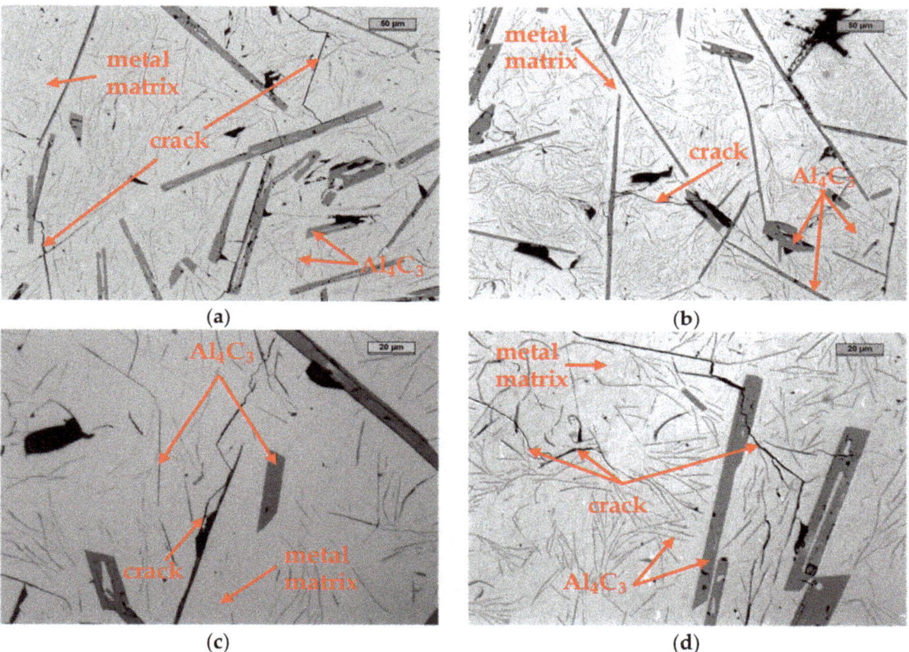


**Figure 8.** Microstructure of specimens of a high-aluminum cast iron with visible fractures of the metal matrix. Photos (**a**–**d**) present the structure of reference samples from high-aluminum cast iron, which were subjected to the phenomenon of spontaneous decomposition.

Figure 8 shows the matrix cracks along the carbide plates, which suggests that the mechanical strength of the carbide was higher compared to the metal matrix. During the crystallization of the alloy and its free contraction, aluminum carbide was able to resist the metal matrix, causing mechanical stress. The largest of them were accumulated on the periphery of the carbide, causing the notch effect, and hence a straight path to the destruction of the metal matrix. In an extreme case, when the bending strength $R_g$ of the carbide was exceeded, it also failed, as shown in Figure 8. During the decohesion of the metallic matrix and the carbide, a micro-gap was formed, into which water vapor from the air could diffuse without any obstacles, thus reacting with the aluminum carbide in accordance with Reaction (1) described in the Introduction to this work.

Subsequently, $Al_4C_3$ began to increase its volume, causing the destruction of the casting in its entire volume, not only in the outer layers. After the formation of micro-cracks, water vapor could rapidly enter the internal structures of the casting, and, as a secondary reaction, hygroscopically interacting with the carbide, this could lead to its rapid destruction.

After analyzing the above observations, it seems advisable to use this type of treatment in order to prevent the decohesion of the metal matrix and $Al_4C_3$ carbide. Therefore, it can be assumed initially that the process of self-decay can be prevented in the following ways, among others:

- Reducing the size of the $Al_4C_3$ carbide precipitates to such values that the metal matrix, which shrinks during crystallization, is able to withstand the stresses caused by the $Al_4C_3$ carbide resisting it;
- Changing the shape of the carbide and reinforcing the strength of the metal matrix;
- Application of an appropriate heat treatment in order to change the shape of $Al_4C_3$ carbides.

In the next part of this publication, the above assumptions are considered. One further issue remains, i.e., the replacement of the $Al_4C_3$ carbide with another carbide with the use of carbide-forming elements, e.g., Ti, V, W. Such treatment may result in the complete elimination of the $Al_4C_3$ carbide by producing another with a changed shape. The compact shape of the new carbide, e.g., TiC, VC, WC, will not cause excessive stress during free shrinkage, contributing to the durability of the casting [5,11].

The experiment shown in [5], which concerns the exchange of the $Al_4C_3$ carbide for a TiC carbide, proves that the hygroscopicity of the $Al_4C_3$ carbide and the prevention of Reaction (1) are crucial in terms of obtaining a material with good mechanical properties. Figure 9 shows the results of these studies.

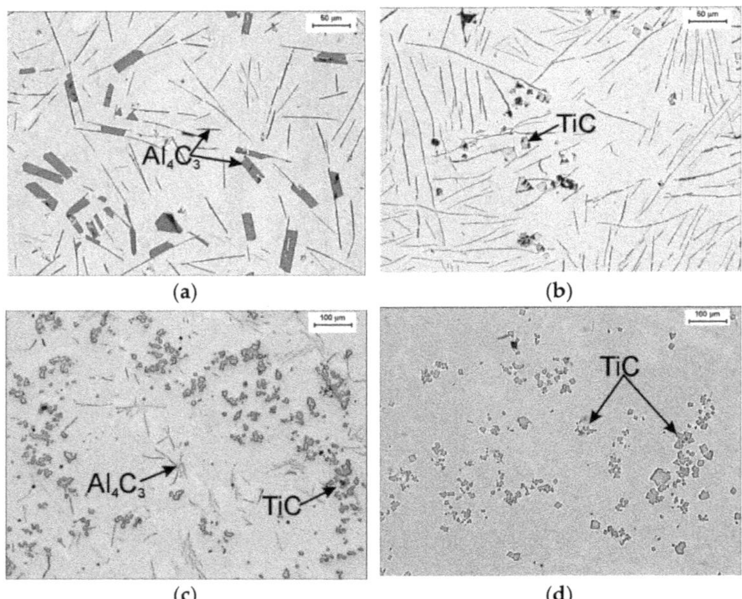


**Figure 9.** The microstructure of high-aluminum cast iron (**a**) and after addition of 1.3 %mass Ti (**b**), 2.7 %mass Ti (**c**), 5.4 %mass Ti (**d**); the structure subjected to spontaneous decomposition with different intensities (**a**–**c**), stable and durable structure (**d**).

Figure 9 shows the structure of a high-aluminum cast iron that underwent self-destruction (spontaneous decomposition) and after Ti was introduced into the cast iron in such an amount that led to the formation of a titanium carbide and the disappearance of the aluminum carbide. It was found that such a material is not subject to the self-destruction mechanism.

## 5. Reducing the Size of the $Al_4C_3$ Carbide Precipitates

In line with the assumptions presented earlier, a special experiment was prepared. Two cast iron samples with the chemical composition specified in Table 1 were subjected to different crystallization times: relatively short, for the cooling rate of 300 °C/s (Figure 10a,b), and long, for the cooling rate of 10 °C/h (Figure 10c,d). It was assumed that, during such different extreme crystallization times, the precipitates of the primary $Al_4C_3$ carbide would have different dimensions.

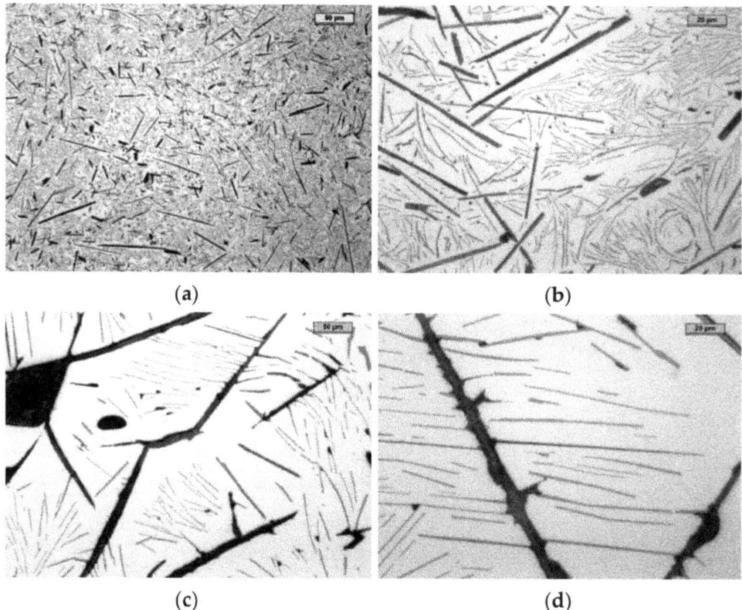

Figure 10. The microstructure of the reference sample cooled at 300 °C/s (a,b) and at 10 °C/h (c,d).

For quick cooling, the first sample was poured into a mold made of copper, with the dimensions of the casting being ∅ = 45 mm in diameter and 4 mm thick, and the second into a sand mold with bentonite. The second cast in the form of a cylinder had the outer dimensions of diameter ∅ = 100 mm and height 100 mm. After the alloy was poured, the mold containing the still molten metal was placed in a temperature-controlled resistance furnace to ensure that it cooled to room temperature.

When analyzing the microstructures of both samples, it could be noticed that (as expected) the precipitation of the $Al_4C_3$ carbide in the case of the sample cooled at a high speed was relatively low. In the case of the sample cooled at a low speed, the $Al_4C_3$ precipitation was 5–10× greater compared to the first sample. The first sample, which cooled quickly, did not disintegrate, while the second sample, which cooled slowly in an oven, disintegrated within one week. This confirms the assumption that one of the reasons for the decomposition is the relatively large longitudinal size of the $Al_4C_3$ carbide, which, by inhibiting free shrinkage, resists the metal matrix, contributing to the decohesion of the matrix–$Al_4C_3$ carbide components.

It can be concluded that a change in the shape of the precipitation, which ultimately will not lead to the effect of the notch, will significantly affect the service life of this high-aluminum cast iron.

The practical use of the described anti-disintegration method is not easy to implement in the case of larger castings, but in the case of casting small details, e.g., in a water-cooled die, it is feasible.

## 6. Change in the Morphology of $Al_4C_3$ Carbide of Cast Iron with the Addition of Large Amounts of Chromium Precipitates and the Effect of Heat Treatment

Among the alloying additives used to improve the ductility of Fe-Al alloys, chromium deserves special mention. There are two possible explanations for its effect on enhancing ductility independent of the environment [13]. One of them is the theory that a Cr additive changes the arrangement of atoms and increases the spacing of screw superdislocations by decreasing the antiphase boundary (APB) energy. The second theory is that Cr addition can alter the surface condition of the sample and the oxidation mechanism, so that the adverse effects of the environment (hydrogen embrittlement) can be mitigated. Therefore the above theory was used to produce a stable material that should not be subject to the process of spontaneous decomposition.

In the experiment, high-aluminum cast iron with the addition of chromium was melted The chemical composition is presented in Table 2. During the melting process, chromium was added to the molten metal. Then, the metal was superheated to the temperature of 1510 °C, after which, at the temperature of 1420 °C, the samples were poured (for metallographic tests) into a metal die previously heated to 300 °C. The die made it possible to obtain four cylindrical samples with a diameter of 18 mm and a length of approx. 100 mm

**Table 2.** Chemical composition of high-aluminum cast iron.

| Element | C | Cr | Si | Mn | Al | S | Fe |
|---|---|---|---|---|---|---|---|
| %mass | 2.12 | 14.20 | 1.03 | 0.30 | 34.33 | 0.01 | remaining |

Two of the four samples were additionally annealed in an oxidizing atmosphere furnace at 950 °C for 24 h and then cooled within the furnace. Then, metallographic specimens were taken from all samples (Figure 11). High-aluminum cast iron samples with the addition of chromium without heat treatment were subject to degradation over a longer period of time. Self-destruction started from one to two years after pouring.

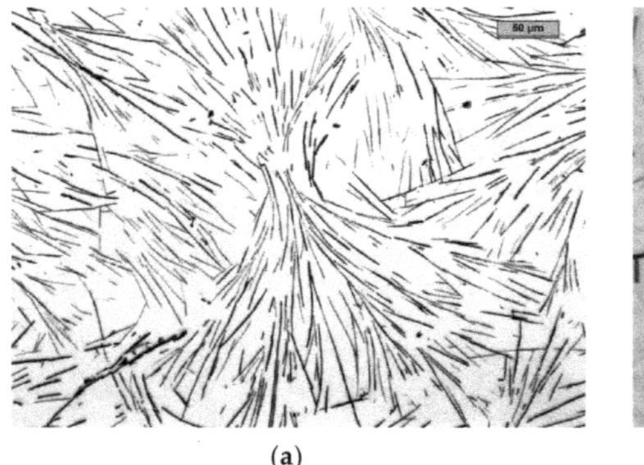


(a)

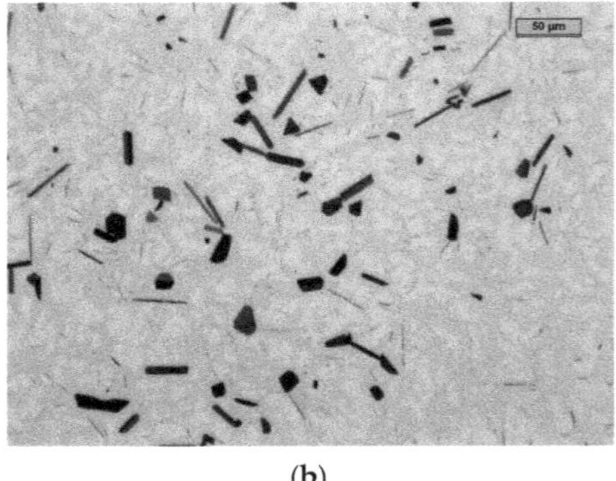


(b)

**Figure 11.** The microstructure of a sample made from high-aluminum cast iron with the addition of chromium, eutectic structure as cast (**a**) and hypereutectic structure after heat treatment (**b**).

The slender, long shapes of eutectic aluminum carbides, almost 100 µm long, took on different shapes and sizes after heat treatment. Cast iron obtained by combining aluminum and chromium obtained a structure composed of aluminum–chromium ferrite reinforced with $Al_4C_3$ carbides. Thus, it had features that combined both chrome cast iron and high-aluminum cast iron. This type of cast iron is unknown in the literature. In other words, we obtained a usable high-aluminum cast iron with much higher chromium content, even exceeding 14%mass. The $Al_4C_3$ carbides after heat treatment are shown in

Figures 11b and 12a. The external appearance of the sample and its disintegration one year after its production are shown in Figure 13.

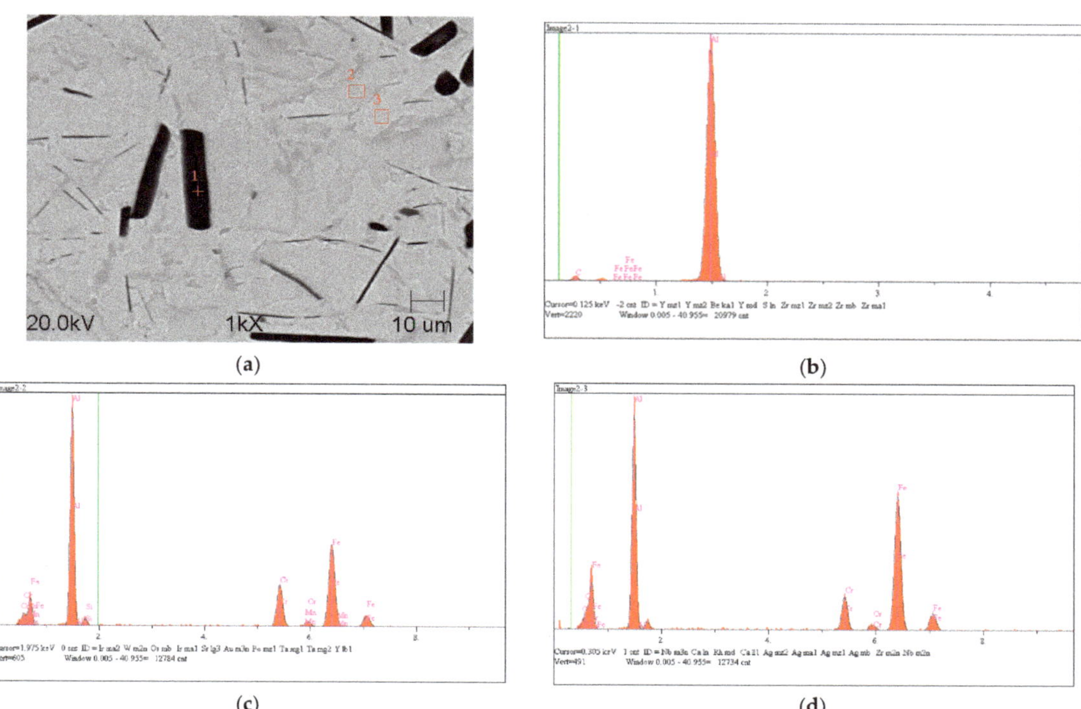


**Figure 12.** SEM microstructure with marked EDS chemical analysis points of a sample made from high-aluminum cast iron after heat treatment (**a**) and EDS X-ray microanalysis of carbide precipitation: No. 1 (**b**), No. 2 (**c**), No. 3 (**d**).

**Figure 13.** The sample without heat-treatment, one year after production—visible surface disintegration of structure.

The metal matrix in the microstructure consisted of two chemically different phases composed of iron, aluminum and chromium. In order to determine the exact composition of the matrix and the visible aluminum carbides, a quantitative analysis of the chemical composition in the energy-dispersive X-ray (EDX) micro area was carried out.

The results of the quantitative analysis of the chemical composition of selected points are presented in Table 3, and the zones of their occurrence are presented in Figure 12b–d. This sample, despite the still existing aluminum carbide $Al_4C_3$, did not disintegrate and did not show the self-destruction mechanism.

**Table 3.** Chemical composition of high-aluminum cast iron.

| Element | C | Al | Cr | Fe | Mn | Si |
|---|---|---|---|---|---|---|
| | %Mass | | | | | |
| Phase No. 2 | 0.00 | 39.18 | 14.03 | 45.06 | 0.00 | 1.73 |
| Phase No. 3 | 0.00 | 31.95 | 8.06 | 55.86 | 0.32 | 1.58 |
| Carbide No.1 | 39.15 | 60.50 | - | - | - | - |

The method developed and described above was patented and published in the patent description PL 222673 B1 [14].

## 7. The Effect of High-Aluminum Cast Iron Modification—A Minor Addition of Boron and Heat Treatment

A modification procedure was used to plasticize the metal matrix. For this, small amounts of boron and bismuth were used. The literature [15–17] shows that boron as a microadditive usually segregates along grain boundaries. For example, it can reduce the segregation of phosphorus at grain boundaries. Then, the hydrogen concentration at the grain boundaries decreases and thus the possibility of intergranular cracking is reduced. Boron segregation increases intergranular cohesion and strengthens grains boundary strength. Boron leads to low sensitivity to hydrogen-induced cracking and, generally, it is applied to improve ductility. Thus, it is an ideal microadditive to be used for castings made of high-aluminum cast iron, which has a tendency to self-destruct. In contrast, the micro-addition of Bi in high-quality grey iron leads to a change in graphite morphology [18]; in the case of high-aluminum cast iron, it was assumed that it would cause the fragmentation of $Al_4C_3$ precipitates. Thus, it was assumed that by adding these elements, the metal matrix would be strong enough not to be damaged by the action of the $Al_4C_3$ carbide plate.

The test melting of cast iron was carried out, the chemical composition of which is shown in Table 4.

**Table 4.** Chemical composition of high-aluminum cast iron.

| Element | C | Si | Mn | Al | S | Bi | B | Fe |
|---|---|---|---|---|---|---|---|---|
| %mass | 2.08 | 1.08 | 0.5 | 36.02 | 0.01 | 0.05 | 0.11 | remaining |

During the melting process, ferro-boron was added to the molten metal composed of high-aluminum cast iron. Then, a metal bath was superheated to a temperature of 1510 °C and, in the final stage, bismuth was added; at a temperature of 1420 °C, samples for metallographic tests were poured into a metal mold preheated to 300 °C. Four cylindrical samples with a diameter of 18 mm and a length of around 100 mm were obtained. Two of the samples were additionally annealed in an oxidizing atmosphere furnace at a temperature of 950 °C for 24 h, and then cooled within the furnace. The fracture for both sets of samples was fine-grained (Figure 14). The sample disintegrated after two months, which proved that the reinforcement of the metal matrix through the modification treatment extended the working time of the casting by 100% compared to the reference casting.

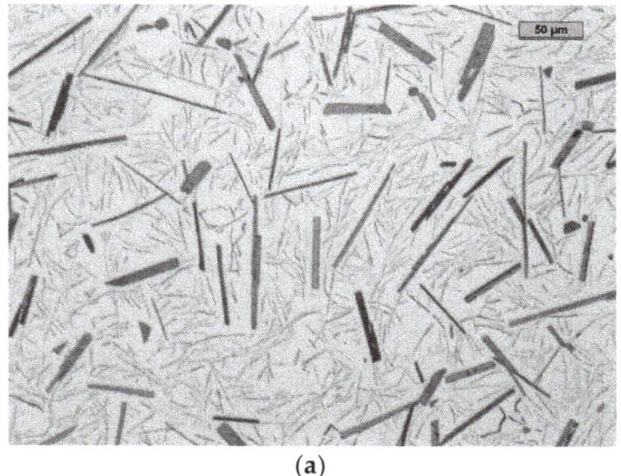


(a)

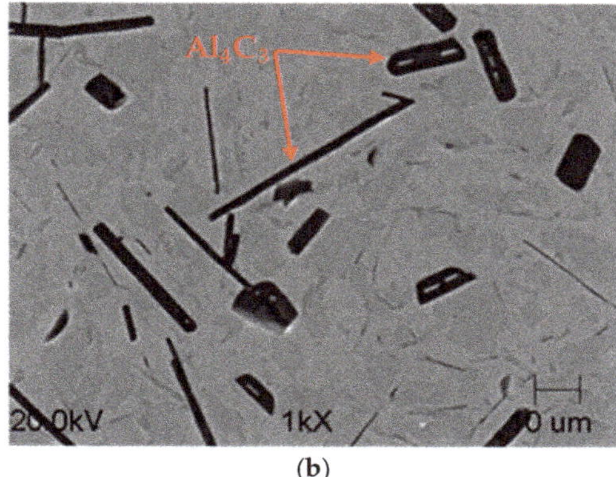


(b)

**Figure 14.** Microstructures of high-aluminum cast iron with the addition of boron and bismuth; hypereutectic structure as cast (**a**) and hypereutectic structure after heat treatment (**b**).

The precipitation of the $Al_4C_3$ carbide in the alloy was not eliminated, as can be seen in Figure 14. This is consistent with the assumption that such a small amount of boron could not completely replace this $Al_4C_3$ carbide with boron carbide. Figure 14 shows $Al_4C_3$ carbide precipitation in the form of plates with a length of approximately 50 µm. If a longer $Al_4C_3$ plate appeared, its thickness was much smaller than that shown in Figure 9 for the reference high-aluminum cast iron. Those carbides that exceeded a thickness of 5 µm were characterized by a length much less than 50 µm. This is key in the stable behavior of the structure of high-aluminum cast iron. After heat treatment, the alloy did not disintegrate. It is a relatively inexpensive casting alloy that may find application in industry. The method developed and described above was patented and published in the patent description PL 222671 B1 [19].

## 8. Summary

To verify the research, several specimens were placed in a specifically made base and 21,000 images were taken for 7 months. Selected images from this set are shown in Figure 15. As can be seen, only three specimens withstood the test of time, i.e., high-aluminum cast iron with the addition of boron and bismuth and heat treatment (B–Bi (HT)), the sample with the addition of titanium (Ti), and the sample named know-how (X).

The research results demonstrate that the reaction of aluminum carbide $Al_4C_3$ with the environment is not the only cause of the mechanism of casting disintegration. These studies provide new data on the emerging stresses in the casting structure. This issue is related to the morphology of aluminum carbide $Al_4C_3$. It was found that one of the causes of casting breakdown is the relatively large longitudinal size of the $Al_4C_3$ carbide, which inhibits free shrinkage and resists the metal matrix by contributing to the decohesion of the metal matrix. The least expensive way to prevent this phenomenon is to modify the high-aluminum cast iron with boron and bismuth and carry out a dedicated heat treatment.

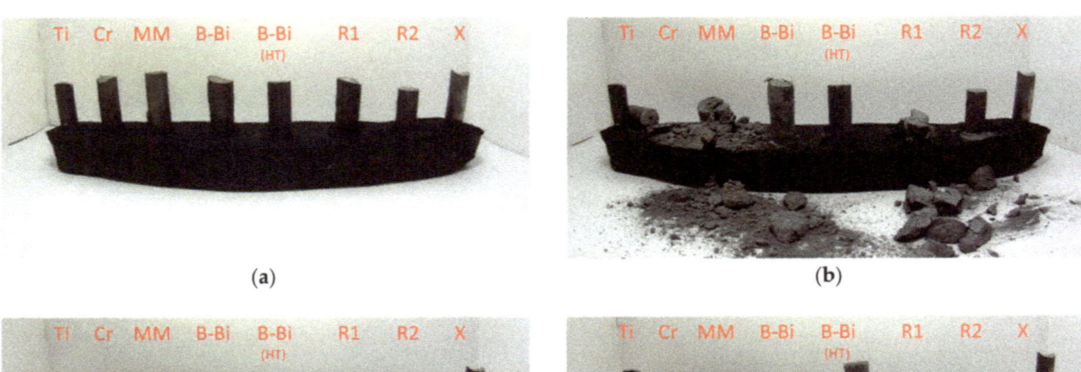


(**a**) (**b**)

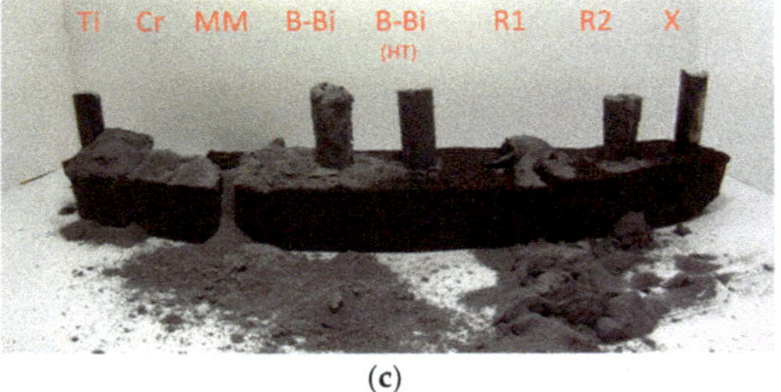


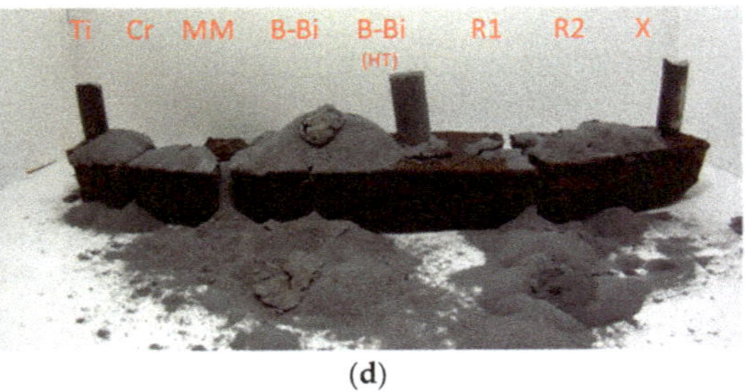


(**c**) (**d**)

**Figure 15.** Cast samples of high-aluminum cast iron with addition of Ti—titanium, Cr—chromium, MM—rare earth elements, B–Bi—boron and bismuth, B–Bi (HT)—boron and bismuth after heat treatment, R1—reference sample (standard cooling rate), R2—reference sample (increased cooling rate) and X—know-how (an attempt will be made to implement into production); observation time: first day (**a**), 74 day (**b**), 148 day (**c**), 216 day (**d**).

## 9. Conclusions

The article defines the mechanism of the spontaneous decay of high-aluminum cast iron castings, which is caused by the large relative size of $Al_4C_3$ carbide precipitates. The mechanism is as follows:

- During the crystallization of the alloy and its free shrinkage, aluminum carbide as a plate resists the metal matrix, causing mechanical stress accumulated mainly on the edges of the carbide, causing the notch effect and detachment of the carbide surface from the matrix. In an extreme case, when the bending strength $R_g$ of the $Al_4C_3$ carbide is exceeded, it is also destroyed.
- During the resulting decohesion of the metal matrix and the $Al_4C_3$ carbide, a micro-gap is formed, into which water vapor contained in the air diffuses, thus reacting with the carbide.
- Subsequently, the precipitation of the $Al_4C_3$ carbide begins to increase its volume causing the destruction of the casting in its entire volume, not only in the outer layers
- After microcracks appear, water vapor enters the internal structures of the casting faster and, as a secondary factor, leads to its rapid destruction.

During this research, methods of counteracting the phenomenon of spontaneous decay were developed. The methods developed include:

- Increasing the speed of cooling. Thanks to this, the dimensions of the $Al_4C_3$ carbide precipitates are so small that the metal matrix, which shrinks during crystallization, is able to overcome the stresses caused by the $Al_4C_3$ carbide precipitation that resists it
- The compact geometric shape of this carbide does not cause excessive stress during free shrinkage, contributing to the durability of the casting
- Changing the shape of the carbide precipitates and strengthening the strength of the metal matrix, e.g., by a modification treatment.

- Heat treatment of high-aluminum cast iron also contributes to changing the shape of carbide precipitates, which makes them durable.

**Author Contributions:** Conceptualization, R.G. and D.K.; Methodology, R.G. and D.K.; Investigations, R.G. and A.S.; Formal analysis, A.S.; Project administration, D.K.; Supervision, D.K.; Resources, A.S.; Visualization, A.S.; Funding acquisition, E.G.; Writing–original draft, R.G.; Writing–review and editing, D.K. and E.G. All authors have read and agreed to the published version of the manuscript.

**Funding:** This research received external funding from project No. 5.72.170.684.

**Institutional Review Board Statement:** Not applicable.

**Informed Consent Statement:** Not applicable.

**Data Availability Statement:** Not applicable.

**Conflicts of Interest:** The authors declare no conflict of interest.

## References

1. Ten, E.B.; Drokin, A.S. High Aluminum Cast Iron Al22D-Advanced Multifunctional Material. *Front. Manuf. Des. Sci.* **2011**, *121–126*, 186–190.
2. Zhumadilova, Z.; Kaldybayeva, S.; Nuruldaeva, G.; Kumar, D. Study of the thermophysical and physical-mechanical properties of high-alloyed aluminum cast iron. *Periódico Tchê Química.* **2019**, *33*, 30–40. [CrossRef]
3. Wojtysiak, A. The mechanism of disintegration Fe-Al-C alloy. *Metalurgia* **1990**, *40*, 43–61.
4. Wojtysiak, A. A Change in the Mechanical Properties of High-Aluminium Cast Iron During Annealing at High Temperatures. *Metalurgia* **1990**, *41*, 81–94.
5. Fraś, E.; Kopyciński, D.; Lopez, H. Processing and properties of high aluminium Fe-Al-C alloys. *Int. J. Cast Metals* **2002**, *15*, 9–15. [CrossRef]
6. Takamori, S.; Osawa, Y.; Halada, K. Aluminum-Alloyed Cast Iron as a Versatile Alloy Materials Transactions. *Spec. Issue Environ. Benign Manuf. Mater. Process. Towar. Dematerialization* **2002**, *2*, 311–314.
7. Takamori, S.; Osawa, Y.; Kakisawa, H.; Minagawa, K.; Halada, K. Aluminum alloyed cast iron as an ecomaterial. *Mater. Sci. Forum* **2003**, *426*, 3305–3310. [CrossRef]
8. Olszówka-Myalska, A. *Magnesium Matrix Composites. Selected Technological Issues*; Wydawnictwo Politechniki Śląskiej w Gliwicach, Monograph: Katowice, Poland, 2017; pp. 1–150.
9. Deeki, S.C.; Sikka, V.K. Nikel and iron aluminides: An overview on properties, processing and applications. *Intermetalics* **1996**, *4*, 357–375.
10. Rapp, R.A.; Zheng, X. Thermodynamic consideration of grain refinement of aluminium alloys by titanium and carbon. *Metall. Trans.* **1991**, *22*, 3071–3075. [CrossRef]
11. Kopyciński, D.; Guzik, E.; Gilewski, R.; Szczęsny, A.; Dorula, J. Analysis of Structure and Abrasion Resistance of the Metal Composite Based on an Intermetallic FeAl Phase with VC and TiC Precipitates. *Arch. Foundry Eng.* **2013**, *13*, 51–54. [CrossRef]
12. Ramirez, B.N.; Schon, C.G. Casting of Iron Aluminides. *J. Phase Equilibria Diffus.* **2017**, *38*, 288–297. [CrossRef]
13. Huang, Y.D.; Yang, W.Y.; Sun, Z.Q. Effect of the alloying element chromium on the room temperature ductility of $Fe_3Al$ intermetallics. *Intermetallics* **2001**, *9*, 119–124. [CrossRef]
14. Guzik, E.; Kopyciński, D.; Szczęsny, A.; Gilewski, R. Method for Producing High-Aluminium Cast Iron with the Addition of Chromium. Polish Patent PL 222673 B1, 9 September 2015.
15. Li, X.; Cheng, Y.; Shen, H.P.; Su, L.C.; Zhang, S.Y. Effect of boron microalloying element on susceptibility to hydrogen embrittlement in high strength mooring chain steel. In Proceedings of the HSLA Steels 2015, Microalloying & Offshore Engineering Steels, Hangzhou, China, 11–13 November 2015; pp. 1211–1218.
16. Yan, B.C.; Zhang, J.; Lou, L.H. Effect of boron additions on the microstructure and transverse properties of a directionally solidified superalloy. *Mater. Sci. Eng. A* **2008**, *474*, 39–47. [CrossRef]
17. Wu, B.; Li, L.; Wu, J.; Wang, Z.; Wang, Y.; Chen, J.; Li, J. Effect of boron addition on the microstructure and stress-rupture properties of directionally solidified superalloys. *Int. J. Miner. Metall. Mater.* **2014**, *21*, 1120–1126. [CrossRef]
18. Haji, M.M. On the Inoculation and Graphite Morphologies of Cast Iron. Ph.D. Thesis, The Royal Institute of Technology (KTH), Stockholm, Sweden, 2014; pp. 1–75.
19. Guzik, E.; Kopyciński, D.; Szczęsny, A.; Gilewski, R. Method for Producing High-Aluminium Cast Iron. Polish Patent PL 222671 B1, 31 August 2016.

*Article*

# The Effect of Cooling Conditions on Martensite Transformation Temperature and Hardness of 15% Cr Chromium Cast Iron

**Mirosław Tupaj [1,*], Antoni Władysław Orłowicz [2], Andrzej Trytek [1], Marek Mróz [2], Grzegorz Wnuk [2] and Anna Janina Dolata [3,*]**

1 Faculty of Mechanics and Technology, Rzeszow University of Technology, Kwiatkowskiego 4, 37-450 Stalowa Wola, Poland; trytek@prz.edu.pl

2 Department of Foundry and Welding, Rzeszow University of Technology, Al. Powstańców Warszawy 12, 35-959 Rzeszów, Poland; zois@prz.edu.pl (A.W.O.); mfmroz@prz.edu.pl (M.M.); gwnuk@prz.edu.pl (G.W.)

3 Department of Advanced Materials and Technologies, Faculty of Materials Engineering, Silesian University of Technology, Krasińskiego 8, 40-019 Katowice, Poland

* Correspondence: mirek@prz.edu.pl (M.T.); anna.dolata@polsl.pl (A.J.D.)

Received: 12 May 2020; Accepted: 16 June 2020; Published: 18 June 2020

**Abstract:** The research reported in the paper concerned the conditions of cooling high-chromium cast iron with about 15% Cr content capable to ensure completeness of transformation of supercooled austenite into martensite in order to obtain high hardness value of the material and thus its high resistance to abrasive wear. For testing, castings were prepared with dimensions 120 mm × 100 mm × 15 mm cast in sand molds in which one of cavity surfaces was reproduced with chills. From the castings, specimens for dilatometric tests were taken with dimensions 4 mm × 4 mm × 16 mm and plates with dimensions 50 mm × 50 mm × 15 mm for heat treatment tests. The dilatometric specimens were cut out from areas subject to interaction with the chill. The austenitizing temperature and time were 1000 °C and 30 min, respectively. Dilatograms of specimens quenched in liquid nitrogen were used to determine martensite transformation start and finish temperatures $T_{Ms}$ and $T_{Mf}$, whereas from dilatograms of specimens quenched in air and in water, only $T_{Ms}$ was red out. To secure completeness of the course of transformation of supercooled austenite into martensite and reveal the transformation finish temperature, it was necessary to continue cooling of specimens in liquid nitrogen. It has been found that $T_{Ms}$ depended strongly on the quenching method whereas $T_{Mf}$ values were similar for each of the adopted cooling conditions. The examined cooling variants were used to develop a heat treatment process allowing to obtain hardness of 68 HRC.

**Keywords:** high-chromium cast iron; austenitizing conditions; cooling conditions; martensite transformation; hardness

---

## 1. Introduction

Resistance to abrasive wear is one of the main criteria in the process of selection materials for construction and operation of machine components and parts [1]. Examples of such elements are replaceable inserts for dies used to form die stampings of refractory materials. The stampings are press molded of an aggregate of $Al_2O_3$-MgO type, containing also $TiO_2$, $SiO_2$, CaO, and $Fe_2O_3$. Particles of the aggregate, characterized by sharp corner edges and hardness values of about 1800 HV0.5, scratch surfaces of die inserts in the course of the press forming process. To date, the inserts are fabricated from tool steels thermally processed to hardness of up to 60–61 HRC. However, the service live of the inserts remains unsatisfactory to manufacturers of refractory materials.

One material used widely for components working in conditions of intensive abrasive wear specific for mining, ceramic, cement, and aggregate processing industries is the high-chromium cast iron [2–5]. Shaping high-chromium cast iron microstructure in order to improve resistance to abrasive wear is oriented at refinement of carbides by increasing the crystallization rate [6–8], improving hardness by introduction of alloying additions [9–11], and applying heat treatment [12,13] aimed at obtaining martensitic matrix [14–16]. However, too high martensite content may lead to fracturing of the matrix in the surface region. On the other hand, during impact loads, iron with a martensitic matrix is better in comparison to iron with the dominant austenitic matrix. The effect of microstructure in high stress abrasion of white cast irons (WCI) have been widely discussed by Heino et al. [17]. They concluded that the austenite-to-martensite ratio and also carbides structure strongly effect on the abrasion wear resistance of WCI specimens.

Unpublished results of present author's tests performed on plate castings of 15% Cr cast iron indicate that quenching them in air or in water from temperature of 1000 °C after austenitizing for 30 min is favorable from the point of view of high hardness values.

Technical literature of the subject lacks results of research concerning the effect of quenching method on martensite transformation start and finish temperature, although such knowledge is necessary for the purpose of developing thermal treatment schedules securing completeness of transformation of supercooled austenite into martensite which would not produce excessive quenching stresses and the resulting hardening cracks.

It is a well-known fact that the reduction of carbon content in austenite results in a decrease of the martensite transformation temperature. The effect of alloying elements on transformations of supercooled austenite in high-chromium cast iron results on their content in austenite. If complete dissolution of carbides was not achieved in the course of austenitizing, then content of carbon and alloying elements in austenite is lower than this revealed by chemical composition analysis of the alloy.

Value of the temperature $T_{Ms}$ can be estimated based on chemical composition with the use of empirical formulas of the type proposed by author of the paper [18]. Relevant relationships are developed for conditions in which complete dissolution of carbides in austenite has occurred. The formulae indicate that the strongest effect on decrease of $T_{Ms}$ value is produced by carbon, and further manganese, nickel, chromium, and molybdenum. Generality of these relationships is not sufficient and for that reason, dilatometric tests are carried out to determine $T_{Ms}$ values more precisely. Dilatometry is a widely used experimental technique for measuring transformations and kinetics of solid-state phase transformations that involve dimensional changes. The course of martensite transformation is accompanied by changes in volume of specimens which are the base on which temperatures and kinetics of phase transformations can be determined with the use of the dilatometric method [19–21].

In view of the fact that carbides are characterized with the higher content of carbon and other alloying elements compared to austenite from which they precipitated, it can be claimed that in case of dissolution of carbides austenite is enriched in these elements. As a result, a decrease of $T_{Ms}$ value is observed. On the other hand, presence of carbides on boundaries of austenite grains hampers their growth by hindering migration of grain boundaries. To break the grain boundary away from a carbide, an amount of energy must be supplied, for instance by heating the alloy up to a sufficiently high temperature. In the case of retention of fine austenite grains, conditions are created for rapid nucleation and development of such diffusional processes as eutectoid or bainitic transformation. It the applied cooling rate enables partial transformation of supercooled austenite into pearlite or bainite, that will be reflected in chemical composition of austenite and thus also in $T_{Ms}$ value. As a result, $T_{Ms}$ value decreases.

The issue of the effect of cooling rate on $T_{Ms}$ in the process of quenching carbide-containing steels was discussed in a number of publications. Authors of [22] have found that in the case of low-alloy high-strength steels, lower values of $T_{Ms}$ corresponded to higher specimen cooling rates. They assumed that at lower values of the cooling rate, precipitation of chromium and molybdenum carbides become possible. That resulted of impoverishment of austenite in content of carbon and alloying elements and ultimately in higher $T_{Ms}$ values. In addition, authors of [23] have found that with increasing rate of cooling 9Cr-1.7W-0.4Mo-Co steel in the quenching treatment, a decrease of $T_{Ms}$ value is observed. The authors suggest that the steel is susceptible to precipitation of carbides in austenite at low cooling rates which is reflected in impoverishment of its content of carbon and other alloying elements. At higher cooling rate values, the process of precipitation of carbides is blocked and austenite is no longer impoverished in carbon and other alloying elements. As a result, lower $T_{Ms}$ values are observed. Authors of papers [24,25] suggest that by deformation of material in the course of austenitizing, it is possible to obtain increased density of lattice imperfections which supports nucleation of martensite and is favorable to increase of $T_{Ms}$ value.

In paper [26] it was demonstrated that plastic deformation of steel at the austenitizing temperature resulted in an increase of $T_{Ms}$ and broadening the $T_{Ms}$–$T_{Mf}$ temperature range. The authors of the study suggest that the effect is connected with an increase of the number of lattice defects that can act as nucleation sites for the start of the martensite transformation.

In the technical literature concerning the issue of high-chromium cast iron heat treatment, numerous research projects were devoted to the possibility to shape microstructure favorable for machinability or high resistance to abrasive wear of castings. Such studies are of great importance to the industrial practice.

Conventional quenching of high-chromium cast iron does not guarantee complete transformation of supercooled austenite into martensite which results in presence of retained austenite in the alloy matrix. It is known that at high chromium content values, retained austenite is stable up to the temperature range of 400–500 °C. Even longer soaking does not result in its transformation but is favorable to precipitation of carbides. That way an impoverishment in carbon and chromium occurs which, in the course of rapid cooling, facilitates transformation of austenite into martensite. In high-alloy steels rich in chromium, such heating and cooling cycle is repeated a number of times in order to secure complete transformation of retained austenite into martensite. Studies on thermal treatment of that type for high-chromium cast iron aimed at increase of its hardness were carried out by authors of papers [27–31]. A common feature of these reports consists of concluding that as a result of precipitation of secondary carbides, impoverishment of austenite in carbon and carbide forming alloying elements (Cr, Mn, Mo) occurs which leads to a rise in the martensite start $T_{Ms}$ temperature. Therefore, the supercooled austenite in the course of quenching and the retained austenite in the course of tempering and rapid cooling can easily be transformed into martensite, resulting in bulk hardness and wear resistance increase.

Studies on development of new heat treatment schedules for high-chromium cast iron were conducted by authors of papers [32–35]. From this point of view, significant are studies on working out TTT (Time – Temperature – Transformation) plots serving as a base for selection of heat treatment parameter values [35].

Authors of [33] demonstrated the possibility to reduce hardness of high-chromium cast iron (14.55% Cr) castings with the use of soaking at subcritical temperatures or with the use of soaking at subcritical temperature preceded with quenching in oil. In the first case, hardness was reduced to the value 38.5 HRC and in the second instance, to 37 HRC whereas the hardness of the material in as-cast condition was 44 HRC. Such treatment was favorable to formation of the matrix microstructure containing "ferrite + grainy carbides". Duration the first heat treatment variant was 31 h and was longer compared to the duration of the second variant which lasted for 17 h. Subcritical soaking temperatures were 650 °C and 750 °C in the first variant of the treatment and 725 °C in the second variant. For a high-chromium cast iron with the same chemical composition (14.55% Cr), the study [35] presented a heating dilatogram from which it followed that the eutectoid transformation start temperature was $TAc_1{}^P$ = 760 °C. Assuming that the rate of heating for which the dilatogram was taken corresponded to the heating rate applied in both of the softening treatment variants, an opportunity appeared for further decrease of hardness at shorter treatment cycle. That can be achieved by increasing the annealing temperature up to the range 725–760 °C.

Authors of [33] reported that in the case of high-chromium cast iron containing 2.7% C, 14.55% Cr, 2.20% Mn, and 0.93% Ni, austenitizing at temperature 950 °C for the period of 2 h secured precipitation of secondary carbides, mainly $M_7C_3$, and the following cooling in oil enabled partial transformation of supercooled austenite into martensite. A factor beneficial for the course of the transformation was a decrease of carbon and chromium content in austenite due to emergence of secondary carbides which led to an increase of $T_{Ms}$ value. The volume fraction of non-transformed austenite was 64.9%. The cast iron with such structure was characterized by hardness of 62 HRC. So high hardness value could be made even higher by means of two methods promoting transformation of retained austenite into martensite. The first method would consist of the application of a one-time or multiple tempering in order to precipitate carbides from retained austenite and force the transformation of austenite into martensite with the use of rapid cooling. In the second method, cryogenic treatment could be used, aimed at transformation of retained austenite into martensite. Such approach to the issue is presented in this paper.

In view of the above, the objective of the study was to acquire a new knowledge in scope of the effect of cooling conditions from the adopted austenitizing temperature on martensite transformation limiting temperatures, and in particular the conditions securing completeness of transformation of supercooled austenite into martensite which is necessary to develop heat treatment processes securing low susceptibility to quenching cracks and high hardness of castings of chromium cast iron containing about 15% Cr.

## 2. Materials and Methods

For the tests, chromium cast iron casting with about 15% content of Cr and dimensions 120 mm × 100 mm × 15 mm were made. Liquid metal was prepared in an induction furnace with about 20 kg charge capacity. The foundry mixture per 10 kg of alloy included: 4 kg of grade L210H21S cast steel; 1.5 kg of special pig iron LS; 0.2 kg of FeCr; 1.0 kg of steel; 0.08 kg of carburizer; 0.05 kg of FeMo; 0.03 kg of FeB; 0.05 kg of FeMn; 0.04 kg of FeSi; and 3 kg of scrap high-chromium cast iron. Analysis of chemical composition was carried out with the use of Q4 Tasman emission spectrometer (Bruker, Kalkar, Germany). Results of analysis are summarized in Table 1.

**Table 1.** Chemical composition of 15% Cr cast iron.

| Element Content (wt.%) | | | | | | | | | | |
|---|---|---|---|---|---|---|---|---|---|---|
| C | Si | Mn | P | S | Cr | Mo | Ni | Cu | B | Fe |
| 3.50 | 1.10 | 0.65 | 0.03 | 0.030 | 15.30 | 0.50 | 0.30 | 0.04 | 0.02 | Bal |

The liquid metal was poured at temperature 1450 °C into two twin-cavity sand molds with horizontally oriented cavities lower surfaces of which were reproduced by steel chills with thickness of 15 mm. Surfaces of chills were covered with black from acetylene-oxygen flame. The molds were provided with knife gates.

After shaking out from molds and cutting gate assemblies off, surfaces of the castings were cleaned in a jet of carborundum particles, smoothed by grinding, and cut into cubes with dimensions 60 mm × 50 mm × 15 mm using Struers Labotom-3 water-cooled metallographic cutter (Struers, Copenhagen, Denmark). Hardness of the cubes on the side of surfaces reproduced by chills was 58 HRC, and on the side of surfaces reproduced by sandmix was 48 HRC. From one side of the cubes, five slices of the material were cut off, each with thickness of 4 mm. Out of the slices, from the regions on the side of surface reproduced by the chill, specimens for dilatometric tests (Ø4 mm × 16 mm) were cut out with the use of BP95d wire erosion machine (ZAP B.P., Końskie, Poland )

Dilatometric tests were carried out on an upgraded dilatometer LS4 (Institute for Ferrous Metallurgy, Gliwice, Poland ) equipped with a computer program for furnace heating and cooling control as well as a program for recording elongation and temperature of specimens. Ni-CrNi thermocouples were used with wire diameter of 0.15 mm. The specimens were tested in protective atmosphere of argon. The specimens were heated at rate $V_{heat}$ = 400 °C/h up to and maintained at the austenitizing temperature of 1000 °C for 30 min and then cooled in compressed air, water, or liquid nitrogen.

The analysis of microstructures was performed using a VEGA 3 scanning electron microscope (SEM, TESCAN, Brno, Czech Republic). All prepared samples were etched in Kalling's reagent for matrix/carbide contrast. The volume fraction of carbide was evaluated with the use of Neophot 2 optical microscope equipped with Videotronic CC20P camera (Videotronic International, Rastatt, Germany) and Multiscan v.08 advanced image analysis system (University of Warsaw, Warsaw, Poland). The analysis was carried out in 10 areas under 500× magnification.

## 3. Results and Discussion

The characteristic microstructure of 15% Cr cast iron in its original state is shown in Figure 1. Microstructure of the cast iron in the as-cast condition is characterized by primary carbides and the matrix containing bainite, the austenite-eutectic carbides eutectic, and martensite. The volume fraction of carbides fell into the range 29.2–30.7%.

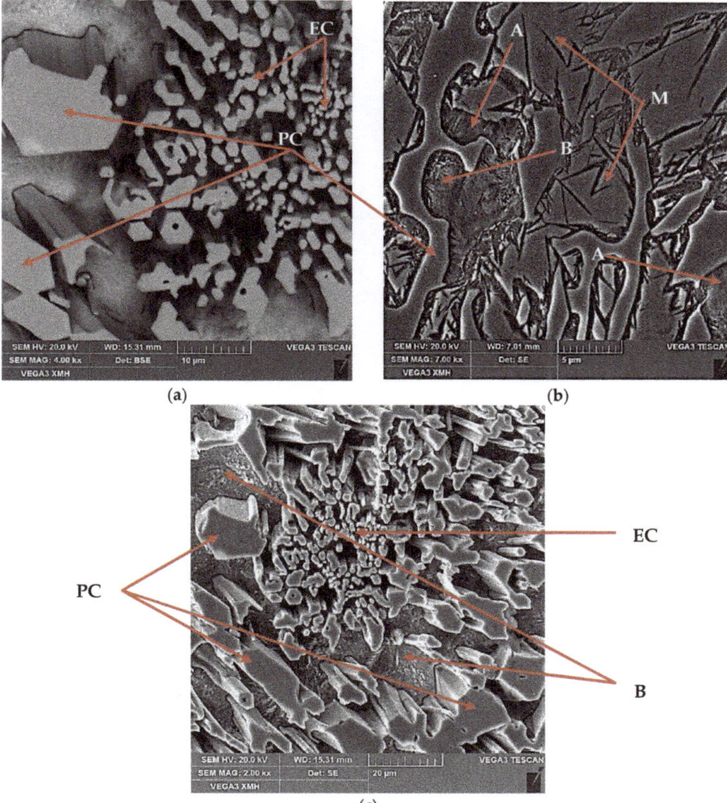


**Figure 1.** SEM microstructure of 15% Cr cast iron specimen as cast condition. Different phases, primary chromium carbide precipitates (PC), eutectic carbide precipitates (EC), martensite (M), bainite (B) and austenite (A) are indicated by arrows: (**a**) after deep etching, (**b**) after light etching, (**c**) after deep etching.

Example dilatograms and selected SEM microstructure images for the different states of the treated samples are presented in Figures 2–4. Martensite transformation start and finish temperatures were determined with the use of the tangent method. The dilatograms contain also average values of the cooling rate in temperature ranges 800–400 °C and 400 °C–TMs.

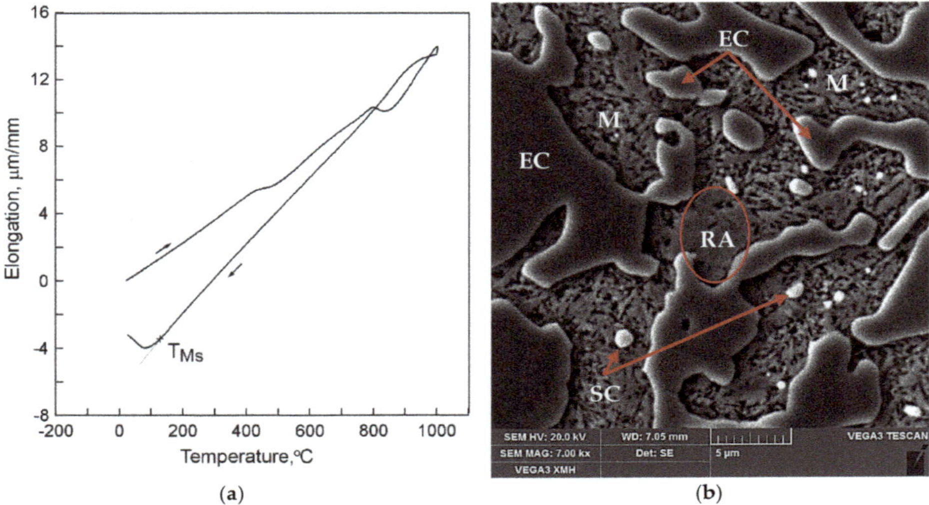


(**a**) (**b**)

**Figure 2.** The 15% Cr cast iron specimen ($V_{heat}$ = 400 °C/h, $T_A$ = 1000 °C, $\tau_A$ = 30 min) cooled in air ($V_{cool\ 800–400}$ = 6.9 °C/s, $V_{cool\ 400–TMs}$ = 1.9 °C/s): (**a**) a dilatogram; (**b**) SEM microstructure, etched section; retained austenite (RA), martensite (M), eutectic carbides (EC), and secondary carbides (SC) have been marked.

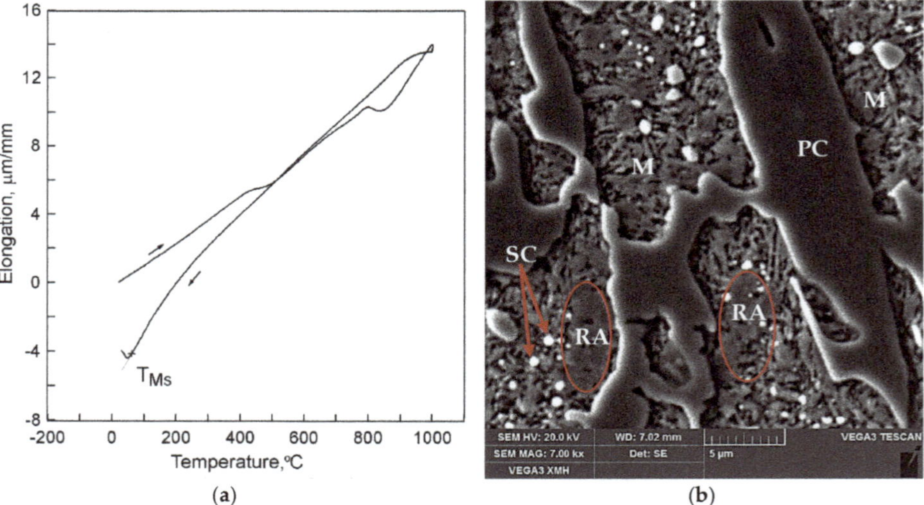


(**a**) (**b**)

**Figure 3.** A dilatogram for 15%Cr cast iron ($V_{heat}$ = 400 °C/h, $T_A$ = 1000 °C, $\tau_A$ = 30 min) cooled in water ($V_{cool\ 800–400}$ = 20.9 °C/s, $V_{cool\ 400–TMs}$ = 3.8 °C/s): (**a**) a dilatogram; (**b**) SEM microstructure, etched section; retained austenite (RA), martensite (M), primary (PC), and secondary carbides (SC) have been marked.

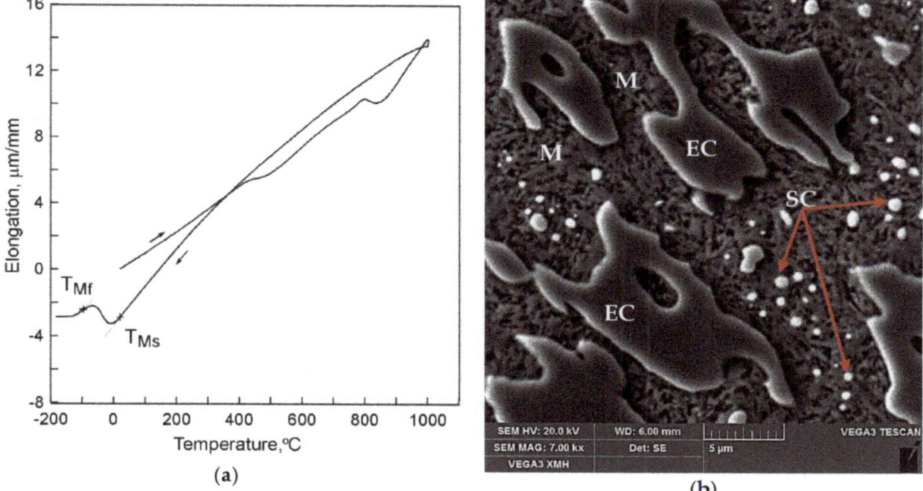


**Figure 4.** A dilatogram for 15% Cr cast iron ($V_{heat}$ = 400 °C/h, $T_A$ = 1000 °C, $\tau_A$ = 30 min) cooled in liquid nitrogen ($V_{cool\ 800–400}$ = 14.8 °C/s, $V_{cool\ 400–TMs}$ = 6.0 °C/s): (**a**) a dilatogram; (**b**) SEM microstructure, etched section; martensite matrix (M) as well as eutecitc (EC), and secondary carbides (SC) have been marked.

It has been found that the course of martensite transformation was accompanied by an increase of cast iron volume which can be noted on dilatograms in the form of inflection in the direction of increasing specimen elongation values with decreasing temperature (Figures 2–4). Once its driving force in the form of decreasing temperature was used up, the transformation of supercooled austenite into martensite was halted which was observed in case of cooling in compressed air (Figure 2a) and in water (Figure 3a). As a result, residual austenite was still visible in the cast iron microstructure. Representative microstructure images are shown in Figures 2b and 3b.

The use of cooling medium with much lower temperature, which was liquid nitrogen (−195.8 °C), resulted in occurrence of a full course of transformation of supercooled austenite into martensite (Figure 4). In the dilatogram of Figure 4a, that is evidenced by occurrence of an inflection in the direction of decreasing specimen elongation values. As can be seen in Figure 4b, the microstructure of the analyzed samples consisted of eutectic carbides (EC) and secondary carbides (SC) surrounded by a martensite matrix (M).

It is worth noting that in the first temperature range 800–400 °C, the cooling rate in liquid nitrogen is lower than the cooling rate in water. That is a result of turbulent evaporation of the medium around the specimen mounted in the dilatometer tube. In the second temperature range 400 °C–$T_{Ms}$, intensity of evaporation of liquid nitrogen decreased significantly which improved effectiveness of heat sinking from specimens. Within that temperature range, the highest cooling rate was observed in the specimen quenched in liquid nitrogen, somewhat lower in the specimen hardened in water, and lowest in the specimen quenched in air.

Results of examination of the effect of medium used to quench 15% Cr cast iron on cooling rate value in temperature ranges 800–400 °C and 400 °C–$T_{Ms}$ and martensite transformation limiting temperatures are given in Table 2.

**Table 2.** The effect of type of medium used to quench 15% Cr cast iron on value of the cooling rate $V_{cool}$ in temperature ranges 800–400 °C and 400 °C–$T_{Ms}$ and on martensite transformation limiting temperature values.

| Cooling Medium | Air | Water | Liquid Nitrogen |
|---|---|---|---|
| Cooling rate (°C/s) | | | |
| in the range 800–400 °C | 6.9 | 20.9 | 14.8 |
| in the range 400 °C–$T_{Ms}$ | 1.9 | 3.8 | 6.0 |
| $T_{Ms}$ (°C) | 128 | 62 | 20 |
| $T_{Mf}$ (°C) | n.a. | n.a. | −97 |

* Heating rate $V_{heat}$ = 400 °C/h; Austenitizing temperature $T_A$ = 1000 °C; Austenitizing time $\tau_A$ = 30 min

The obtained results indicate that cooling conditions have an effect on the martensite transformation start temperature. Hysteresis of the martensite transformation start temperature is characterized by decreasing together with increasing cooling rate in the second temperature range 400 °C–$T_{Ms}$. For cooling conditions either in air, water, or liquid nitrogen, the martensite transformation start temperature assumes positive values, from 128 °C for air, through 62 °C for water, to 20 °C for liquid nitrogen. Only the use of liquid nitrogen secures completeness of transformation of supercooled austenite into martensite.

Examination of the effect of the cooling medium type (compressed air, water, liquid nitrogen) used in the process of quenching plate specimens on their hardness values were carried out for the same cooling rate values and conditions of austenitizing as those adopted in dilatometric tests. In the course of heating and austenitizing in Nabertherm N/61H furnace, specimens were dusted with dry quartz sand. Hardness was evaluated with the use of Rockwell hardness testing machine in scale C. Results of testing hardness of the specimens on the side reproduced by chill are summarized in Table 3.

**Table 3.** Hardness values for 15% Cr cast iron plate specimens on the side reproduced by chill, cooled in compressed air, water, or liquid nitrogen from temperature 1000 °C after austenitizing for 30 min.

| Cooling Medium | Air | Water | Liquid Nitrogen |
|---|---|---|---|
| HRC hardness | 63 | 64 | 68 |

* Heating rate $V_{heat}$ = 400 °C/h; Austenitizing temperature $T_A$ = 1000 °C; Austenitizing time $\tau_A$ = 30 min

The obtained results indicate that the use of liquid nitrogen as the cooling medium resulted in the temperature decrease deep enough to secure completeness of martensite transformation in the group of tested specimens. It turned out that cooling in water allowed to achieve higher specimen hardness values compared to those observed in specimens cooled in compressed air. The results suggest that the degree of transformation of supercooled austenite into martensite obtained in the course of cooling the specimens in water is higher compared to the degree of transformation of supercooled austenite into martensite in specimens cooled in compressed air.

Further dilatometric tests included two stages of cooling. After 1 h, the specimens cooled in compressed air were, without dismounting from dilatometer, subject to further cooling in liquid nitrogen (deep cryogenic treatment). The same dilatometric testing procedure was applied to specimens cooled in water. Example dilatograms taken in the course of the tests are presented in Figure 5.

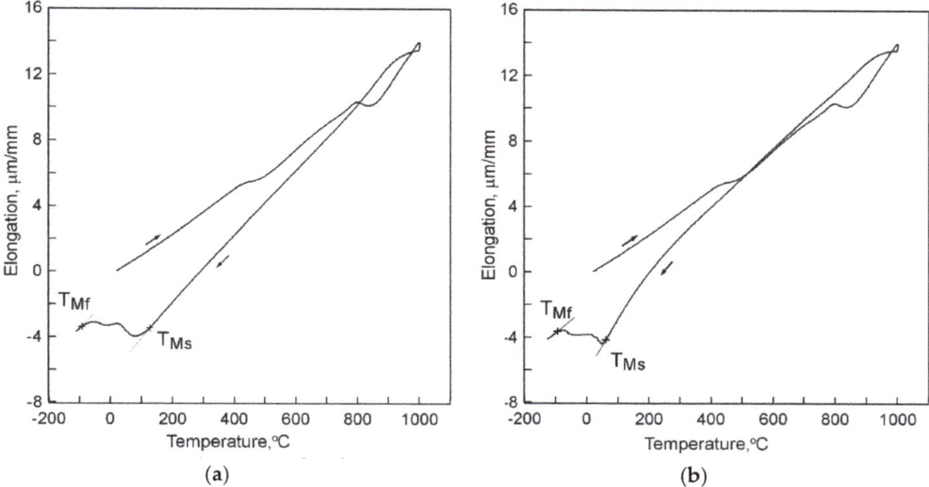


**Figure 5.** A dilatograms of 15% Cr cast iron ($V_{heat}$ = 400 °C/h, $T_A$ = 1000 °C, $\tau_A$ = 30 min) cooled: (**a**) in air and after 1 h, in liquid nitrogen; (**b**) in water and after 1 h, in liquid nitrogen.

Analysis of the obtained dilatograms indicates that application of the two-stage procedure of cooling to dilatometric specimens triggered resumption of transformation of supercooled austenite into martensite (Figure 6).

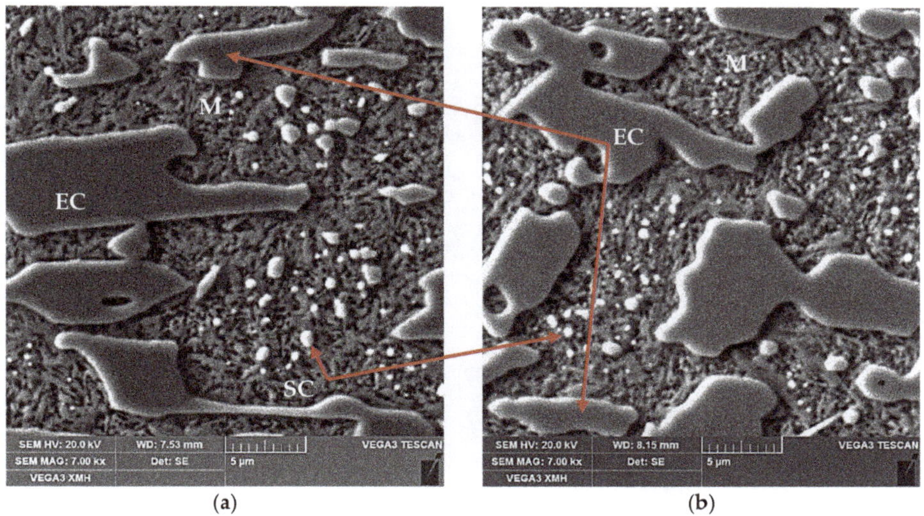


**Figure 6.** The SEM microstructure images of 15% Cr cast iron ($V_{heat}$ = 400 °C/h, $T_A$ = 1000 °C, $\tau_A$ = 30 min) cooled: (**a**) in air and after 1 h, in liquid nitrogen; (**b**) in water and after 1 h, in liquid nitrogen.

The obtained dilatograms illustrate changes in specimen dimensions in the course of temperature decreasing. The nature of the changes is however different than in the case of direct cooling the specimens in liquid nitrogen (Figure 4). The last resolute inflection of the plot line in direction corresponding to decrease of specimen dimensions evidences occurrence of a complete course of supercooled austenite transformation into martensite and allows to determine the martensite transformation finish temperature $T_{Mf}$. The obtained dilatograms evidence the possibility to trigger transformation of supercooled austenite into martensite as a result of immersing specimens in a cooling medium with sufficiently low temperature. The microstructure of the tested samples consisted of martensite (M), eutectic carbides (EC), and secondary carbides (SC), whereas no residual austenite (RA) was observed (Figure 6). On the other hand, it can be assumed that the course of transformation of supercooled austenite into martensite is less violent compared to the variant with direct cooling in liquid nitrogen. This should result in lower level of quenching stresses and thus less susceptibility of castings to quenching cracks.

Results of the research on the effect of two-stage quenching on martensite transformation temperatures in 15% Cr cast iron are given in Table 4.

**Table 4.** Results of the research on determination of martensite transformation limiting temperatures in 15% Cr cast iron after two-stage cooling, first in compressed air and then in liquid nitrogen and first in water and then in liquid nitrogen.

| Cooling Medium | First Compressed Air, then Liquid Nitrogen | First Water, then Liquid Nitrogen |
|---|---|---|
| $T_{Ms}$ (°C) | 128 | 62 |
| $T_{Mf}$ (°C) | −93 | −95 |

* Heating rate $V_{heat}$ = 400 °C/h; Austenitizing temperature $T_A$ = 1000 °C; Austenitizing time $\tau_A$ = 30 min

The obtained results indicate that the use of a medium with significantly lower temperature such as liquid nitrogen (−195.8 °C) in the second stage of cooling allowed to finish off the process of transformation of supercooled austenite still remaining in specimens cooled in air or in water into martensite.

When two-stage cooling was applied in order to ensure completeness of martensite transformation, $T_{Mf}$ values in the variant with compressed air and in the variant with water (−93 °C and −95 °C, respectively) may be considered comparable.

Results of examination of the effect of two-stage cooling first in compressed air and then in liquid nitrogen or first in water and then in liquid nitrogen on hardness values of plate specimens on their sides reproduced with chills are given in Table 5. The plate specimens, protected from oxidation by dusting with a dry quartz sand sprinkle, were austenitized at temperature 1000 °C for 30 min in Nabertherm N/61H furnace.

**Table 5.** Results of hardness measurements for plate specimens of 15% Cr cast iron austenitized at 1000 °C for 30 min, then cooled in compressed air and further in liquid nitrogen or in compressed air and further in liquid nitrogen. Measurements taken on surface reproduced by chill.

| Cooling Medium | First Compressed Air, then Liquid Nitrogen | First Water, then Liquid Nitrogen |
|---|---|---|
| Plate hardness (HRC) | 68 | 68 |

* Heating rate $V_{heat}$ = 400 °C/h; Austenitizing temperature $T_A$ = 1000 °C; Austenitizing time $\tau_A$ = 30 min

The obtained results indicate that under the same conditions of austenitizing, application of diversified processes of cooling securing complete transformation of supercooled austenite into martensite allowed to obtain the same high hardness values of 15% Cr cast iron.

The austenite, in view of its supersaturation with carbon and alloying elements, is thermodynamically unstable. For that reason, it is susceptible to precipitation of carbides out of it. Elevation of alloy temperature favors the process energetically. Precipitation of carbides in high-chromium cast iron leads to impoverishment of austenite in carbon, chromium, and other elements included in composition of chromium carbides. Authors of the paper [35] demonstrated that precipitations of chromium carbides from austenite in a cast iron containing 14.55% Cr proceeded fastest at the temperature of 950 °C. The process started as early as after 10 s and ended within the period of 160 min. The austenite impoverished in carbon and other alloying elements, transforms into martensite in the course of rapid cooling. The austenite-martensite transformation temperature is the higher, the less carbon and carbide formers are contained in austenite. The process of precipitation of chromium carbides requires diffusive migration of atoms making up the carbides to thermodynamically favorable locations such as grain boundary defects, crystalline lattice defects, glide bands, or austenite regions adjacent to carbides in which plastic deformations could occur as a result of different linear expansion coefficient values. The observed higher values of $T_{Ms}$ for conditions of quenching in air can be explained so that the period of keeping specimens in the range of high temperatures turned out to be sufficiently long for the process of precipitation of carbides and change in chemical composition of austenite. In the conditions of cooling in liquid nitrogen, the period of maintaining specimens in the range of high temperatures was too short for the process of precipitation of carbides in the course of cooling which resulted in lower $T_{Ms}$ values.

## 4. Conclusions

The obtained results of the research on the effect of variant cooling processes applied to 15% Cr cast iron specimens from the austenitizing temperature of 1000 °C after the austenitizing time of 30 min on martensite transformation temperatures and hardness values indicate that:

- the cooling rate, especially in the temperature range 400 °C–$T_{Ms}$, is decisive for martensite transformation start temperatures;
- cooling in air or in water does not secure completeness of transformation of supercooled austenite into martensite;
- the martensite transformation start temperature $T_{Ms}$ is sensitive to the cooling rate and amounts to 128 °C for cooling in air, 62 °C for cooling in water, and 20 °C for cooling in liquid nitrogen;
- in case of cooling in air or in water, it is possible to force completeness of the course of transformation of supercooled austenite into martensite, however this requires application of an additional operation of cooling in a medium with temperature lower than the martensite transformation finish temperatures (−93 °C and −95 °C, respectively);
- application of diversified variant cooling processes guaranteeing completeness of transformation of supercooled austenite into martensite allowed to obtain the hardness value of 68 HRC in plate specimens on their side reproduced by chills.

**Author Contributions:** Conceptualization, M.T., A.W.O., A.T., M.M., and A.J.D.; methodology, M.T., A.W.O., A.T., and M.M.; software, G.W.; validation, M.T., A.W.O., A.T., and M.M.; formal analysis, M.T., A.W.O., A.T., and M.M..; investigation, G.W.; resources, M.T., A.W.O., A.T., and M.M.; data curation, G.W.; writing—original draft preparation, M.T., A.W.O., and A.J.D.; writing—review and editing, M.T., A.W.O., and A.J.D.; visualization, M.T. and A.J.D.; supervision, A.W.O. All authors have read and agreed to the published version of the manuscript.

**Funding:** This research received no external funding.

**Conflicts of Interest:** The authors declare no conflict of interest. The funders had no role in the design of the study; in the collection, analyses, or interpretation of data; in the writing of the manuscript, or in the decision to publish the results".

## References

1. Hutchings, I.M.; Shipway, P. Design and selection of materials for tribological applications. In *Tribology—Friction and Wear of Engineering Materials*, 2nd ed.; Elsevier Science: London, UK, 2017; Volume 8, pp. 283–302.
2. Atapek, S.H.; Polat, S.A. Study of wear of high-chromium cast iron under dry friction. *Metal Sci. Heat Treat.* **2013**, *55*, 181–183. [CrossRef]

3. Gasan, H.; Erturk, F. Effects of a destabilization heat treatment on the microstructure and abrasive wear behavior of high-chromium white cast iron investigated using different characterization techniques. *Metall. Mater. Trans.* **2013**, *44A*, 4993–5005. [CrossRef]
4. Gu, J.; Xiao, P.; Song, J.; Li, Z.; Lu, R. Sintering of a hypoeutectic high chromium cast iron as well as its microstructure and properties. *J. Alloys Compd.* **2018**, *740*, 485–491. [CrossRef]
5. Pokusová, M.; Brúsilová, A.; Šooš, L.; Berta, I. Abrasion Wear Behavior of High-chromium Cast Iron. *Arch. Foundry Eng.* **2015**, *15*, 17–22. [CrossRef]
6. Coronado, J.J.; Sinatora, A. Abrasive wear study of white cast iron with different solidification rates. *Wear* **2009**, *267*, 2116–2121. [CrossRef]
7. Gromczyk, M.; Kondracki, M.; Studnicki, A.; Szajnar, J. Stereological Analysis of Carbides in Hypoeutectic Chromium Cast Iron. *Arch. Foundry Eng.* **2015**, *15*, 17–22. [CrossRef]
8. Orłowicz, W.; Trytek, A. Effect of rapid solidification on sliding wear of iron castings. *Wear* **2003**, *254*, 154–163. [CrossRef]
9. Scandian, C.; Boher, C.; Mello, J.D.B.; Rezai-Aria, F. Effect of molybdenum and chromium contents in sliding wear of high-chromium white cast iron: The relationship between microstructure and wear. *Wear* **2009**, *267*, 401–408. [CrossRef]
10. Studnicki, A.; Dojka, R.; Gromczyk, M.; Kondracki, M. Influence of Titanium on Crystallization and Wear Resistance of High Chromium Cast Iron. *Arch. Foundry Eng.* **2016**, *16*, 117–123. [CrossRef]
11. Kopyciński, D.; Piasny, S.; Kawalec, M.; Madizhanova, A. The Abrasive Wear Resistance of Chromium Cast Iron. *Arch. Foundry Eng.* **2014**, *14*, 63–66. [CrossRef]
12. Kopyciński, D.; Guzik, E.; Siekaniec, D.; Szczęsny, A. Analysis of the High Chromium Cast Iron Microstructure after the Heat Treatment. *Arch. Foundry Eng.* **2014**, *14*, 43–46. [CrossRef]
13. Gonzalez-Pociño, A.; Alvarez-Antolin, F.; Asensio-Lozano, J. Influence of Thermal Parameters Related to Destabilization Treatments on Erosive Wear Resistance and Microstructural Variation of White Cast Iron Containing 18% Cr. Application of Design of Experiments and Rietveld Structural Analysis. *Materials* **2019**, *12*, 3252. [CrossRef] [PubMed]
14. Coronado, J.J.; Gómez, A.; Sinatora, A. Tempering temperature effects on abrasive wear of mottled cast iron. *Wear* **2009**, *267*, 2070–2076. [CrossRef]
15. Hadji, A.; Bouhamla, K.; Maouche, H. Improving Wear Properties of High-Chromium Cast Iron by Manganese Alloying. *Int. J. Metalcast.* **2016**, *10*, 43–55. [CrossRef]
16. Bedolla-Jacuinde, A.; Guerra, F.; Mejia, I.; Vera, U. Niobium Additions to a 15%Cr–3%C White Iron and its Efects on the Microstructure and on Abrasive Wear Behavior. *Metals* **2019**, *9*, 1321. [CrossRef]
17. Heino, V.; Kallio, M.; Valtonen, K.; Kuokkala, V.T. The role of microstructure in high stress abrasion of white cast irons. *Wear* **2017**, *388–389*, 119–125. [CrossRef]
18. Andrews, K.W. Empirical Formulae for the Calculation of Some Transformation Temperatures. *J. Iron Steel Inst.* **1965**, *203*, 721–727.
19. Yang, H.S.; Bhadeshia, H.K.D.H. Uncertainties in Dilatometric Determination of Martensite Start Temperature. *Mater. Sci. Technol.* **2007**, *23*, 556–560. [CrossRef]
20. Warke, V.S.; Sisson, R.D., Jr.; Makhlouf, M.M. A Model for Converting Dilatometric Strain Measurements to the Fraction of Phase Formed During the Transformation of Austenite to Martensite in Powder Metallurgy Steels. *Metall. Mater. Trans. A* **2009**, *40A*, 569–572. [CrossRef]
21. Gomez, M.; Medina, S.F.; Caruana, G. Modelling of Phase Transformation Kinetics by Correction of Dilatometry Results for a Ferritic Nb-microalloyed Steel. *ISIJ Int.* **2003**, *43*, 1228–1237. [CrossRef]
22. De Souzaa, S.d.; Moreiraa, P.S.; de Fariaa, G.L. Austenitizing Temperature and Cooling Rate Effects on the Martensitic Transformation in a Microalloyed-Steel. *Mater. Res.* **2020**, *23*, 1–9.
23. Gao, Q.; Wang, C.; Qu, F.; Wang, Y.; Qiao, Z. Martensite transformation kinetics in 9Cr-1.7W-0.4Mo-Co ferrite steel. *J. Alloys Compd.* **2014**, *610*, 322–330. [CrossRef]
24. Durlu, T.N. Effects of High Austenitizing Temperature and Austenite Deformation on Formation of Martensite in Fe-Ni-C Alloys. *J. Mater. Sci.* **2001**, *36*, 5665–5671. [CrossRef]
25. Sastri, A.S.; West, D.R.F. Effect of Austenitizing Conditions on the Kinetics of Martensite Formation in Certain Medium-Alloy Steels. *J. Iron Steel Inst.* **1965**, *203*, 138–149.

26. Alvarado-Meza, M.A.; García-Sanchez, E.; Covarrubias-Alvarado, O.; Salinas-Rodriguez, A.; Guerrero-Mata, M.P.; Colás, R. Effect of the High-Temperature Deformation on the Ms Temperature in a Low C Martensitic Stainless Steel. *J. Mater. Eng. Perform.* **2013**, *22*, 34–350. [CrossRef]
27. Efremenko, V.; Shimizu, K.; Chabak, Y. Effect of Destabilizing Heat Treatment on Solid-State Phase Transformation in High-Chromium Cast Irons. *Metall. Mater. Trans. A* **2013**, *44A*, 5434–5446. [CrossRef]
28. Powell, G.L.F.; Laird, G.J., II. Structure, nucleation, growth and morphology of secondary carbides in high chromium and Cr-Ni white cast irons. *Mater. Sci.* **1992**, *27*, 29–35. [CrossRef]
29. Maratray, F. Choice of appropriate compositions for chromium-molybdenum white irons. *Trans. AFS* **1971**, *79*, 121–124.
30. Wiengmoon, A.; Chairuangsri, T.; Pearce, J.T.H. A Microstructural Study of Destabilised 30wt%Cr-2.3wt%C High Chromium Cast Iron. *Iron Steel Inst. Jpn. Int.* **2004**, *44*, 396–403. [CrossRef]
31. Laird, G., II; Powell, G.L.F. Solidification and Solid-State Transformation Mechanisms in Si Alloyed High-Chromium White Cast Irons. *Metall. Trans. A* **1993**, *24A*, 981–988. [CrossRef]
32. Guitar, M.A.; Suárez, S.; Prat, O.; Guigou, M.D.; Gari, V.; Pereira, G.; Mücklich, F. High Chromium Cast Irons: Destabilized-Subcritical Secondary Carbide Precipitation and Its Effect on Hardness and Wear Properties. *J. Mater. Eng. Perform.* **2018**, *27*, 3877–3885. [CrossRef]
33. Efremenko, V.G.; Wu, K.M.; Chabak, Y.U.G.; Shimizu, K.; Isayev, O.B.; Kudin, V.V. Alternative Heat Treatments for Complex-Alloyed High-Cr Cast Iron Before Machining. *Metall. Mater. Trans. A* **2018**, *49A*, 3430–3440. [CrossRef]
34. Karantzalis, A.E.; Lekatou, A.; Mavros, H. Microstructural Modifications of As-Cast High-Chromium White Iron by Heat Treatment. *J. Mater. Eng. Perform.* **2009**, *18*, 174–181. [CrossRef]
35. Efremenko, V.G.; Chabak, Y.U.G.; Brykov, M.N. Kinetic Parameters of Secondary Carbide Precipitation in High-Cr White Iron Alloyed by Mn-Ni-Mo-V Complex. *J. Mater. Eng. Perform.* **2013**, *22*, 1378–1385. [CrossRef]

*Article*

# Dehydroxylation of Perlite and Vermiculite: Impact on Improving the Knock-Out Properties of Moulding and Core Sand with an Inorganic Binder

**Artur Bobrowski [1,*], Karolina Kaczmarska [1], Maciej Sitarz [2], Dariusz Drożyński [1], Magdalena Leśniak [2], Beata Grabowska [1] and Daniel Nowak [3]**

1 Faculty of Foundry Engineering, AGH—University of Science and Technology, Reymonta 23, 30059 Krakow, Poland; karolina.kaczmarska@agh.edu.pl (K.K.); dd@agh.edu.pl (D.D.); beata.grabowska@agh.edu.pl (B.G.)
2 Faculty of Materials Science and Ceramics, AGH—University of Science and Technology, Mickiewicza 30, 30059 Krakow, Poland; msitarz@agh.edu.pl (M.S.); mlesniak@agh.edu.pl (M.L.)
3 Faculty of Mechanical Engineering, Wroclaw University of Science and Technology, 27 Wybrzeże Wyspiańskiego, 50370 Wrocław, Poland; daniel.nowak@pwr.edu.pl
* Correspondence: arturb@agh.edu.pl

**Abstract:** The article presents the results of research aimed at examining the type of swelling material introduced into moulding or core sand to improve their knock-out properties. Tests on Slovak perlite ore (three grain sizes), Hungarian perlite ore and ground vermiculite (South Africa) were carried out. For this purpose, thermal and structural analyses (FTIR—Fourier Transform Infrared Spectroscopy), a chemical composition test (XRF-X-Ray Fluorescence), phase analysis (XRD—X-Ray Diffraction), and scanning electron microscopy (SEM—Scanning Electron Microscope) as well as final strength tests of moulding sands with the addition of perlite ore and vermiculite were carried out. The results of thermal studies were related to IR (Infrared Spectroscopy) spectra and XRD diffractograms. It has been shown that the water content in the pearlite ore is almost three times lower than in vermiculite, but the process of its removal is different. Moreover, the chemical composition of the perlite ore, in particular the alkali content and its grain size, may influence its structure. The phenomena of expansion (perlite) and peeling (vermiculite) have a positive effect on the reduction of the final sand strength and eliminate technological inconveniences (poor knocking out) that significantly limit the wide use of moulding sands with inorganic binders.

**Keywords:** aluminosilicate; perlite; vermiculite; dehydroxylation; thermal analysis; FTIR; XRD; XRF; SEM; moulding sand; inorganic binder

**Citation:** Bobrowski, A.; Kaczmarska, K.; Sitarz, M.; Drożyński, D.; Leśniak, M.; Grabowska, B.; Nowak, D. Dehydroxylation of Perlite and Vermiculite: Impact on Improving the Knock-Out Properties of Moulding and Core Sand with an Inorganic Binder. *Materials* **2021**, *14*, 2946. https://doi.org/10.3390/ma14112946

Academic Editor: Tomasz Wróbel

Received: 22 March 2021
Accepted: 26 May 2021
Published: 29 May 2021

## 1. Introduction

The current state of knowledge indicates that many attempts have been made to eliminate the main technological disadvantage, which is poor knock-out property, that limits the wide use of moulding sands with inorganic binders in a foundry. Taking into consideration the increasing requirements and limitations of environmental protection regulations, it can be predicted that the competitiveness of this type of binder in relation to organic binders will increase. Their wide use is supported by the possibility of obtaining good technological and mechanical properties, the purchase price, much lower than that of organic binders, as well as a small amount of harmful products of thermal decomposition (in the form of gases) generated when pouring molds with a liquid casting alloy and lower costs of waste disposal [1,2]. In addition, the foundries also bear costs related to environmental fees, the amount of which strongly depends on the quantity and quality of generated waste, including gas products. The assessment of the harmfulness of moulding sand does not only concern emissions to the atmosphere. What should also be remembered are the work environment and the frequent exposure of workers to large amounts of harmful substances. To increase the competitiveness and interest of the foundries in inorganic binders, it seems necessary to eliminate or significantly reduce poor knock-out properties. They are the result of the so-called second maximum strength resulting from exposure to a high temperature [3,4]. Many attempts have been made to improve the knock-out properties with the use of various types of materials or by modifying the structure of inorganic binders, e.g., by introducing metal oxides. All these measures were aimed at reducing the final strength of the moulding sand and thus led to an improvement in the knock-out properties [5–13].

Knowing the characteristic feature of commonly available materials of mineral origin (perlite ore, vermiculite), which is the increase in volume under the influence of temperature (expansion and exfoliation phenomena), associated with the high speed dehydroxylation reaction [14–18], it was assumed that their introduction into moulding or core sand would reduce the final sand strength and would reduce the technological inconvenience associated with poor knock-out properties. The kinetic energy of the swelling additive located between the grains of the grain matrix will disrupt the continuity of the hardened binder layer and contribute to loosening of the hardened sand and lowering the final strength. Thanks to this, the knock-out will also be improved. Because the minerals belong to the group of inorganic materials, they will not emit harmful gaseous products during heating. In this way, it becomes possible to preserve the basic advantage of sands with inorganic binders—they will remain environmentally friendly.

Based on a literature review, it was found that Greek perlite (Milos Island) is the best studied perlite deposit in Europe [19–21]. The paper [21] compares the chemical compositions of Greek, Italian, Hungarian, Chinese, and Turkish perlites and describes the impact of heat treatment processes on changes in IR spectra. However, there is no reference to Slovak perlite and the relationship between the chemical composition and phase changes under the influence of temperature and the results of FTIR structural studies. In the publication [22] on the Slovak perlite ore from the Lehôtka deposit near Brehmi, it was shown that it mainly contains biotite, albite, quartz, and smectite, but the influence of the ore grain size on any changes in the phase composition or changes taking place in the pearlite ore under the influence of temperature (dehydroxylation, chemical composition, phase transitions) have not been described. No detailed studies were found for thermal, structural, and phase analyses for perlite ore from Hungarian deposits beyond the chemical composition listed in [21].

In the works [23–25], thermal properties of natural vermiculite were determined. The publication [23,24] concerned the Santa Olalla vermiculite deposit (Huelva, Spain), and tests on samples before and after grinding in a vibratory mill were carried out. In turn, in the publication [25], the authors, referring to the obtained research results, stated that, in their opinion, this process was not sufficiently explained, and therefore a detailed characterization of the structural evolution during the dehydration of high-purity raw vermiculite from the Chinese deposit in Hebei Province using the XRD method in-situ was carried out.

Earlier studies of South African vermiculite from the Palabora deposit showed that it is not pure, as the content of potassium oxide ($K_2O$) is higher than 0.35% [26] and contains relatively high levels of iron [27]. It is considered to be one of the most environmentally friendly due to the fact that it is free of asbestos-like fibers and free crystalline silica [28], which is extremely important for the preparation of moulding sand.

In the literature, one can find work on reducing the final strength of moulding sand with an inorganic binder with the use of mineral additives, where the sand was introduced with an additive named Glassex, which is expanded perlite [4,7,10]. A significant gap is the lack of data on the effect of perlite ore with a different grain size and type of deposit on the ability to reduce the final strength of moulding and core sand, and thus improve its knock-out properties. There is also no comparison on the difference between the final strength of moulding sands with an inorganic binder if, at the stage of their preparation, perlite or vermiculite ore is added.

According to the authors, the results contained in the publication supplement the knowledge available so far.

## 2. Materials and Methods

### 2.1. General Characteristics of Materials

Perlite ore from Slovak deposits (SP1–SP3 samples; Perlit AF Sp. Z o.o., Kazimierz Biskupi, Poland) of various grain sizes, Hungarian perlite ore (HP sample; Perlit AF Sp. Z o.o., Kazimierz Biskupi, Poland), and South African vermiculite (V; Vermiculite Poland, Ełk, Poland) were used for the tests. The moulding sands for determination of the tensile strength ($R_m^{tk}$) according to the following protocol was carried out:

- Moulding sand without additives (G)—quartz sand form Szczakowa mine, Sibelco Poland (Bukowno, Poland) (97.28%), inorganic Geopol® binder form Sand Team S.r.o, Holubice, Czech Republic) (2.43%), ester hardener SA72 form Sand Team S.r.o, Holubice, Czech Republic (0.29%);
- Moulding sand with perlite ore (SP1–SP3 and HP)—quartz sand form Szczakowa mine, Sibelco Poland (Bukowno, Poland) (95.42%), inorganic Geopol® binder form Sand Team S.r.o, Holubice, Czech Republic (2.38%), ester hardener SA72 form Sand Team S.r.o, Holubice, Czech Republic (0.29%), perlite ore (1.91%);
- Moulding sand with vermiculite (V)—quartz sand form Szczakowa mine, Sibelco Poland (Bukowno, Poland) (96.34%), inorganic Geopol® binder form Sand Team S.r.o, Holubice, Czech Republic (2.41%), ester hardener SA72 form Sand Team S.r.o, Holubice, Czech Republic (0.29%), vermiculite (0.96%).

#### 2.1.1. Perlite Ore

Perlite ore is a transformed effusive rock built of volcanic glass, whose name comes from the French "pearl" (in German "Perlstein"), and is connected with the ball (pearl) form. This form is a result of the stresses caused by the rapid cooling of the glaze. Perlite ore also has a characteristic color and gloss [16,29,30]. Chemically, the perlite is a metastable, amorphous, hydrated potassium-sodium aluminosilicate with a cryptocrystalline structure. In the structure of perlite, apart from chemically bound water, the content of which usually ranges from 2.0 to 5.0% by volume, there may also be quartz inclusions, plagioclase, biotite, and secondary minerals such as montmorillonite or zeolite [30,31]. Deposits of the Slovak perlite are found in tertiary rocks in volcanic areas in the central and eastern part of Slovakia. They are a part of a complex of rocks made of tuffs, rhyolites, and andesites. The resources of these deposits are estimated at over 30 million tons [14]. The tested samples came from the Lehôtka deposit. In turn, Hungarian perlite belongs to the group of Upper Miocene rocks of the rhyolite-rhyodacite type. It was created as a result of undersea volcanic activity on the edge of a subduction zone, with very acidic viscous lava and pyroclastic materials. This ore contains 90–95% of amorphous volcanic glaze, 5.0 to 6.0% of crystalline components, mainly quartz, plagioclase, biotite, and 2.5–3.5% of water [14,32]. Despite comparable prices of these raw materials, perlite from the Hungarian deposits differs from the Slovak perlite ore in terms of properties of the obtained expanded perlite. The Slovak perlite is characterized by a creamier color, higher density, and higher mechanical strength. On the other hand, the Hungarian perlite ore is characterized by a very fine graining and lower mechanical strength [14].

#### 2.1.2. Vermiculite

The name "vermiculite" comes from the Latin word vermiculus and means "worm". Thus, it refers to the characteristic appearance of the mineral after its rapid heating. Then, the conversion of its internal layers occurs due to the separation of the inter-packet water. The evaporation of the water results in a 15- to 30-fold increase in volume. This process is called exfoliation, i.e., the separation of layers (plates) in grains, and is carried out at a temperature of about 900–1000 °C. Its effect is to increase the volume while reducing the apparent and bulk density [33–37]. In the case of some vermiculites, the beginning of the exfoliation process can be observed at a temperature of 300 °C [36]. As shown in previous studies [34], as a result of rapid heating to 100 °C, vermiculite loses about 50% of the interlayer water, but at this temperature, the layer separation process (exfoliation) does not occur. The vermiculite ((Mg, Fe, Al)$_3$(Al, Si)$_4$O$_{10}$(OH)$_2$·4H$_2$O) belongs to the group of clay minerals, which belong to the group of 2:1 type phyllosilicate packets, where the octahedron layer is closed between two tetrahedron layers, vertices facing each other [34,37–41]. Vermiculite is formed as a result of the hydrolysis and weathering of magma packet rocks, hydrothermal action, seepage of groundwater or a combination of these three factors, from dark types of biotite, hydrobiotite, and phlogopite belonging to the mica group [34–36,42]. Vermiculite is very rarely present as a pure component, and most often it forms mosaic-like mixed-pack structures. Depending on the deposit and its impurities, its color ranges from yellow-golden, brown to olive [34,43–45].

### 2.2. Characterization of Methods

The following devices and research methodology were used in the research:

- The perlite ore samples did not require preparation. In turn, vermiculite was subjected to the process of grinding (comminuting) in a ball mill in order to obtain a fine-grained fraction, because only the material prepared in this way is suitable for use in moulding sands.
- Sieve analysis on a standard set with a mesh size from 0.056 to 1.6 mm (average of two measurements), in accordance with the Polish Standard (PN–85/H–11001) was carried out.
- Simultaneous TG/DTG/DTA (Thermal Gravimetry/Differential Thermal Gravimetry/Differential Thermal Analysis) thermal examinations of the tested materials were performed with use of the Thermal Analyzer produced by Jota (Kraków, Poland). The temperature range of test was 20–1000 °C, and the heating rate was 10 °C/min in an air atmosphere in alumina pans.
- Structural studies were carried out by means of infrared spectroscopy (FTIR) by a transmission technique consisting of preparing pellets in potassium bromide (KBr) using an Digilab Excalibur FTS 3000 (Bio Rad, Hercules, CA, USA) spectrometer with a standard DTGS detector. The spectra were recorded in the Resolution Pro (Varian/Agilent Technologies, Santa Clara, CA, USA) program in the specific infrared range (4000–400 cm$^{-1}$), with a resolution of 4 cm$^{-1}$. In order to obtain the appropriate quality of the spectra, the samples were ground in the form of fine powder in agate mortars and then mixed with KBr in the ratio 1:100. Test material samples were obtained as a result of rapid heating in a silite furnace at a temperature in the range of 100–1000 °C. The residence time of the sample in the oven was 10 min in each case.
- The chemical composition of the investigated materials was determined by the X-ray Fluorescence Spectrometry (XRF) method, using a WD-XRF Axios Max spectrometer with Rh 4 kW PANalytical lamp (Malvern Panalytical, Malvern, UK).
- To analyze the phase composition of the materials, a Philips/Panalytical X'Pert Pro MD powder diffractometer (Malvern Panalytical, Malvern, UK) using Cu K$\alpha$1 radiation was used. Standard Bragg-Brentano geometry with a θ–2θ setup was applied (0.008° step size and 5–90° 2θ range). Highscore Plus 3.0 software with a database (Version PDF-4+2021 powder diffraction database, 2021, International Centre for Diffraction Data, Newtown Square, Pennsylvania, USA) was used to determine the positions of the observed peaks and assign them to the appropriate phases.

- Moulding sands with an inorganic geopolymer binder were prepared in a roller mixer. First, sand without additives (without perlite ore and vermiculite) was prepared, which was the reference point. Then, the moulding sands with additives were prepared. The mixing time was 5 min in each case. The loose components, i.e., the grain matrix (quartz sand) and the perlite ore/ground vermiculite (mixing time 1 min) were mixed first, then the SA72 hardener (Sand Team S.r.o, Holubice, Czech Republic) (mixing time 1 min) was added, and the (Geopol binder Sand Team S.r.o, Holubice, Czech Republic) (mixing time 3 min) was added in the last stage.
- Standard fittings for determining the tensile strength were made of the prepared sands. A portion of the moulding sand was placed in a special matrix, coupled with a device for a vibratory concentration of samples (LUZ-2e apparatus manufactured by Multiserw Morek, Brzeźnica, Poland). The hardening process in the open air, under laboratory conditions (temperature 20–21 °C, relative humidity 30–35%) was carried out. After hardening the moulding sand (within 24 h), the samples were heated together in the furnace to a temperature in the range of 100–1000 °C, with a heating rate of 10 °C/min (the same as in the case of the thermal analysis). After the sand samples and the furnace had cooled down to the ambient temperature, the tensile strength ($R_m^{tk}$) was determined using the LRu-2 moulding sand strength tester (Multiserw-Morek, Brzeźnica, Poland). Moulding sand without additives (as a reference point) (G), sands with the addition of the Slovak perlite ore of various grain sizes (SP1–SP3), the Hungary perlite ore (HP), and crushed vermiculite (ground) (V) were also made.
- Microscopic images of moulding sands with additives were made with the use of a HITACHI TM-3000. A Hitachi TM3000 scanning electron microscope (SEM) (Hitachi High-Tech Co., LTD, Tokyo, Japan) was used in this study to investigate the surface morphology of specimens, with a 30 nm resolution, a charge reduction mode of 5 or 15 kV voltage, and 15× to 30,000× magnification capabilities. For high-resolution imaging, the samples were carbon sprayed using a Quorum Techologies Q150T high vacuum sputter (Lewes, UK).

## 3. Results and Discussion

### 3.1. Sieve Analysis

In the first stage of the research, sieve analysis of materials was carried out. Tables 1 and 2 show their grain size composition and a comparison of the basic grain size parameters.

**Table 1.** Grain size analysis of Slovak perlite ore (SP1–SP3), Hungarian perlite ore (HP), ground vermiculite (V) and quartz sand (QS—matrix of moulding sands).

| Mesh Size Sieve | Material | | | | | |
|---|---|---|---|---|---|---|
| | **SP1** | **SP2** | **SP3** | **HP** | **V** | **QS** |
| 1.600 | 0.00 | 0.00 | 0.00 | 0.00 | 0.00 | 0.00 |
| 0.800 | 0.00 | 0.00 | **0.95** | 0.00 | 0.00 | 0.00 |
| 0.630 | 0.00 | 0.00 | **1.58** | 1.84 | 0.00 | 0.48 |
| 0.400 | 0.00 | **48.16** | **29.98** | **30.08** | 0.06 | 7.19 |
| 0.320 | 0.00 | **30.21** | 40.10 | **20.62** | 0.13 | **18.63** |
| 0.200 | 0.00 | **11.58** | 25.94 | **29.09** | 3.36 | **50.21** |
| 0.160 | 6.52 | 4.38 | 0.31 | 7.93 | 9.83 | **13.10** |
| 0.100 | 9.29 | 4.10 | 0.38 | 7.36 | 22.54 | 9.71 |
| 0.071 | **9.35** | 0.67 | 0.20 | 1.22 | **13.34** | 0.56 |
| 0.056 | **22.03** | 0.33 | 0.20 | 0.53 | **3.66** | 0.07 |
| Bottom | **52.81** | 0.57 | 0.36 | 1.33 | **47.08** | 0.05 |
| Sum | 100.00 | 100.00 | 100.00 | 100.00 | 100.00 | 100.00 |

**Table 2.** Comparison of grain size distribution of tested materials.

| Parameter | Unit | Material | | | | | |
|---|---|---|---|---|---|---|---|
| | | SP1 | SP2 | SP3 | HP | V | QS |
| Average grain size, $d_L$ | mm | 0.08 | 0.30 | 0.31 | 0.24 | 0.06 | 0.23 |
| Average grain size, $D_{50}$ | mm | 0.10 | 0.40 | 0.36 | 0.33 | 0.07 | 0.26 |
| The mesh size sieve where the main faction has gathered | - | 0.071/ 0.056/ bottom | 0.40/ 0.32/ 0.20 | 0.40/ 0.32/ 0.20 | 0.40/ 0.32/ 0.20 | 0.071/ 0.056/ bottom | 0.32/ 0.20/ 0.16 |
| Main fraction, $F_g$ | % | 92.61 | 89.95 | 96.02 | 79.79 | 82.96 | 81.94 |
| Distribution factor, $S_0$ | - | 1.27 | 1.20 | 1.19 | 1.38 | 1.93 | 1.27 |
| Inclination indicator, $S_k$ | - | 1.04 | 1.04 | 1.05 | 0.93 | 0.98 | 1.00 |
| Homogeneity degree, $GG$ | % | 61.00 | 68.00 | 74.00 | 50.00 | 21.00 | 68.00 |
| Grain number, $L$ | - | 150.52 | 42.90 | 40.76 | 52.27 | 21.86 | 55.27 |

### 3.2. Thermal Analysis

Figures 1–3 show the results of the thermal analysis for the Slovak perlite ore of various grain size (SP1–SP3), and Figure 4 shows the results of the determination for the Hungarian perlite ore (HP). In turn, the results of the analysis for ground vermiculite (V) are shown in Figure 5.

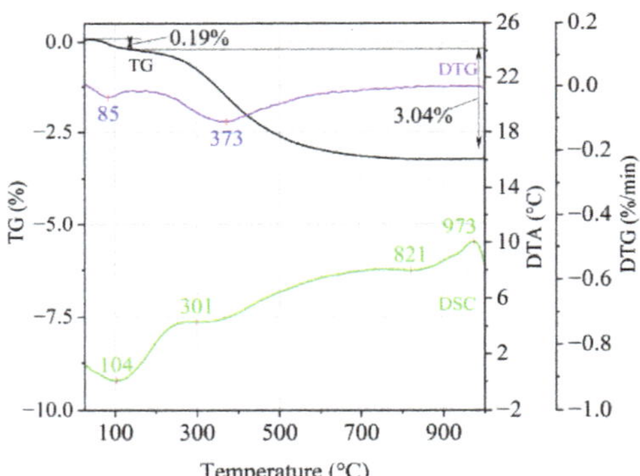


**Figure 1.** Thermal analysis curves of Slovak perlite ore (SP1).

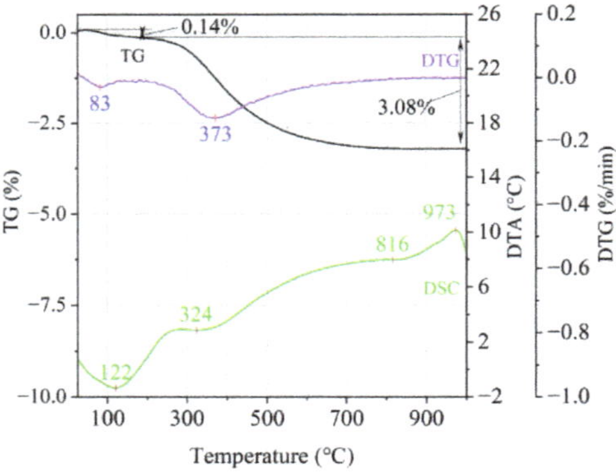


**Figure 2.** Thermal analysis curves of Slovak perlite ore (SP2).

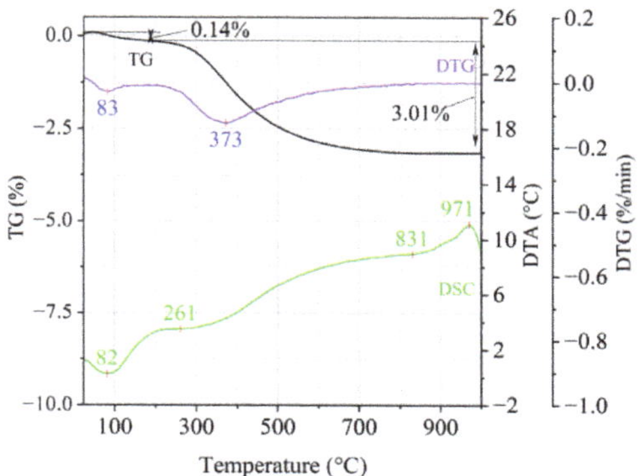


**Figure 3.** Thermal analysis curves of Slovak perlite ore (SP3).

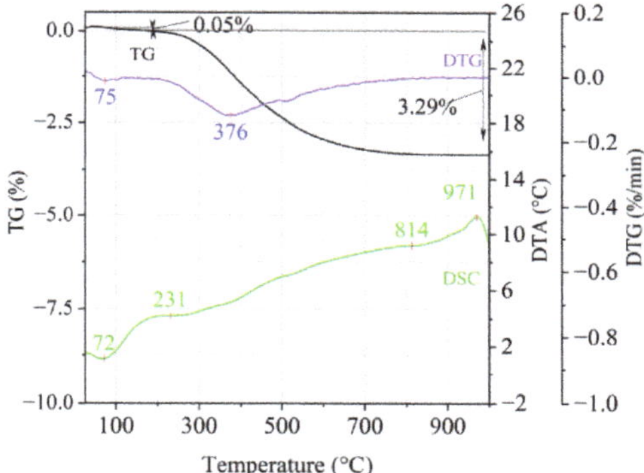


**Figure 4.** Thermal analysis curves of Hungarian perlite ore (HP).

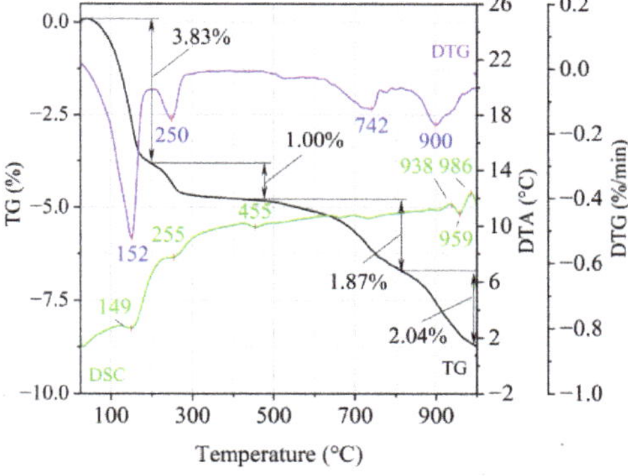


**Figure 5.** Thermal analysis curves of ground vermiculite (V).

The endothermic effect with a maximum at a temperature of approx. 100 °C indicates the ongoing dehydration process (weight loss up to 190 °C). Slovak perlite ore then loses a small amount of adsorbed water (SP1—0.19%, SP2—0.14%, SP3—0.14%), while Hungarian perlite loses only 0.05%. The temperature of maximum weight loss for SP1–SP3 was approx. 85 °C and for perlite it was 75 °C. As a result of further heating from 190 °C to 1000 °C, about 3% weight loss was noted for the Slovak pearlite ore samples. Slightly greater weight loss was recorded for the Hungarian pearlite ore sample, amounting to 3.29% (Figure 4). The dehydroxylation process takes place over a wide temperature range, which proves the presence of several types of water (free, structural) in the perlite ore structure. The gradual heating is therefore not conducive to the rapid dehydroxylation reaction of the perlite ore associated with a significant increase in volume, as part of the "energy charge" accumulated in the material structure is released in stages. No significant relationship was found between the grain size of Slovak perlite ore (with a different main fraction) and the amount of water accumulated in its structure.

On the TG-DTG curves obtained for the ground vermiculite sample (Figure 5), it is possible to indicate four stages of weight loss. On the DTA curve there are four endothermic and two exothermic effects. At the temperature of 195 °C, there is a first loss of weight about 3.83% (maximum weight loss rate at 152 °C) and it should be related to the reaction of removing the water adsorbed on the surface of the material (endothermic effect at 149 °C). A second weight loss was recorded at a temperature range 195–430 °C, which is related to the reaction of removing the inter-packet water and water associated with the cation exchange (endothermic effect with maximum at 255 °C). The beginning of the dehydroxylation process, characterized by a slow rate associated with the release of OH groups bound in the vermiculite structure, occurs at the third and fourth stage of weight loss at temperatures above 500 °C. The third weight loss (1.87%) occurs in the range of 500–800 °C with a maximum weight loss rate at 742 °C. The fourth, in the range of approx. 850–1000 °C with a maximum weight loss rate at 900 °C, is associated with a weight loss of approx. 2.04%. The appearance of exothermic thermal effects around 940 °C and 980 °C indicates a change in the initial structure of the vermiculite and the crystallization of a new phase. Between them, an endothermic heat effect (approx. 959 °C) was recorded, which may indicate dehydroxylation of the residual hydroxyl groups. These changes inspired the authors to conduct infrared (FTIR) structural research and phase composition analysis (XRD).

### 3.3. X-ray Fluorescence Analysis (XRF)

In order to determine the chemical composition of the starting materials and to properly interpret the results of X-ray diffraction (XRD), tests were performed using X-ray fluorescence spectroscopy (XRF). Tables 3 and 4 show the elemental and oxide compositions respectively.

**Table 3.** Elemental composition of the tested materials.

| Component | SP1 | SP2 | SP3 | HP | V |
|---|---|---|---|---|---|
| O | 47.6 | 47.35 | 48.63 | 48.31 | 39.60 |
| Si | 33.18 | 33.18 | 33.88 | 34.58 | 15.96 |
| Al | 6.45 | 6.14 | 7.69 | 6.32 | 3.47 |
| K | 5.89 | 6.66 | 3.78 | 5.13 | 7.04 |
| Fe | 2.72 | 2.65 | 1.43 | 2.05 | 20.90 |
| Ca | 1.73 | 1.68 | 0.95 | 1.35 | 1.25 |
| Na | 1.26 | 1.36 | 2.81 | 1.61 | 0.11 |
| Ti | 0.32 | 0.21 | 0.12 | 0.11 | 1.89 |
| Mg | 0.30 | 0.16 | 0.37 | 0.08 | 8.61 |
| Others | 1.09 | 0.61 | 0.34 | 0.46 | 1.17 |

**Table 4.** Oxide composition of the tested materials.

| Component | SP1 | SP2 | SP3 | HP | V |
|---|---|---|---|---|---|
| Na2O | 1.70 | 1.84 | 3.79 | 2.18 | 0.14 |
| MgO | 0.49 | 0.27 | 0.61 | 0.14 | 14.27 |
| Al2O3 | 12.20 | 11.60 | 14.53 | 11.94 | 6.55 |
| SiO2 | 70.99 | 70.99 | 72.47 | 73.99 | 34.13 |
| K2O | 7.09 | 8.03 | 4.55 | 6.18 | 8.48 |
| CaO | 2.43 | 2.35 | 1.33 | 1.89 | 1.75 |
| Fe2O3 | 3.89 | 3.80 | 2.04 | 2.93 | 29.88 |
| TiO2 | 0.53 | 0.50 | 0.21 | 0.18 | 3.16 |
| Others | 0.68 | 0.62 | 0.47 | 0.57 | 1.64 |

Based on the analysis of the chemical composition, it was found that the size of the Slovak perlite ore fraction influences the content of some oxides. As indicated in the work [46], the $K_2O/Na_2O$ ratio is particularly important from the point of view of the expansion susceptibility of perlite ore. As shown in Table 4, the $K_2O/Na_2O$ ratio for the smallest fraction (SP1) is 4.17, for the average 4.36, and for the coarsest 1.20. On the other hand, for the Hungarian perlite ore, which, according to the grain composition analysis, has the main fraction collected on the same sieves as the medium-grained perlite (SP2), the $K_2O/Na_2O$ ratio is 2.83. It should be noted that the content of $K_2O$, CaO, $Fe_2O_3$, and $TiO_2$ oxides decreases with the increase in the grain size of Slovak pearlite, while the share of $SiO_2$, $Na_2O$, and MgO increases. Hungarian perlite has the lowest MgO and $TiO_2$ contents. Turkish perlite [47] has a similar content of $SiO_2$ (approx. 71%) and $Al_2O_3$ (13%), but it contains fewer $Fe_2O_3$ inclusions (approx. 1.6%). Compared to Macedonian perlite [48], which is considered rich in $K_2O$ and $Na_2O$ oxides (4.21% and 3.56%, respectively), Slovak and Hungarian perlite in particular has an even higher $K_2O$ content, even above 8%, in the case of the medium fraction (SP2) and above 7%—the smallest fraction (SP1).

Ground vermiculite (V) is characterized by a high content of $SiO_2$ and $Fe_2O_3$ (34.12% and 29.88%, respectively). It also has a high $K_2O$ content (approx. 8.5%), which is consistent with the literature [37], but the MgO content is still higher and amounts to 14.27%. Compared to the Iranian vermiculite from the province of Gilan [49], which has a similar MgO content, a significant difference should be noted in the content of $Al_2O_3$—Iranian 15.70% in relation to 6.55 and $Fe_2O_3$—12.90% to 29.88%. In the Iranian vermiculite, the content of the remaining components is: $K_2O$ = 0.97%, CaO = 5.70%, $Na_2O$ = 0.16%. The $SiO_2$ content is at a similar level.

### 3.4. Structural Analysis (FTIR)

Figures 6–10 show the IR spectra of the tested materials exposed to the temperature in the range of 100–1000 °C (prepared in an oven).

Based on the IR spectra of Slovak perlite samples, rapidly heated in a furnace at a fixed temperature, it can be concluded that the water removal process runs over a wide temperature range. Initially, there is a visible weakening of the intensity of the band in the wavenumber range 3650–3400 $cm^{-1}$ and the band occurring around 1640 $cm^{-1}$. The dehydroxylation process associated with the removal of OH groups takes place up to the temperature of about 800 °C. For the sample of Slovak perlite with the smallest grain size (SP1), the primary structure of the material decomposed at 1000 °C (Figure 11). The remaining samples, both Slovak and Hungarian perlite ores, retain their structure, but a change in the half-width of the bands can be indicated as the temperature increases. The analysis of the spectra of vermiculite (V) shows that at 500 °C, water is removed from its structure, which is manifested in the disappearance of the characteristic bands within the wavenumbers 3440 $cm^{-1}$ (stretching vibrations νOH) and 1643 $cm^{-1}$ (deformation vibrations δOH). Additionally, at a temperature of 900 °C, changes in its structure begin, initially manifested by the appearance of two bands: at 878 $cm^{-1}$ and 839 $cm^{-1}$, along with an increase in their intensity in the spectra of the sample exposed to the temperature of 1000 °C and a clear shift of the maximum intensity of the main band from 1002 to 1032 $cm^{-1}$, characteristic of Si–O asymmetric stretching vibrations [50,51]. According to literature data [52] at 900 °C the illite decomposes and enstatite appears, and when heated to 1000 °C, enstatite decomposes and forsterite appears. Such a course is indicated by the appearance of an intense band at 878 $cm^{-1}$, characteristic of forsterite [53].

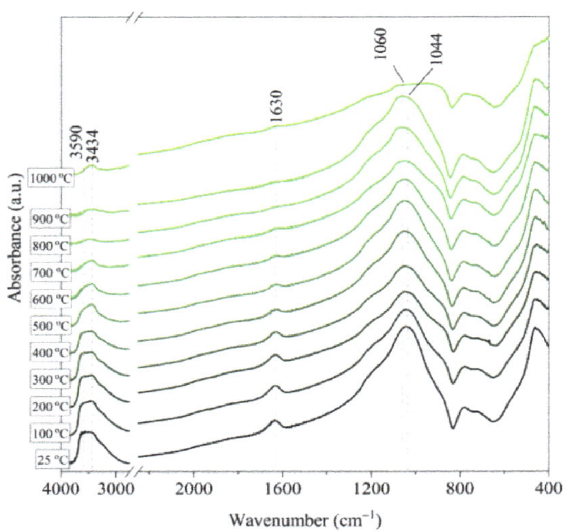


**Figure 6.** IR spectra of Slovak perlite ore (SP1) after annealing at a temperature in the range of 100–1000 °C.

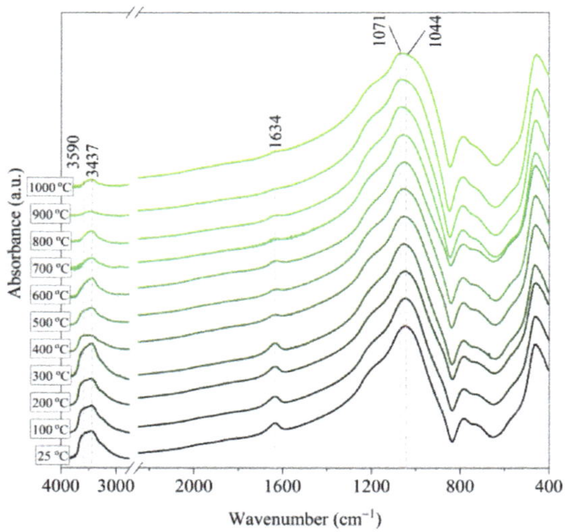


**Figure 7.** IR spectra of Slovak perlite ore (SP2) after annealing at a temperature in the range of 100–1000 °C.

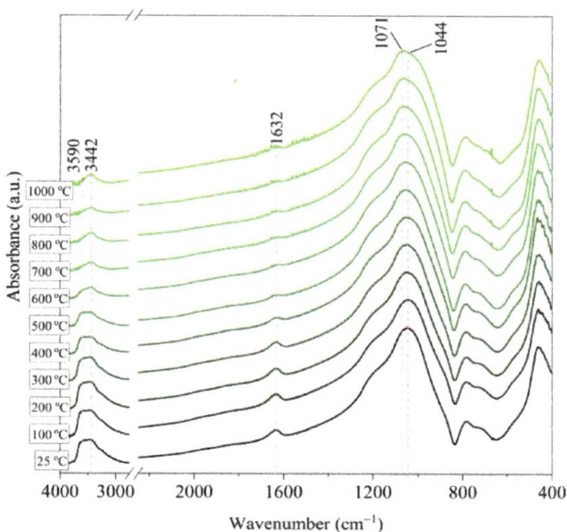


**Figure 8.** IR spectra of Slovak perlite ore (SP3) after annealing at a temperature in the range of 100–1000 °C.

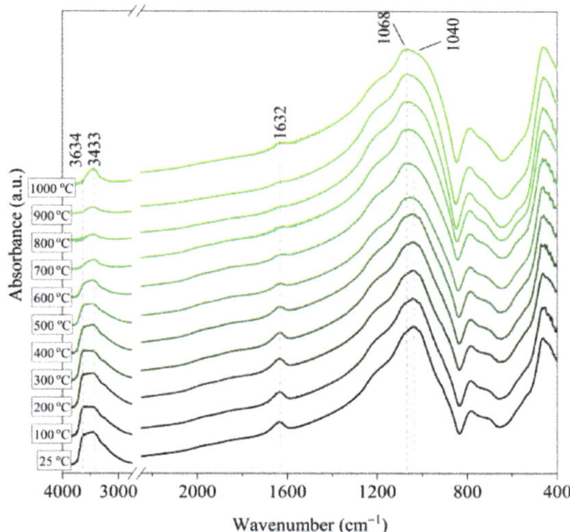


**Figure 9.** IR spectra of Hungarian perlite ore (HP) after annealing at a temperature in the range of 100–1000 °C.

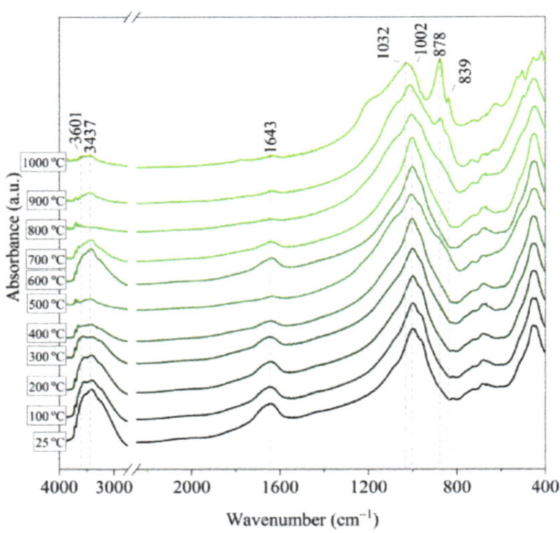


**Figure 10.** IR spectra of vermiculite (V) after annealing at a temperature in the range of 100–1000 °C

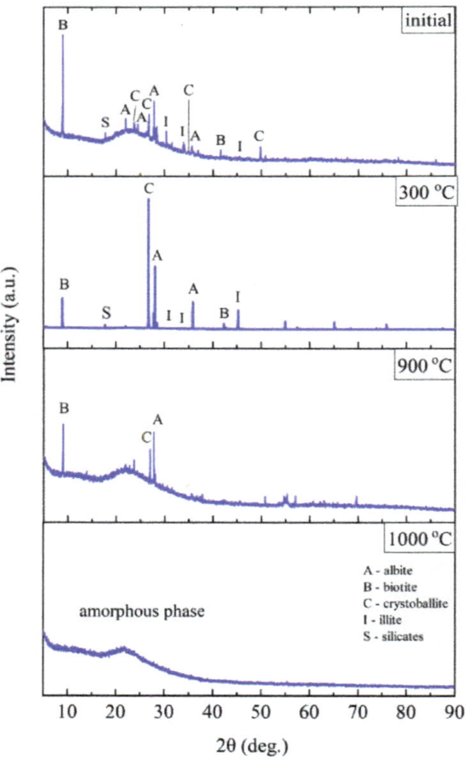


**Figure 11.** XRD diffractograms for Slovak perlite ore SP1.

### 3.5. Analysis of the Change in the Phase Composition (XRD) under the Influence of Temperature

Greek perlite (Milos, Tsigrado, and Trachilas fields) was investigated in the publication [20]. XRD analysis of the starting perlite revealed the presence of glassy phase, quartz, feldspar K, plagioclase, and mica. Small amounts of ilmenite, zirconium, and spinel were also identified, but only in the Tsigrado pearlite sample, which were removed in the expansion process. In turn, the publication [19] showed that Greek perlite consists mainly of silicate glass (amorphous) and additives in the form of biotite, quartz, and feldspar. In the case of Macedonian perlite [48], XRD analysis showed the presence of a large amount of aluminosilicate amorphous and small amounts of crystalline, represented mainly by feldspar, quartz, and cristobalite. Feldspar is represented as K-plagioclase, Na-feldspar, and microcline. The less pronounced presence of $SiO_2$ polymorphs represented as α–quartz and cristobalite was also indicated. Compared to the original perlite, the expanded one showed a marked increase in the amount of cristobalite present.

The phase composition tests were carried out at the characteristic points obtained in the thermal analysis and in the FTIR structural tests. Figures 11–14 show the diffraction patterns for the tested samples of Slovak perlite (SP1–SP3) and Hungarian (HP) ore. Figure 15 shows the diffractograms for ground vermiculite (V).

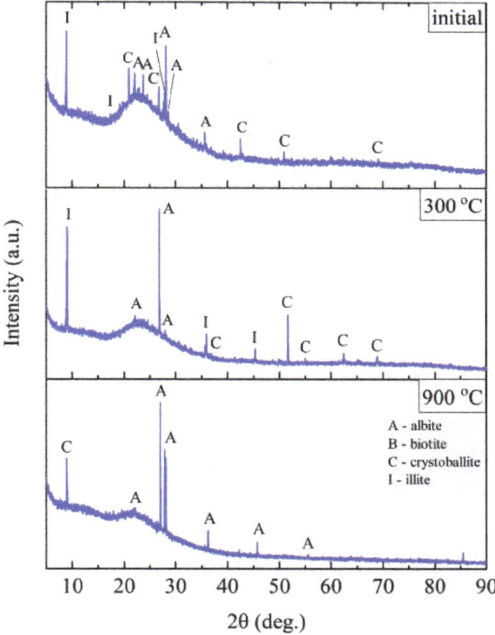


**Figure 12.** XRD diffractograms for Slovak perlite ore SP2.

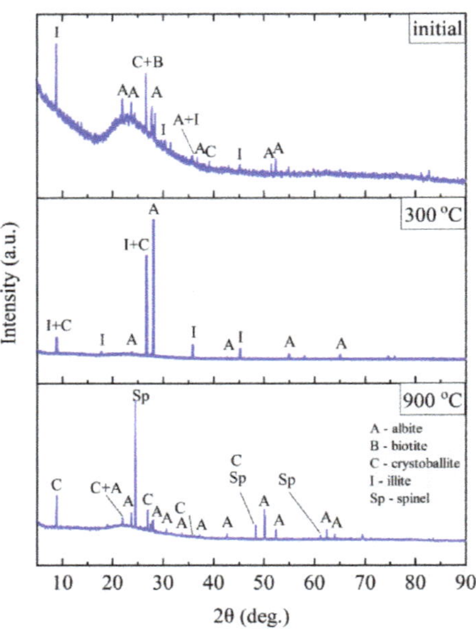


**Figure 13.** XRD diffractograms for Slovak perlite ore SP3.

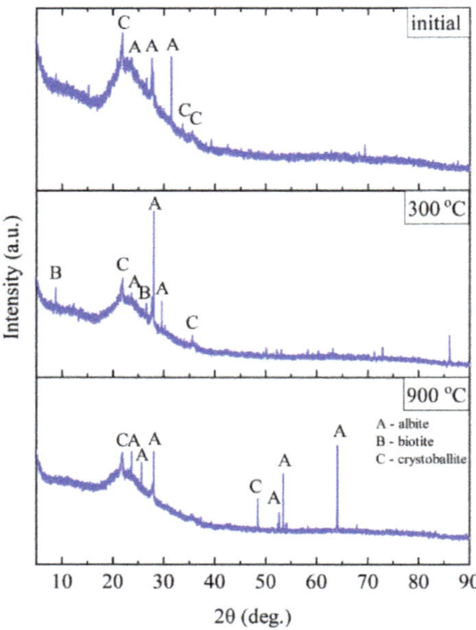


**Figure 14.** XRD diffractograms for Hungarian perlite ore HP.

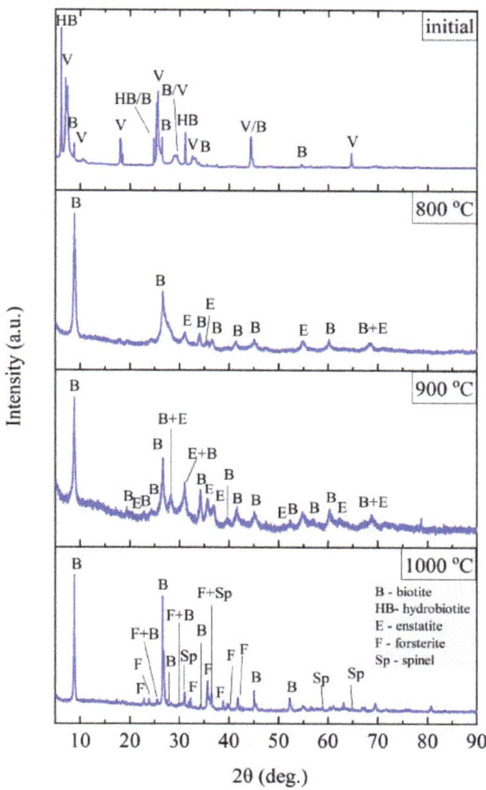


**Figure 15.** XRD diffractograms for vermiculite (V).

The finest fraction of Slovak perlite (SP1) in the initial state consists mainly of the amorphous phase—this is indicated by the presence of the so-called amorphous halo in the range of 2θ = 18–30°, with visible inclusions of the crystalline phase represented by cristobalite (C), albite (A), illite (I), biotite (B), and calcium and aluminum silicates (S)—Figure 11. At the temperature above 300 °C, no changes in the phase composition were found, while above the temperature of 900 °C, reflections from the crystalline phases disappear, and the sample consists only of the amorphous phase, which confirms the conclusions of the spectroscopic studies (Figure 6) regarding the disintegration of the structure. The mean fraction (SP2) and the coarsest fraction (SP3) have a phase composition similar to that of the SP1 sample. In addition to the amorphous halo, reflections related to the presence of albite (A), cristobalite (C), and illite (I) were identified, but no biotite (B) was found, which was probably screened entirely to the finest fraction (SP1). At a temperature above 300 °C, the phase composition of the SP2 and SP3 samples also did not change, and at a temperature above 900 °C, two phases were identified in the SP2 perlite: albite (A) and cristobalite (C), while in the SP3 sample there was a clear increase in the amount of the amorphous phase and the formation of a new phase—spinel (Sp), most likely as a result of illite (I) decomposition [54].

The Hungarian perlite (HP) is the most homogeneous in terms of phase composition. The presence of the amorphous phase (amorphous halo), cristobalite (C), and albite (A) was recorded. During heating, at a temperature above 300 °C, a decrease in the intensity of the amorphous halo associated with the amorphous phase is visible, and at the same time, an increase in the intensity of reflections from cristobalite (C). Above 900 °C, the presence of cristobalite (C) and albite (A) was found.

Three main phases were distinguished in the sample of the starting vermiculite after grinding: vermiculite, hydrobiotite, and biotite. As a result of heating the sample to the temperature of 700 °C, the phase composition does not change. Only above 800 °C do the reflections from vermiculite disappear, which proves the decomposition of layered minerals, and on the diffractograms obtained for samples annealed at 800 and 900 °C, two phases were distinguished—biotite and enstatite. Due to the clear changes in the structure noticed in the MIR (Mid-infrared) spectrum at 1000 °C (Figure 10), a phase analysis was performed for the sample annealed at this temperature. The results of the phase analysis (Figure 15) are consistent with the conclusions presented on the basis of spectroscopic studies. The clear shift in the position of the band associated with the Si–O stretching vibration can be related to the appearance of iron-aluminum spinel, in which aluminum is present in coordination 6 with the formation of forsterite.

### 3.6. Investigation of the Final Strength ($R_m^{tk}$) of Moulding Sands with Mineral Additives

The knock-out tests determine the susceptibility of the moulding sand to fragmentation under the influence of mechanical impact and removal from the mold (moulding sand) or casting (core sand) after the casting has cooled down to the knock-out temperature. Assessment of the knock-out properties of the moulding sand was carried out by determining the final strength, which allows capturing the first and second maximum strength (the first at a temperature of 200–300 °C, the second in the range of 800–900 °C). The research was aimed at showing the effect of the type of material on the obtained final strength depending on the annealing temperature. The figure shows the results of the determination for moulding sand without additives (G), sand with the addition of the Slovak perlite ore (SP1–SP3), the Hungary perlite ore (HP), and the ground vermiculite (V).

Figure 16 shows that the moulding sand with a geopolymeric binder without additives (G) shows an increase in strength at 200 °C (first hardening).

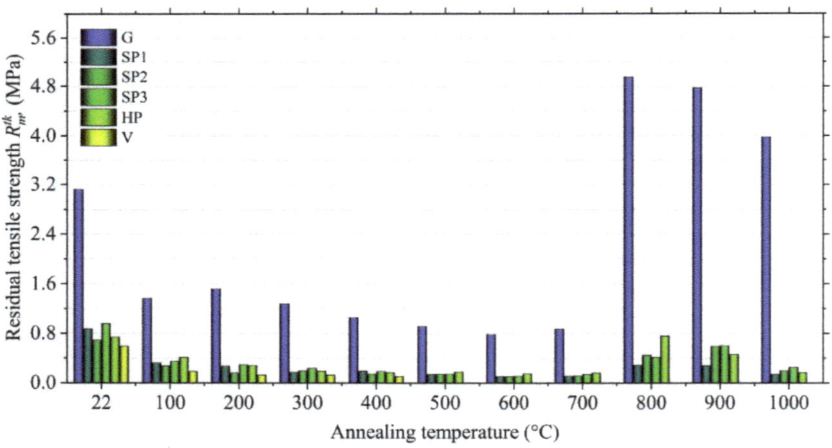


**Figure 16.** Final tensile strength ($R_m^{tk}$) of moulding sands with Geopol® binder, depending on the type of additive.

The second maximum strength occurs at the temperature of 800 °C and is 4.95 MPa. The introduction of perlite ore into the moulding sand, regardless of the fraction size, has a positive effect on the reduction of the final strength, especially at higher temperature. The final tensile strength ($R_m^{tk}$) with the addition of SP2 and SP3 Slovak perlite ore at the temperature of 800 °C decreases to about 0.4 MPa. Even better results are obtained by the moulding sand with the addition of the finest fraction (SP1), for which the final tensile strength value $R_m^{tk}$ = 0.29 MPa was obtained. In the case of the use of Hungarian perlite ore (HP), a visible decrease in final strength was also noted compared to the moulding sand without additives (G), but the strength at 800 °C is twice as high as in the case of moulding sand with the addition of Slovak pearlite ore of similar grain size (SP2). A better effect is obtained at a higher temperature, i.e., 900 °C, which, according to the authors, should be associated with a generally higher content of water accumulated in the structure of Hungarian perlite (3.29% according to thermal analysis) released at a higher temperature. The desired effect of lowering the final strength obtained by introducing vermiculite into the moulding sand is achieved at a much lower temperature than in the case of perlite ore. The moulding sand with vermiculite (1.0 mass parts.), already at the temperature of 500 °C, has a final tensile strength less than 0.01 MPa. This additive effect is due to the lower dehydroxylation temperature.

An important aspect from the technological point of view is the selection of the fraction of additives depending on the size of the matrix of the moulding sand. Proper adjustment of the size of the additive allows for better filling of voids between grains (pores). The greater degree of fragmentation means that they are able to be distributed throughout the entire volume of the moulding sand, which, when heated, gives the sand gives better results in terms of the number of microcracks and translates into a lower value of final strength.

The chemical composition of individual fractions may also influence the swelling process. The finer fraction of Slovak perlite is characterized by a higher $K_2O/Na_2O$ ratio. Perhaps it is an additional factor determining the reduction of the final strength of the moulding sand with SP1 ore.

Additionally, positive results are obtained with a lower share of vermiculite in the moulding sand (1.0 mass parts). As shown in previous studies [55], vermiculite added to the moulding sand in the initial form (nor ground) causes its destruction to an excessive degree.

### *3.7. SEM Microscopic Imaging*

The microscopic tests were carried out to illustrate the mechanism of action of swelling additives in moulding sands with an inorganic binder, i.e., to capture microcracks resulting from increasing the volume of additives located in the inter-grain pores of the quartz matrix. Figures 17 and 18 show the images made with the use of an electron microscope (SEM) for selected samples of moulding sands subjected to temperature influence: with the addition of Slovak perlite (SP1) ore and ground vermiculite (V).

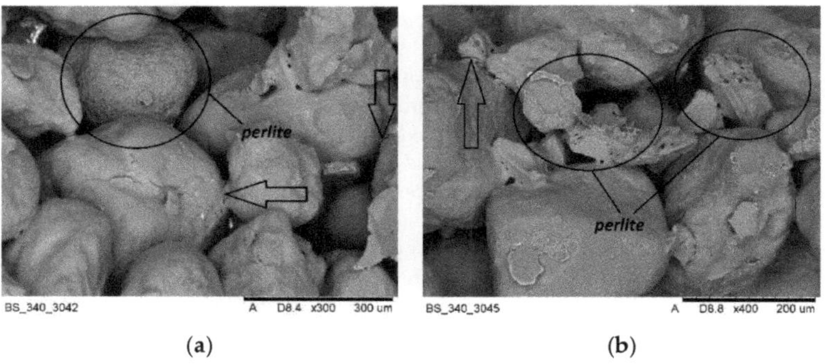


**Figure 17.** SEM imaging of moulding sand with SP1 perlite ore after annealing at 800 °C: (**a**) magnification 300×, (**b**) magnification 400×.

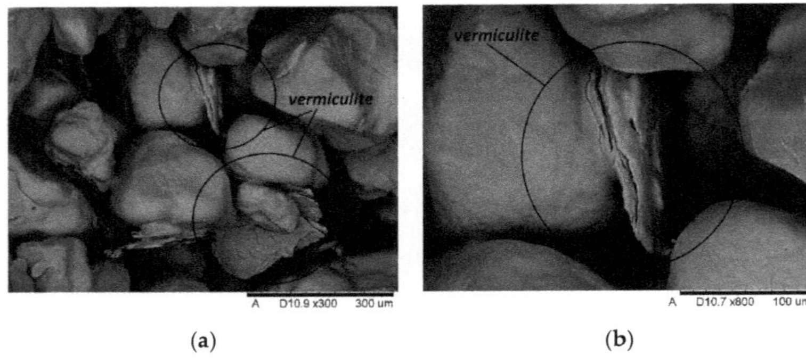


(a) (b)

**Figure 18.** SEM imaging of moulding sand with vermiculite (V) after annealing at 800 °C: (**a**) magnification 300×; (**b**) magnification 800×.

The SEM photos presented above show that the additives introduced during the preparation of the moulding sand (perlite ore, vermiculite) are located between the grains of the moulding sand matrix. Under the influence of the high temperature of the casting alloy, the mineral additives swell, and the energy released in this process is transferred to the adjacent matrix grains, covered with a layer of binder, causing microcracks. The broken binder continuity reduces the final strength of the moulding sand, and thus contributes to the elimination of the technological inconvenience of poor knock-out. As a result, the efficiency of castings production is improved and energy and labor costs are reduced, without harming the natural environment.

## 4. Conclusions

Based on the research, the following conclusions were drawn:

- on the basis of the conducted thermal analysis, no significant correlation was found between the grading of the Slovak perlite ore (SP1–SP3) and the amount of water accumulated in its structure;
- the chemical composition, in particular the alkali ratio $K_2O/Na_2O$, and the size of the fraction may have a significant impact on the expandability of the pearlite ore. In the case of the finest fraction at a temperature above 900 °C, it is possible to create an amorphous phase in the entire sample volume;
- in the finest fraction of Slovak perlite (SP1) at the temperature above 300 °C, no changes in the phase composition were found, while above the temperature of 900 °C, reflections from the crystalline phases disappear, and the sample consists only of the amorphous phase.
- at temperatures above 900 °C, two phases were identified in the SP2 perlite: albite and cristobalite, while in the SP3 sample, the formation of a new phase—spinel was found, most likely as a result of illite decomposition.
- Hungarian perlite is characterized by greater stability in terms of phase composition during heating in the tested temperature range. Above 900 °C, the presence of cristobalite and albite was found;
- vermiculite has a much higher water content than perlite ore—regardless of grain size and origin. The formation of the spatial structure characteristic of vermiculite, where the layers separate and there is a significant increase in volume, is observed at a temperature of approx. 500 °C;
- it was shown that the structure of ground vermiculite does not change until a temperature of about 700 °C. Only above 800 °C do the reflections from vermiculite disappear, which proves the decomposition of layered minerals, and the diffractograms obtained for samples annealed at 800 and 900 °C distinguish two phases—biotite and enstatite;

- in the IR spectrum obtained for a sample of vermiculite annealed at the temperature of 1000 °C, a clear shift in the band position related to the Si–O stretching vibration was shown, which was interpreted as the appearance of an iron-aluminum spinel in which aluminum is in 6 coordination with the formation of forsterite;
- the processes of expanding perlite ore and exfoliating vermiculite have a positive effect on reducing the final strength of moulding and core sand with inorganic binders used in the foundry, eliminating technological inconveniences—poor susceptibility to knocking out. However, in the considered variants, the best effects of lowering the final strength of moulding sands with an inorganic geopolymer binder were obtained in the case of introducing the SP1 Slovak perlite ore in the finest fraction, which correlates well with the conclusion presented above.

**Author Contributions:** Conceptualization, A.B., K.K. and D.D.; methodology, A.B., K.K., D.D., M.S., B.G. and D.N.; investigation, A.B., K.K., D.D., M.L. and D.N.; data analysis, A.B., K.K., M.S., M.L.; writing—original draft preparation, A.B. and K.K.; writing—review and editing, A.B., K.K., M.S., B.G. All authors have read and agreed to the published version of the manuscript.

**Funding:** This research received no external funding.

**Institutional Review Board Statement:** Not applicable.

**Informed Consent Statement:** Not applicable.

**Data Availability Statement:** The data are contained within the article and/or are available on request from the corresponding author.

**Conflicts of Interest:** The authors declare no conflict of interest.

## References


1. Bobrowski, A.; Holtzer, M.; Żymankowska–Kumon, S.; Dańko, R. Harmfulness assessment of moulding sands with a geopolymer binder and a new hardener, in an aspect of the emission of substances from the BTEX group. *Arch. Metall. Mater.* **2015**, *60*, 341–344. [CrossRef]
2. Drożyński, D.; Bobrowski, A.; Holtzer, M. Influence of the reclaim addition on properties of moulding sands with the Geopol binder. *Arch. Foundry Eng.* **2015**, *15*, 138–142. [CrossRef]
3. Lewandowski, J.L. *Materials for Foundry Molds*; Publishing House Akapit: Krakow, Poland, 1997.
4. Major-Gabryś, K.; Dobosz, S.M. The influence of the Glassex additive on technological and knock-out properties of the moulding sands with hydrated sodium silicate and new ester hardeners. *Metall. Foundry Eng.* **2011**, *37*, 33–40. [CrossRef]
5. Izdebska-Szanda, I.; Baliński, A. New generation of ecological silicate binders. *Procedia Eng.* **2011**, *10*, 887–893. [CrossRef]
6. Baliński, A. Influence of the modification method of hydrated sodium silicate on the effectiveness of changes in the final strength of moulding sand. *Trans. Foundry Res. Inst.* **2017**, *57*, 161–168. [CrossRef]
7. Dobosz, S.M.; Major–Gabryś, K. The mechanism of improving the knock-out properties of moulding sands with water glass. *Arch. Foundry Eng.* **2008**, *8*, 37–42.
8. Dobosz, S.M.; Major–Gabryś, K. Self-hardening moulding sands with water glass and new ester hardener. *Inżynieria Mater.* **2006**, *27*, 576–579.
9. Hutera, B.; Stypuła, B.; Kmita, A.; Bobrowski, A.; Drożyński, D.; Hajos, M. Effect of metal oxides nanoparticles on the tensile strength properties of foundry moulding sands with water glass. *Giessereiforschung* **2012**, *64*, 14–18.
10. Dobosz, S.M.; Major–Gabryś, K. Glassex—A new additive improving the knock-out properties of moulding sands with water glass. *Archives of Foundry.* **2004**, *4*, 63–68.
11. Wang, J.; Fan, Z.; Wang, H.; Dong, X.; Huang, N. An improved sodium silicate binder modified by ultra-fine powder materials. *China Foundry* **2007**, *4*, 026–030.
12. Bobrowski, A.; Stypuła, B.; Hutera, B.; Kmita, A.; Drożyński, D.; Starowicz, M. FTIR spectroscopy of water glass—The binder moulding modified by ZnO nanoparticles. *Metalurgija* **2012**, *51*, 477–480.
13. Bobrowski, A.; Kmita, A.; Starowicz, M.; Stypuła, B.; Hutera, B. Effect of Magnesium Oxide Nanoparticleson Water Glass Structure. *Arch. Foundry Eng.* **2012**, *12*, 9–12. [CrossRef]
14. Burkowicz, A. Expanded perlite—A thermal insulation material little known in Poland. *Sci. J. Inst. Miner. Energy Econ. Pol. Acad. Sci.* **2016**, *96*, 7–22.
15. Pichór, W.; Janiec, A. Thermal stability of expanded perlite modified by mullite. *Ceram. Int.* **2009**, *35*, 527–530. [CrossRef]
16. Samar, M.; Saxena, S. Study of chemical and physical properties of perlite and its application in India. *Int. J. Sci. Technol. Manag.* **2016**, *5*, 70–80.

17. Żelazowska, E.; Pichniarczyk, P.; Najduchowska, M. Light glass crystalline aggregates from waste materials for the building materials industry. *Mater. Ceram.* **2014**, *66*, 321–330.
18. Żelazowska, E.; Pichniarczyk, P.; Sacha, S.; Zawiła, J.; Rybicka–Łada, J.; Marczewska, A. Obtaining glass-crystalline materials based on perlite ore and expanded perlite waste. *Mater. Ceram.* **2012**, *64*, 411–416.
19. Kaufhold, S.; Reese, A.; Schwiebacher, W.; Dohrmann, R.; Grathoff, G.H.; Warr, L.N.; Halisch, M.; Müller, C.; Schwarz–Schampera U.; Ufer, K. Porosity and distribution of water in perlite from the island of Milos, Greece. *SpringerPlus* **2014**, *3*, 598. [CrossRef] [PubMed]
20. Tsikouras, B.; Passa, K.S.; Iliopoulos, I.; Katagas, C. Micorstructural Control on perlite expansibility and geochemical balance with a novel application of Isocon analysis: An example from Milos Island Perlite (Greece). *Materials* **2016**, *6*, 80. [CrossRef]
21. Roulia, M.; Chassapis, K.; Kapoutsis, J.A.; Kamitsos, E.I.; Savvidis, T. Influence of thermal treatment on the water release and the glassy structure of perlite. *J. Mater. Sci.* **2006**, *41*, 5870–5881. [CrossRef]
22. Osacky, M.; Uhlik, P.; Kuchta, L. *Experimental Alteration of Volcanic Glass;* Acta Mineralogica–Petrographica, Abstract Series 5 Szeged, Hungary, 2006; Volume 86.
23. Balek, V.; Perez-Rodriguez, J.L.; Perez–Maqueda, L.A.; Subrt, J.; Poyato, J. Thermalbehaviour of ground vermiculite. *J. Therm Anzlysis Calorim.* **2007**, *88*, 819–823. [CrossRef]
24. Poyato, L.; Perez–Maqueda, L.A.; Justo, A.; Balek, V. Emanation thermal analysis of natural and chemically-modified vermiculite *Clays Clay Miner.* **2002**, *50*, 791–798. [CrossRef]
25. Lingya, M.; Xiaoli, S.; Yunfei, X.; Jingming, W.; Xiaoliang, L.; Jianxi, Z.; Hongping, H. The structural change of vermiculite during dehydration processes: A real-timein-situXRD method. *Appl. Clay Sci.* **2019**, *183*, 1–7. [CrossRef]
26. Marcos, C.; Arango, Y.C.; Rodríguez, I. X-ray diffraction studies of the thermal behaviour of commercial vermiculites. *Appl. Clay Sci.* **2009**, *42*, 368–378. [CrossRef]
27. Marcos, C.; Rodríguez, I. Expansion behaviour of commercial vermiculites at 1000 °C. *Appl. Clay Sci.* **2010**, *48*, 492–498. [CrossRef]
28. Eaves, D. Sampling and analysis of crude vermiculite samples for possible asbestiform fibre and quartz content. *IOM Consult Rep.* **2006**, *924*, 1130.
29. Bagdassarov, N.; Ritter, F.; Yane, Y. Kinetics of perlite glasses degassing TG and DSC analysis. *Glass Sci. Technol.* **1999**, *72*, 277–290
30. Austin, G.S.; Barker, J.M. Commercial perlite deposits of New Mexico and North America. In Proceedings of the New Mexico Geological Society 49th AnnualFall Field Conference, Socorro, Mexico, 19–22 October 1998; pp. 271–277.
31. Kozhukhova, N.I.; Chizhov, R.V.; Zhernovsky, I.V.; Strokova, V.V. Structure formation of geopolymer perlite binder vs. type of alkali activating agent. *ARPN J. Eng. Appl. Sci.* **2016**, *11*, 12275–12281.
32. Herskovitch, D.; Lin, I.J. Upgrading of raw perlite by a dry magnetic technique. *Magn. Electr. Sep.* **1996**, *7*, 145–161. [CrossRef]
33. Goodall, R.; Williams, C.; Fernie, J.A.; Clyne, T.W. Thermal Expansion and Stiffness Characteristics of a Highly Porous, Fire Resistant Composite Material. Available online: https://www.researchgate.net/publication/228871362_Thermal_Expansion_and_Stiffness_Characteristics_of_a_Highly_Porous_Fire-Resistant_Composite_Material (accessed on 25 March 2021).
34. Hillier, S.; Marwa, E.M.; Rice, C.M. On the mechanism of exfoliation of vermiculite. *Clay Miner.* **2013**, *48*, 563–582. [CrossRef]
35. Koksal, F.; Gencel, O.; Kaya, M. Combined effect of silica fume and expanded vermiculite on properties of lightweight mortars a ambient and elevated temperatures. *Constr. Build. Mater.* **2015**, *88*, 175–187. [CrossRef]
36. Suvorov, S.A.; Skurikhin, V.V. Vermiculite—A promising material for high-temperature heat insulators. *Refract. Ind. Ceram.* **2003** *44*, 186–187. [CrossRef]
37. Valášková, M.; Martynková, G. Vermiculite: Structural Properties and Examples of the Use. *Clay Miner. Nat. Their Characterisation Modyfication Appl.* **2012**, 209–238. [CrossRef]
38. Rebilasová, S.; Peikertová, P.; Gröplová, K.; Neuwirthová, L. The influence of mechanical treatment of vermiculite on preparation of the composites vermiculite/$TiO_2$. In Proceedings of the NANOCON 2011 Conference Proceedings, Brno, Czech Republic 21–23 September 2011.
39. Gailhanou, H.; Blanc, P.; Rogez, J.; Mikaelian, G.; Horiuchi, K.; Yamamura, Y.; Saito, K.; Kawaji, H.; Warmont, F.; Grenèche, J.M et al. Thermodynamic properties of saponite, nontronite, and vermiculite derived from calorimetric measurements. *Am. Mineral* **2013**, *98*, 1834–1847. [CrossRef]
40. Handke, M. *Crystallochemistry of Silicates*, 2nd ed.; University Scientific and Technical Publishers of AGH: Krakow, Poland, 2008
41. Campos, A.; Moreno, S.; Molina, R. Characterization of vermiculite by XRD and spectroscopic techniques. *Earth Sci. Res. J.* **2009** *13*, 108–118.
42. Abidi, S.; Nait-Ali, B.; Joliff YFavotto, C. Impact of perlite, vermiculite and cement on the thermal conductivity of a plaster composite material: Experimental and numerical approaches. *Compos. Part B* **2015**, *68*, 392–400. [CrossRef]
43. Derkowski, A.; Drits, V.A.; McCarty, D.K. Nature of rehydroxylation in dioctahedral 2:1 layer clay minerals. *Am. Mineral.* **2012** *97*, 610–629. [CrossRef]
44. Derkowski, A.; Drits, V.A.; McCarty, D.K. Rehydration in a dehydrated-dehydroxylated smectite in environment of low water vapor content. *Am. Mineral.* **2012**, *97*, 110–127. [CrossRef]
45. Harmon, R.S.; Wörner, G.; Goldsmith, S.T.; Harmon, B.A.; Gardner, C.B.; Lyons, W.B.; Ogden FLPribil, M.J.; Long, D.T.; Kern, Z Fórizs, I. Linking silicate weathering to riverine geochemistry—A case study from a mountainous tropical setting in west-central Panama. *Geol. Soc. Am. Bull.* **2016**, *128*, 1780–1812. [CrossRef]

46. Varga, P.; Uhlik, P.; Lexa, J.; Šurka, J.; Bizovská, V.; Hudec, P.; Pálková, H. The influence of porosity on the release of water from perlite glass by thermal treatment. *Mon. Chem. Chem. Mon.* **2019**, *150*, 1025–1040. [CrossRef]
47. Erdoğan, S.T. Properties of Ground Perlite Geopolymer Mortars. *J. Mater. Civil Eng.* **2015**, *27*, 1–10. [CrossRef]
48. Rekaa, A.A.; Pavlovski, B.; Lisichkov, K.; Jashari, A.; Boev, B.; Boev, I.; Lazarova, M.; Eskizeybek, V.; Oral, A.; Jovanovski, G.; et al. Chemical, mineralogical and structural features of native and expanded perlite from Macedonia. *Geol. Croat.* **2019**, *72*, 215–221. [CrossRef]
49. Mazoomi, F.; Jalali, M. Effect of vermiculite, nanoclay and zeolite on ammonium transport through saturated sandy loam soil: Columns experiments and modeling approaches. *Catena* **2019**, *176*, 170–180. [CrossRef]
50. Handke, M.; Sitarz, M.; Długoń, E. Amorphous SiCxOy coatings from ladder-like polysilsesquioxanes. *J. Mol. Struct.* **2011**, *993*, 193–197. [CrossRef]
51. Sitarz, M.; Jastrzębski, W.; Jeleń, P.; Długoń, E.; Gawęda, M. Preparation and structural studies of black glasses based on ladder-like silsesquioxanes. *Spectrochim. Acta Part A* **2014**, *132*, 884–888. [CrossRef]
52. Pérez–Maqueda, L.A.; Balek, V.; Payato, J.; Pérez–Rodriquez, J.L.; Šubrt, J.; Bountsewa, I.M.; Beckman, I.N.; Málek, Z. Stdy of natural and ion exchanged vermiculite by emanation thermal analysis, TG, DTA and XRD. *J. Therm. Anal. Calorim.* **2003**, *71*, 715–726. [CrossRef]
53. Handke, M. Vibration spectroscopy of silicates and the character of Si–O bonding in silicates. *Sci. J. AGH* **1984**, *48*.
54. De Araujo, J.H.; Da Silva, N.F. Thermal Decomposition of Illite. *Mater. Res.* **2004**, *7*, 359–361. [CrossRef]
55. Bobrowski, A. *The Phenomenon of Dehydroxylation of Selected Mineral Materials from the Aluminosilicates Group as the Determinant Factor of the Knock-Out Improvement of Moulding and Core Sands with Inorganic Binder*; Archives of Foundry Engineering; Polish Academy of Sciences: Gliwice, Poland, 2018.

materials

MDPI

*Article*

# Influence of Al/Ti Ratio and Ta Concentration on the As-Cast Microstructure, Phase Composition, and Phase Transformation Temperatures of Lost-Wax Ni-Based Superalloy Castings

Małgorzata Grudzień-Rakoczy [1,*], Łukasz Rakoczy [2], Rafał Cygan [3], Konrad Chrzan [1], Ondrej Milkovič [4,5] and Zenon Pirowski [1]

1 Łukasiewicz Research Network-Krakow Institute of Technology, Zakopianska 73, 30-418 Krakow, Poland; konrad.chrzan@kit.lukasiewicz.gov.pl (K.C.); zenon.pirowski@kit.lukasiewicz.gov.pl (Z.P.)
2 Faculty of Metals Engineering and Industrial Computer Science, AGH University of Science and Technology, al. Mickiewicza 30, 30-059 Krakow, Poland; lrakoczy@agh.edu.pl
3 Consolidated Precision Products Corporation, Investment Casting Division, Hetmańska 120, 35-078 Rzeszow, Poland; rafal.cygan@cppcorp.com
4 Institute of Materials Research, Slovak Academy of Sciences, Watsonova 47, 040 01 Košice, Slovakia; omilkovic@saske.sk
5 Institute of Experimental Physics, Slovak Academy of Sciences, Watsonova 47, 040 01 Košice, Slovakia
* Correspondence: malgorzata.grudzien@kit.lukasiewicz.gov.pl

**Abstract:** The as-cast microstructure, alloying element segregation, solidification behavior, and thermal stability of model superalloys based on Inconel 740 with various Al/Ti ratios (0.7, 1.5, 3.4) and Ta (2.0, 3.0, 4.0 wt%) concentrations were investigated via ThermoCalc simulations, scanning and transmission electron microscopy, energy-dispersive X-ray spectroscopy, dilatometry, and differential scanning calorimetry. The solidification of the superalloys began with the formation of primary $\gamma$ dendrites, followed by MC carbides. The type of subsequently formed phases depended on the superalloys' initial Al/Ti ratio and Ta concentration. The results obtained from solidification simulations were compared to the obtained microstructures. For all castings, the dendritic regions consisted of fine $\gamma'$ precipitates, with their size mainly depending on the initial Al/Ti ratio, whereas in the interdendritic spaces, (Nb, Ta, Ti)C carbides and Nb-rich Laves phase precipitates were present. In high Al/Ti ratio superalloys, β-NiAl precipitates, strengthened by $\eta$ and $\alpha$-Cr phases, were observed. Based on dilatometric results, the dissolution of $\gamma'$ precipitates was accompanied by a substantial increase in the coefficient of thermal expansion. The end of the dilatation effect took place around the $\gamma'$ solvus temperature, as determined via calorimetry. Moreover, the bulk solidus temperature was preceded by the dissolution of the Laves phase, which may be accompanied by local melting.

**Keywords:** Inconel 740; phase transformation; investment casting; superalloy; solidification

itation: Grudzień-Rakoczy, M.; akoczy, Ł.; Cygan, R.; Chrzan, K.; ilkovič, O.; Pirowski, Z. Influence f Al/Ti Ratio and Ta Concentration n the As-Cast Microstructure, Phase omposition, and Phase ransformation Temperatures of ost-Wax Ni-Based Superalloy astings. *Materials* **2022**, *15*, 3296. ttps://doi.org/10.3390/ a15093296

cademic Editors: Alexander Churyumov and Guozheng Quan

eceived: 4 March 2022
ccepted: 28 April 2022
ublished: 4 May 2022

ublisher's Note: MDPI stays neutral ith regard to jurisdictional claims in ublished maps and institutional affil- tions.



## 1. Introduction

Ni-based superalloys are a class of materials extensively used for hot components in industrial power plants and aerospace engines, due to their excellent creep strength, oxidation, and hot-corrosion resistance at elevated service temperatures [1]. Advanced ultra-supercritical (A-USC) power plant designs are expected to have efficiencies of coal-fired power plants above 35 pct. With component service temperatures approaching 760 °C, Ni-based superalloys must replace ferritic-martensitic, high-strength steels, and austenitic stainless steels to fulfill the long-term, high-temperature strength requirements in these pressurized cycles [2,3]. The IN740 Ni-based superalloy has been studied within the scope of European projects leading to COM-TES700 test loop exposures at the Esbjerg (Sweden) and Scholven (Germany) power plants [4]. IN740 was developed on the basis of the Nimonic 263 superalloy chemical composition [5]. Compared to Nimonic 263, Cr and Mo concentrations in IN740 were optimized to achieve improved sulfidation resistance, whereas Nb was added to increase high-temperature strength [6]. The US A-USC Consortium applies higher service temperatures (760 °C) that require steam transfer pipes and boiler tubing to be produced from precipitation-strengthened alloys with a higher volume fraction of intermetallic phase precipitates. Up to 725 °C, IN740 exhibits good microstructural stability during long-term annealing. At 760 °C, however, three main limitations appear, i.e., the high coarsening rate of the $\gamma'$ phase, plate-like $\eta$-phase formation, and high Si-containing brittle G-phase formation, which lead to a decrease in strength and ductility [7–9]. Modifying the IN740 Ni-based superalloy's chemical composition has been the subject of several research studies. Shin et al. [10] investigated the influence of Mo (0.5, 3.0, 6.0 wt.%) on the $\gamma'$ precipitates' coarsening kinetics in IN740. The authors found that the coarsening rate of the $\gamma'$ precipitates decreased with increasing Mo, and the activation energies for coarsening were 245 kJ/mol, 261 kJ/mol, and 278 kJ/mol for 0.5, 3.0, and 6.0 wt.% Mo, respectively. Abbasi et al. [11] investigated the influence of Al and Ti concentrations on IN740's microstructure stability and oxidation resistance. It was shown that an increase in Al concentration (up to 2.4 wt.%), with low Ti contents, leads to grain refinement, improves oxidation resistance, and reduces the number of voids in the vicinity of the superalloy/oxide layer boundary. The opposite effect, however, was found for increased Ti concentrations. The result was an increased mass gain after oxidation and numerous voids at the superalloy/oxide layer boundary, due to $TiO_2$ precipitates not constituting a strong enough barrier for Cr and O ions. The most popular and commercially used IN740 superalloy modification is the IN740H superalloy, which satisfies the service requirements for A-USC applications [12]. Presently, IN740H is widely used in A-USC power plants in wrought form as boiler components. However, large complex components in boilers and other casings in gas turbines and valve chests additionally require thick-wall castings. Using the material in its cast variant would be highly beneficial in terms of size, geometry, and complexity range [13,14]. Detrois et al. [15,16] have indicated possible ductility losses during creep in castings, with a chemical composition in line with IN740′s composition. They suggested that excessive carbides at the grain boundaries are responsible for the effect. Despite the presence of Cr-rich $M_{23}C_6$ carbides, which impede grain-boundary sliding by restricting deformation at the boundaries, the local chemistry can change and subsequently lead to the formation of cracking initiation sites. In this work, IN740 superalloy chemical composition modifications consisted of adding Al (increase in Al/Ti ratio) as the main $\gamma'$-former and introducing Ta, which is widely known as a strong MC-carbide-former and limits the formation of $M_{23}C_6$ carbides [17,18]. The reported order of most to least stable MC carbide is TaC > NbC > TiC > VC, where Nb is approximately equivalent to tantalum in stabilizing MC carbides [19]. As the production of superalloys in the form of large castings is desirable, the process was carried out in an argon gas atmosphere, rather than in vacuum. The casting of Ni-based superalloys presents unique challenges, as the microstructure segregates during casting. The casting process of Ni-based superalloys with a high refractory element content, is usually followed by solution heat treatment which can also be challenging [20]. As the diffusivity of the alloying elements is low and the diffusion path from dendrite cores to interdendritic spaces is long, it is difficult to homogenize the chemical composition via heat treatment. Furthermore, due to the remaining eutectic phases in the interdendritic spaces, the local solidus temperature is lower than the bulk solidus temperature [21]. If the solutioning temperature is not carefully selected incipient melting will take place. In this sense, the solutioning of Ni-based superalloys by means of conventional processing requires a long time. Well-designed heating scenarios will be needed to mitigate the risk of incipient melting and achieve as much chemical homogenization as possible [22,23]. Analyzing the material's solidification behavior, phase composition, and stability is necessary to prepare the best procedure to dissolve those as much as possible during the solution heat treatment, without risk of incipient melting of the superalloy. The results from this work could provide significant insight into composition optimization and solutioning heat treatment control of novel Ni-based superalloys.

## 2. Materials and Methods

### 2.1. Materials

For the purposes of this work, nine Ni-based superalloys, with increased Al concentration and Ta addition compared to the standard IN740, were cast. The castings were manufactured via lost-wax casting. First, wax models, combined with a gating system, were prepared (Figure 1).

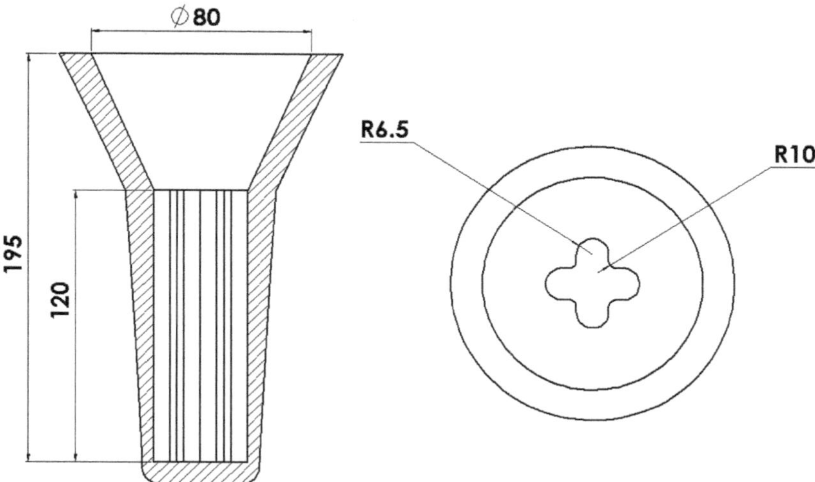


**Figure 1.** The geometry of the shell mold used for the castings preparation.

The chemical composition modification procedures for the nine cast Ni-based superalloys are presented in Table 1. A 3 kg IN740 ingot was inductively melted in a corundum crucible for 35 min in an argon atmosphere. Next, the alloying additions of Al (rods with diameter 6 mm and length 100 mm, 99.99% purity) and Ta (disks ∅10 × 0.2 mm, 99.9% purity) were added. The melted superalloy was then deoxygenated twice using a Ni-Mg compound (before melt-pouring). The pouring temperature was in the range of 1500–1550 °C, and was controlled by S-type thermocouples (PtRh10-Pt), with a ∅ 0.8 mm wire used for temperature measurements. For more detailed instructions on casting procedures and shell mold fabrication, please consult our previous paper [24]. The chemical compositions of all investigated materials, given in Table 2, were analyzed using a spark optical emission spectrometer.

**Table 1.** IN740 modification procedures, wt%.

| Ta Concentration | Al/Ti Concentration Ratio | | |
|---|---|---|---|
| | **Low: 0.7 (±0.10)** | **Medium: 1.5 (±0.15)** | **High: 3.4 (±0.20)** |
| Low: 2.0 (±0.10) | A1 | A4 | A7 |
| Medium: 3.0 (±0.15) | A2 | A5 | A8 |
| High: 4.0 (±0.20) | A3 | A6 | A9 |

**Table 2.** A1–A9 superalloy chemical composition, wt%.

| Alloy | Cr | Co | Al | Ti | Nb | Fe | Mo | Mn | Si | C | Ta | Ni |
|---|---|---|---|---|---|---|---|---|---|---|---|---|
| A1 | 22.34 | 19.96 | 1.30 | 1.72 | 1.80 | 0.62 | 0.43 | 0.30 | 0.59 | 0.03 | 2.00 | |
| A2 | 22.46 | 19.91 | 1.28 | 1.78 | 1.92 | 0.61 | 0.43 | 0.30 | 0.60 | 0.02 | 3.00 | |
| A3 | 22.41 | 19.91 | 1.31 | 1.68 | 1.88 | 0.62 | 0.44 | 0.30 | 0.61 | 0.02 | 4.00 | |
| A4 | 22.84 | 19.71 | 2.80 | 1.82 | 1.92 | 0.63 | 0.42 | 0.29 | 0.59 | 0.03 | 2.00 | |
| A5 | 22.74 | 19.69 | 2.82 | 1.76 | 1.80 | 0.64 | 0.43 | 0.30 | 0.60 | 0.03 | 3.00 | Balance |
| A6 | 22.28 | 19.85 | 2.75 | 1.83 | 1.94 | 0.56 | 0.43 | 0.22 | 0.56 | 0.02 | 4.00 | |
| A7 | 21.76 | 18.85 | 6.06 | 1.81 | 1.88 | 0.63 | 0.42 | 0.29 | 0.60 | 0.03 | 2.00 | |
| A8 | 21.20 | 19.05 | 6.34 | 1.76 | 1.82 | 0.61 | 0.40 | 0.28 | 0.56 | 0.03 | 3.00 | |
| A9 | 21.43 | 18.65 | 6.12 | 1.84 | 1.86 | 0.62 | 0.42 | 0.29 | 0.60 | 0.03 | 4.00 | |

### 2.2. Methods

The Thermo-Calc software (version 2021a, Stockholm, Sweden), with the TCNI10 database, was used to analyze the solidification process using the Scheil model. The Scheil model considers the segregation of solute elements during solidification and assumes the absence of back-diffusion during calculations. Complete solidification is assumed with less than 1% liquid present in the calculation, and generally, the solidification range is quite broad [25]. In this work, the model was used to predict the type of primary phases in the as-cast superalloys. The superalloys were prepared into 10 × 10 × 5 mm samples and their surface was mechanically polished. The samples were then subjected to X-ray diffraction (XRD) to identify the formed phases. Measurements were made in Bragg–Brentano geometry using a Philips X'Pert Pro MRD diffractometer, (Philips, Amsterdam, The Netherlands) equipped with Cu radiation (λ = 1.54 Å). Measurements were made at room temperature in the range of 2θ = 30–100°, with a step size of 0.01671°.

For scanning electron microscopy (SEM) characterization, specimens (10 × 10 × 5 mm) were cut from the middle of each casting and mounted in resin, ground, polished, and electrochemically etched in 10% oxalic acid. Back-scattered electron (BSE) imaging and energy-dispersive X-ray spectroscopy (EDX) were performed using 20 kV accelerating voltage, set on a Phenom XL (Phenom-World, Eindhoven, The Netherlands) microscope, equipped with an EDX detector. Semi-quantitative chemical analyses were carried out using the ZAF correction. Observations of $\gamma'$ precipitates in dendritic regions were performed using a high-resolution Merlin Gemini II microscope (ZEISS, Oberkochen, Germany), equipped with a Schottky field-emission gun (Bruker, Billerica, MA, USA). The images were obtained using secondary electrons (SE).

Quantitative analysis of the $\gamma'$ precipitates and carbides (manually contoured on images) was carried out using ImageJ (National Institutes and the Laboratory for Optical and Computational Instrumentation, University of Wisconsin, Madison, WI, USA). The measurements were conducted in five places on each casting, which included more than 200 $\gamma'$ precipitates (SEM-SE, magnification ×100,000) and at least 50 carbide precipitates (SEM-BSE, magnification ×10,000). Next, the precipitates' area (A) and perimeter (P) were determined. The size of $\gamma'$ precipitates in the dendritic regions and carbides in the interdendritic spaces were represented as the equivalent diameter (Equation (1)), while the circularity of $\gamma'$ precipitates was calculated to evaluate their morphological differences (Equation (2)).

$$\overline{\phi} = \sqrt{\frac{4A}{\pi}} \tag{1}$$

$$\zeta = \frac{4\pi A}{P^2} \tag{2}$$

Samples A3, A6, and A9 were investigated in more detail via TEM using bright field (BF), selected area electron diffraction (SAED), and high-angle annular dark-field scanning transmission electron microscopy (HAADF-STEM) methods. Samples were first mechanically ground down to a thickness of about 0.05 mm, after which 3 mm discs were punched and dimpled using a Gatan dimple grinder (Gatan, Inc., Pleasanton, CA, USA) on both sides. The final step was thinning using an $Ar^+$ ion beam (PIPS of Gatan) at low angles. Before loading into the microscope, the thin foils were plasma-cleaned (NanoMill 1040 Fischione) to remove surface contaminations. Observations of samples A3 and A6 were performed using a Tecnai G2 20 TWIN (FEI) microscope. Analysis of the A9 superalloy was carried out with the Cs-corrected probe FEI Titan$^3$ G2 60–300 TEM with the ChemiSTEM system (Thermo Fisher Scientific, Eindhoven, The Netherlands). The accelerating voltage was 300 kV. The STEM-EDX data were acquired at 300 kV and were then subjected to analysis using the Esprit software (Bruker, v1.9) (Bruker, Billerica, MA, USA), in which the standardless Cliff–Lorimer quantification method was applied. The JEMS software (JEMS-SWISS, Jongny, Switzerland) was applied to fit the diffraction peaks. To analyze the phase transformations in the elevated temperatures, dilatometry and differential scanning calorimetry were carried out. The dilatometry measurements were carried out with a NETZSCH DIL 402C/4/G (Netzsch, Selb, Germany) device on samples $\phi 6 \times 25$ mm up to melting point (under argon shielding) with a heating rate of 0.08 °C/s. Differential scanning calorimetry was performed using a Pegasus DSC 404C/3/G (Netzsch, Selb, Germany) device on samples weighing about 40 mg in the range 25–1480 °C with a heating rate of 10 °C/min.

## 3. Results

### 3.1. Solidification Simulation via Thermo-Calc

Figure 2 presents the solidification process predictions of the A1–A9 superalloys, based on Scheil's model. The solidification process of low Al/Ti ratio superalloys, with Ta additions in the range of 2.0 to 4.0 wt%, is predicted to begin with $\gamma$ phase crystal formation at 1369 °C (A1), 1364 °C (A2), and 1355 °C (A3), followed by MC carbide precipitation at 1289 °C (A1), 1266 °C (A2), and 1260 °C (A3). Upon further temperature decrease, and at the stage where the mass fraction of the solid exceeds 0.9, the Laves phase is expected. Finally, the Scheil model predicts the formation of the G phase in the final stage of solidification. In the case of superalloys A4, A5, and A6 (medium Al/Ti ratio), solidification with $\gamma$ phase formation is estimated to begin at 1355 °C, 1349 °C, and 1341 °C, respectively, whereas MC carbides are expected to precipitate at 1294 °C, 1288 °C, and 1267 °C. Additionally, superalloys A3 and A4 share the same liquidus temperature. The data show that for constant Ta content, an increase in Al/Ti ratio leads to an increase in the MC carbide precipitation temperature. At 1106 °C, the precipitation of the $\eta$ phase is predicted in superalloy A6. The lack of $\eta$ phase in the A4 and A5 superalloys indicates that sufficiently high Ta concentration at the medium Al/Ti ratio is required for its formation. All superalloys are projected to contain Laves phase precipitations. Similarly to superalloys with a low Al/Ti ratio, the model predicts G phase formation in the final solidification stage for medium Al/Ti ratios. The results of the Scheil solidification simulation of superalloys A7, A8, and A9 indicate that their liquidus temperature is the lowest of all tested superalloys, i.e., 1328 °C, 1320 °C, and 1313 °C, respectively. As in the case of superalloys with low and medium Al/Ti ratios, an increase in Ta concentration is accompanied by a decrease in the liquidus temperature. The MC carbides are speculated to precipitate at higher temperatures, when the mass fraction of the liquid phase is higher than that of the solid phase. At 1263 °C, the $\beta$-NiAl phase is expected to form, regardless of the Ta concentration. With further cooling, the Laves phase should precipitate at 1139 °C (A7), 1141 °C (A8), and 1145 °C (A9). The $\eta$ phase is estimated to precipitate at 1093 °C in superalloy A7, whereas for samples A8 and A9, these temperatures are expected to be slightly higher, at 1100 °C and 1101 °C, respectively. The last phase predicted to form from the liquid is the $\sigma$ phase. Unlike in castings A1–A6, no G phase is expected.

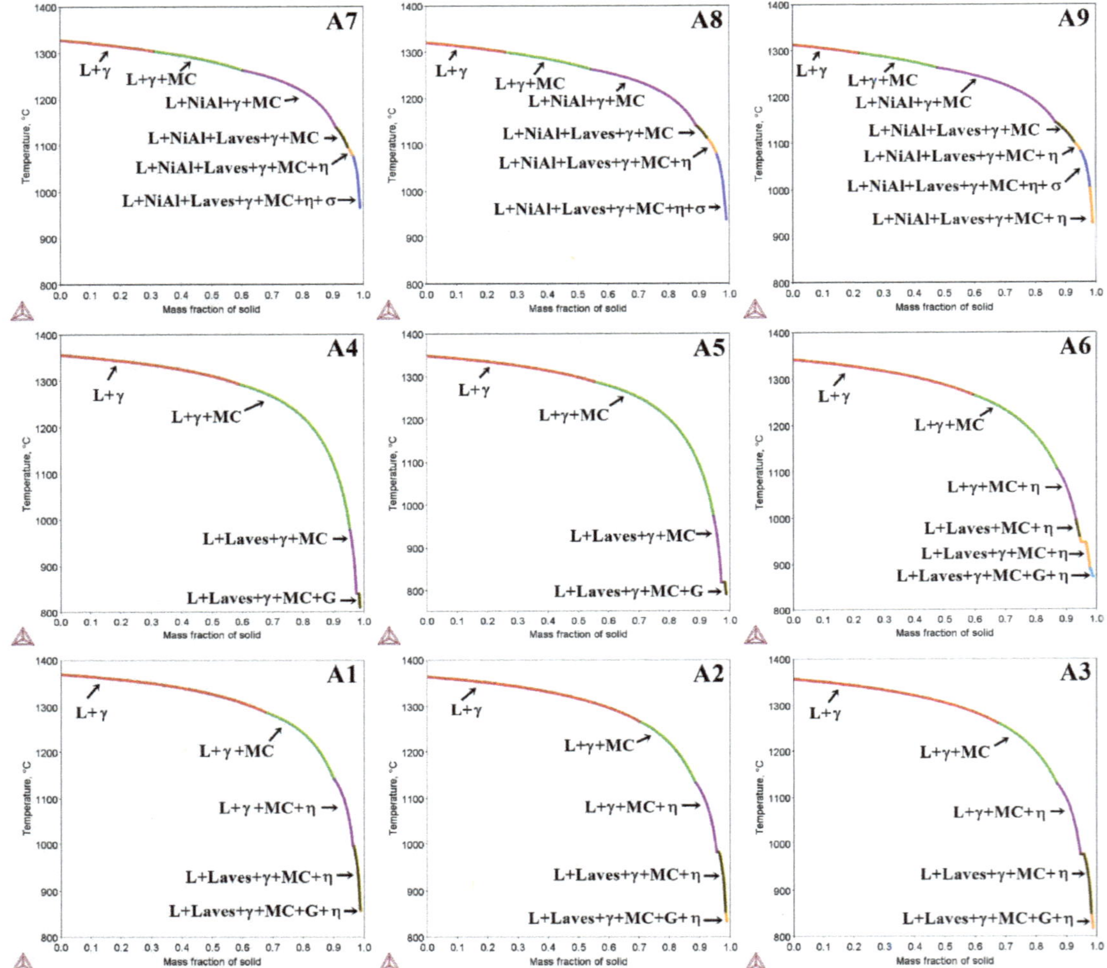


**Figure 2.** Scheil solidification simulation of the (A1–A9) superalloys.

### 3.2. Analysis of the Microstructure of the A1–A9 Superalloys in As-Cast Condition

The XRD patterns of the as-cast superalloys are presented in Figure 3. Peaks of four phases were identified, namely the γ matrix, γ′ intermetallic phase, β-NiAl, and MC carbides.

The microstructures of the as-cast A1–A9 superalloys are shown in Figure 4. The dendritic regions are characterized by a relatively homogenous microstructure, whereas precipitates with bright phase contrast can be found in the interdendritic regions. The highest percentage of precipitates in the interdendritic spaces is observed in the A7–A9 superalloys. Local fine precipitates with a black contrast are visible in all castings.

The dendritic regions, regardless of the casting variant, are predominantly composed of very fine precipitates with a spherical or near-cubic morphology (Figure 5). The finest precipitates were observed in variants A1–A3 (low Al/Ti ratio), while the largest were observed in samples with the highest Al/Ti ratio. These precipitates were subjected to stereological parameter analyses, and the results are summarized in Table 3.

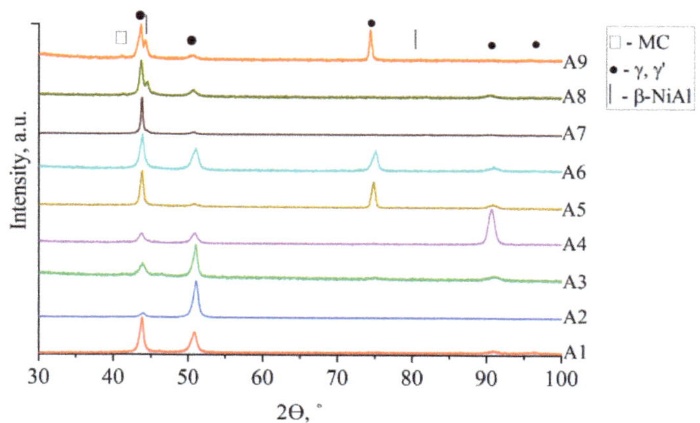


**Figure 3.** XRD spectra of the as-cast A1–A9 superalloys.

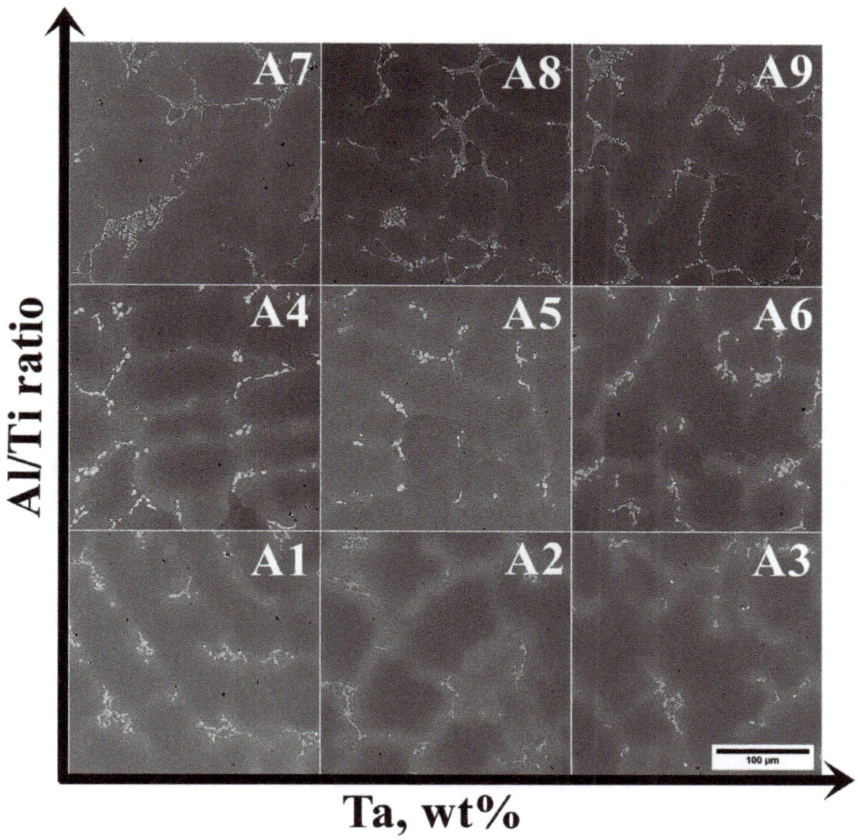


**Figure 4.** Dendritic structure of the as-cast (A1–A9) superalloys, SEM-BSE.

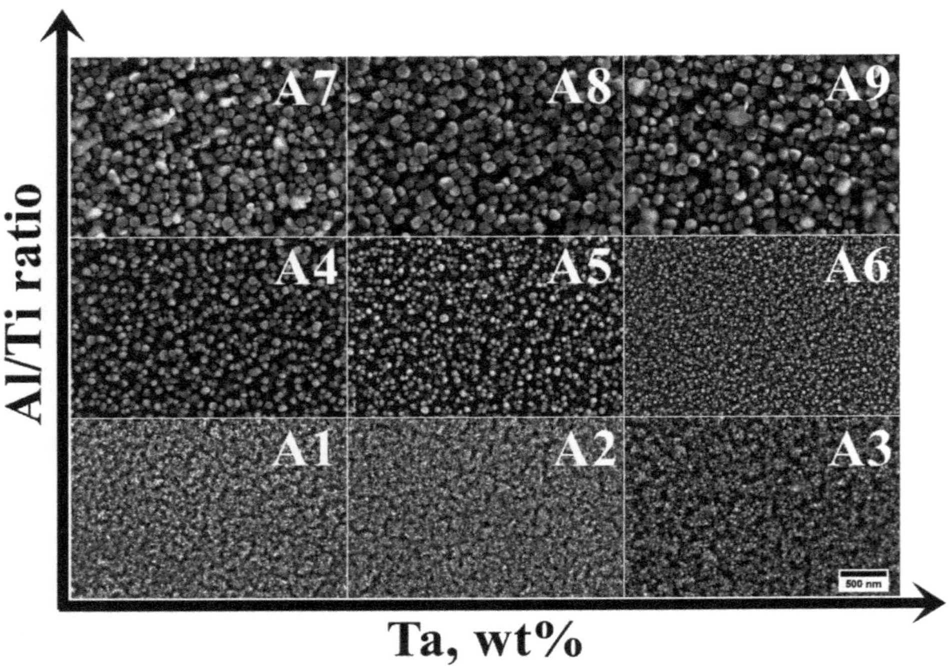


**Figure 5.** Precipitates in the dendritic regions of the (A1–A9) castings, SEM-SE.

**Table 3.** $\gamma'$ precipitates size in the dendritic regions.

| Parameter/Casting | A1 | A2 | A3 | A4 | A5 | A6 | A7 | A8 | A9 |
|---|---|---|---|---|---|---|---|---|---|
| Mean size of the $\gamma'$ precipitates, nm | 26 (±3) | 31 (±4) | 46 (±6) | 99 (±10) | 76 (±9) | 41 (±4) | 110 (±16) | 112 (±25) | 134 (±18) |
| Circularity, $\xi$ | 0.96 (±0.04) | 0.93 (±0.03) | 0.93 (±0.02) | 0.90 (±0.02) | 0.91 (±0.04) | 0.94 (±0.02) | 0.89 (±0.02) | 0.88 (±0.03) | 0.87 (±0.03) |

In superalloys with low and high Al/Ti ratios, the mean size of the $\gamma'$ precipitates increases with Ta concentration. The mean diameter increases from 26 nm to 46 nm in the first group, while in the second group, it increases from 110 nm to 134 nm. In superalloys with a medium Al/Ti ratio, a decrease in the mean size of the $\gamma'$ precipitates from 99 nm to 41 nm with increasing Ta is observed.

The precipitates in the dendrites were analyzed by TEM and SAED (Figure 6). It was confirmed that the $\gamma$ matrix is strengthened by coherent $\gamma'$ precipitates. Since the lattice parameters of the disordered matrix and ordered $\gamma'$ precipitates are very similar, the electron diffraction patterns exhibit some common maxima, e.g., {200} reflections. Additionally, {100} superlattice reflections, originating only from $\gamma'$ precipitates, are also present. The crystallographic relationship is as follows: {100} $\gamma$//{100} $\gamma'$; <010> $\gamma$//<010> $\gamma'$, which is referred to as the cube-cube orientation relationship.

In the interdendritic spaces, the microstructure of the castings is more complex. Numerous precipitates of various distributions, geometries, sizes, and phase contrasts are observed (Figure 7).

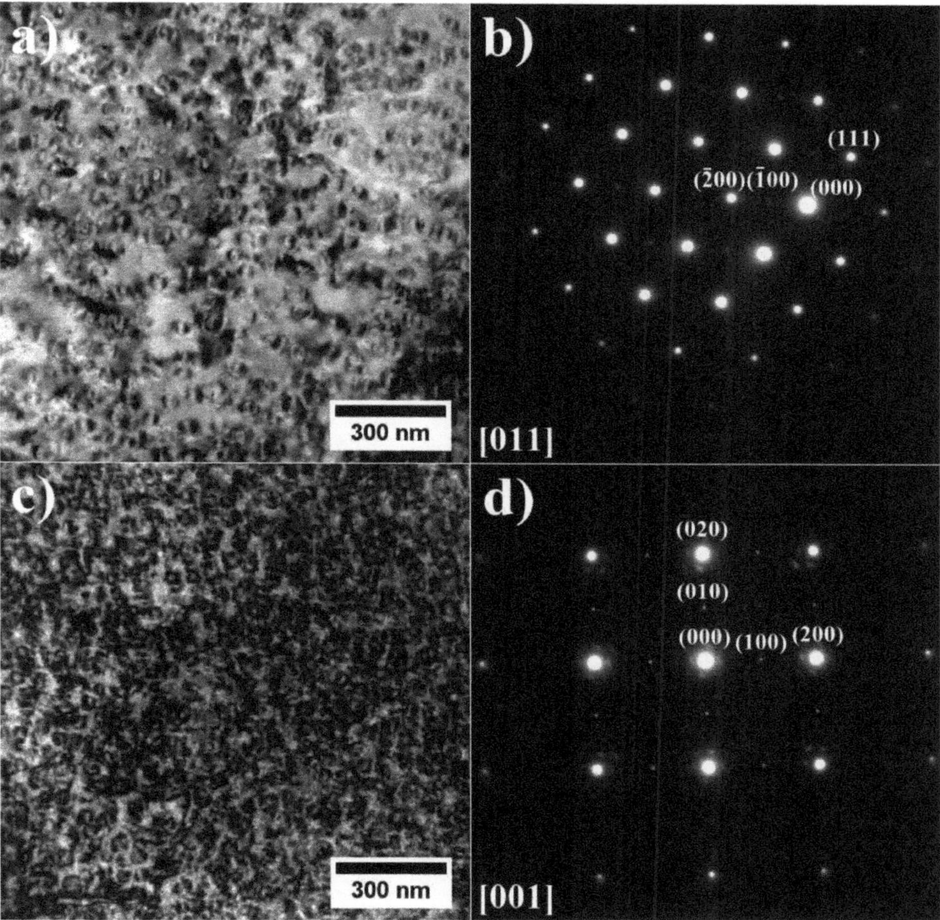


**Figure 6.** Morphology of the $\gamma'$ precipitates in the dendritic region and corresponding SAED patterns: (**a**,**b**) A3; (**c**,**d**) A6, TEM-BF.

MC carbides (bright precipitates: high Z-number) are considered to be the second most important precipitates, after $\gamma'$ precipitates, in strengthening polycrystalline Ni-based superalloys. Semi-quantitative SEM-EDX analyses were carried out and it was observed that the carbides are composed mainly of three elements: Nb, Ti, and Ta (Figure 8). Based on Ta/Nb atomic concentration relationship calculations, it can be observed that the increase in Ta concentration independently of the Al/Ti ratio leads to the partial replacement of Nb in the MC carbides composition (Figure 8a). It looks visible with increasing Ta contents in the superalloy. A similar effect is observed for the Ta/Ti relationship (Figure 8b), where an increase in Ta contents led to an increase in the Ta/Ti ratio in MC carbides. Based on these results, the main carbide-alloying element was found to be Nb. Whether Ta or Ti is second in order depends on the initial Ta concentration in the superalloy. In superalloys A1–A6, Ti is the second most concentrated element in the MC carbides, while in superalloys A7–A9, Ta takes over.

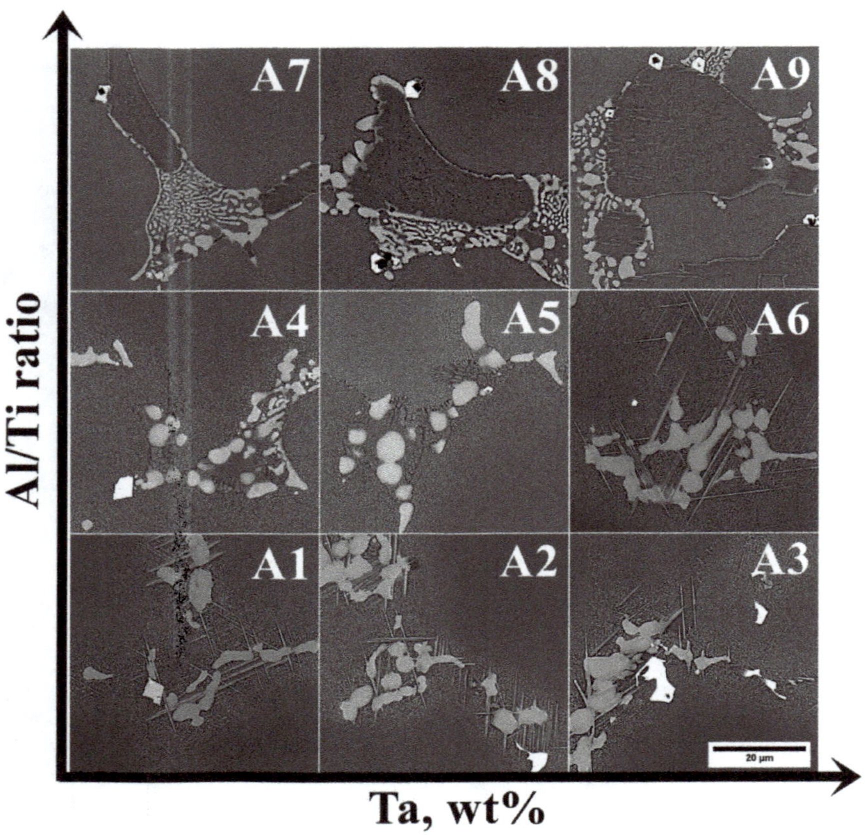


**Figure 7.** Morphology of precipitates in the interdendritic spaces of the (A1–A9) castings, SEM-BSE.

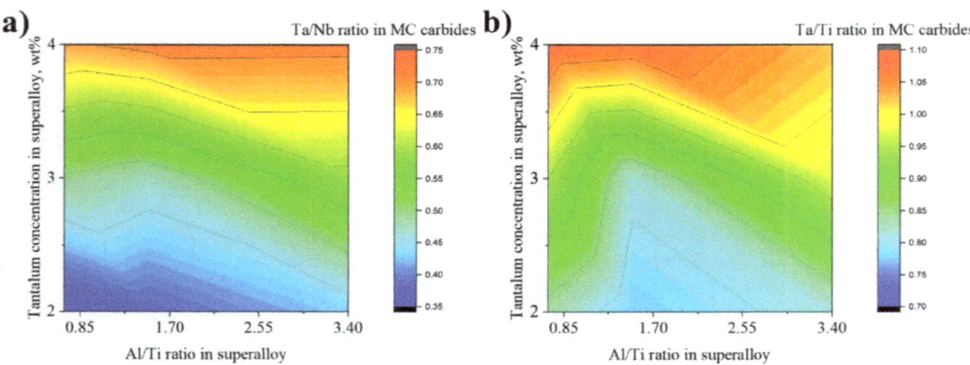


**Figure 8.** The concentration relationship in the MC carbides: (**a**) Ta/Nb; (**b**) Ta/Ti, calculated based on the semi-quantitative SEM-EDX measurements.

Based on SEM-EDX results, it was shown that the addition of Ta strongly influences the relationship between the main elements that form MC carbides. The carbides were subjected to image analysis to check whether the chemical composition of the superalloy

(Ta addition) influences the carbides' size. The results are shown in Table 4. The mean size of the carbides is between 2.6 and 4.5 μm. The results indicate that there are no significant differences or relationships between the chemical composition of the carbide and its size. The carbides in all castings have a regular blocky-shaped morphology. Their non-complex shape indicates that they were formed during solidification, when a relatively high fraction of the liquid phase is present [26]. This is in line with thermodynamic simulation results, as the MC carbides are the second phase to appear directly after the γ matrix in all castings.

**Table 4.** Mean size of the MC carbides in A1–A9 superalloy castings.

| Superalloy | Mean Size, μm | Standard Deviation, μm |
|---|---|---|
| A1 | 3.1 | 1.2 |
| A2 | 3.8 | 1.2 |
| A3 | 3.1 | 1.6 |
| A4 | 3.5 | 1.5 |
| A5 | 3.0 | 0.7 |
| A6 | 2.7 | 0.7 |
| A7 | 4.5 | 1.3 |
| A8 | 2.6 | 0.9 |
| A9 | 3.1 | 0.9 |

In the interdendritic spaces of all castings, there are precipitates with grey phase contrast and regular morphologies. Their amount in the microstructure is much higher than that of MC carbides. Based on SEM-EDX analysis of these precipitates, it was shown that they are rich in Ni and Nb. This unique chemical composition suggests that the Laves phase is formed with a chemical formula based on the $A_2B$ intermetallic compound. Position A is usually occupied by Ni, Cr, and Co, while B is occupied by Nb, Ti, and Ta. The atomic concentration relationship (Ni, Co, Cr)/(Nb, Ta, Ti) was calculated and is presented in Figure 9. This relation is within the 2.28–2.83 range. A more clear relationship is observed between the precipitates' Ta/Nb ratio and the superalloys' initial Ta concentration, where the increase of the latter is accompanied by an increase in the former. For Inconel 718, Baeslack et al. [27] observed that if Ta replaces Nb, Ta-rich MC carbides and Ta-rich Laves phase form. Based on microstructure observations, it was shown that the Laves phase quantities do not significantly differ between the standard superalloy and the Ta-modified one. Simulated weld HAZ experiments carried out in the Gleeble 1500 system indicated, however, a higher Laves phase liquation temperature for the Ta-modified sample (1225 °C) compared to the standard Inconel 718 alloy (1175 °C).

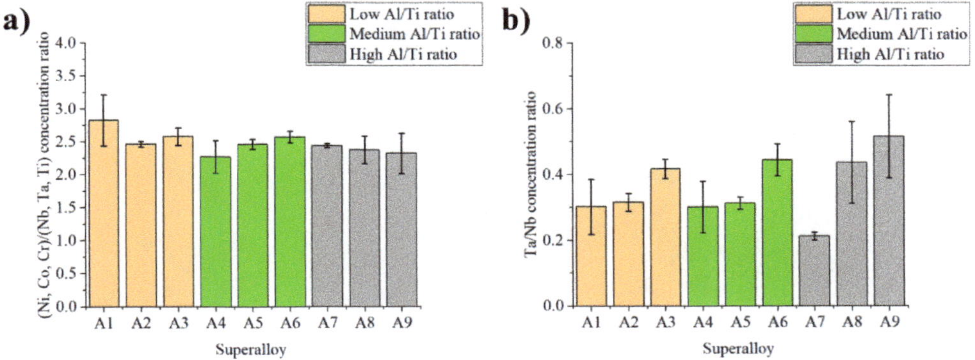


**Figure 9.** The atomic concentration relationship in the Laves phase precipitates in the interdendritic spaces: (**a**) (Ni, Co, Cr)/(Nb, Ta, Ti); (**b**) Ta/Nb, calculated based on semi-quantitative SEM-EDX measurements.

The SAED pattern, shown in Figure 10, indicates that the crystal structure of Nb-rich precipitates is hexagonal ($P6_3/mmc$), with nominal lattice parameters of a = 4.824 Å and c = 15.826 Å, consistent with the crystal structure of the Laves phase (Strukturbericht notation C36) [28]. The detected Laves phase seems to be the product of a non-equilibrium eutectic reaction $L \rightarrow \gamma$ + Laves occurring at the end of superalloy solidification, as predicted by Scheil solidification simulations. The solidification of superalloys begins with the formation of Nb-lean gamma dendrites which leads to enrichment of the liquid phase in Nb. Cieślak [29] stated that the solidification of Alloy 718 (and other Nb-bearing superalloys) is dominated by the segregation behavior of Nb and the precipitation of Nb-rich eutectic constituents in the form of $\gamma$/Laves and $\gamma$/MC (NbC). Radhakrishna [30] indicated that Laves phase formation requires a local niobium concentration of 10–30 wt%. The Laves phase is detrimental to the material's mechanical properties, affecting structural integrity, and can lead to premature failure of critical components during service [30]. Their amount should be significantly decreased during solution heat treatment to avoid weak-zone microstructures between the Laves phase and the matrix interface. Additionally, plate-like precipitates are visible at the edges of the Laves phases. It has been confirmed by FIB-SEM reconstruction tomography of the as-cast 718Plus superalloy that the Laves phase/$\gamma$ interfaces are favorable for $\eta$ phase formation [31].

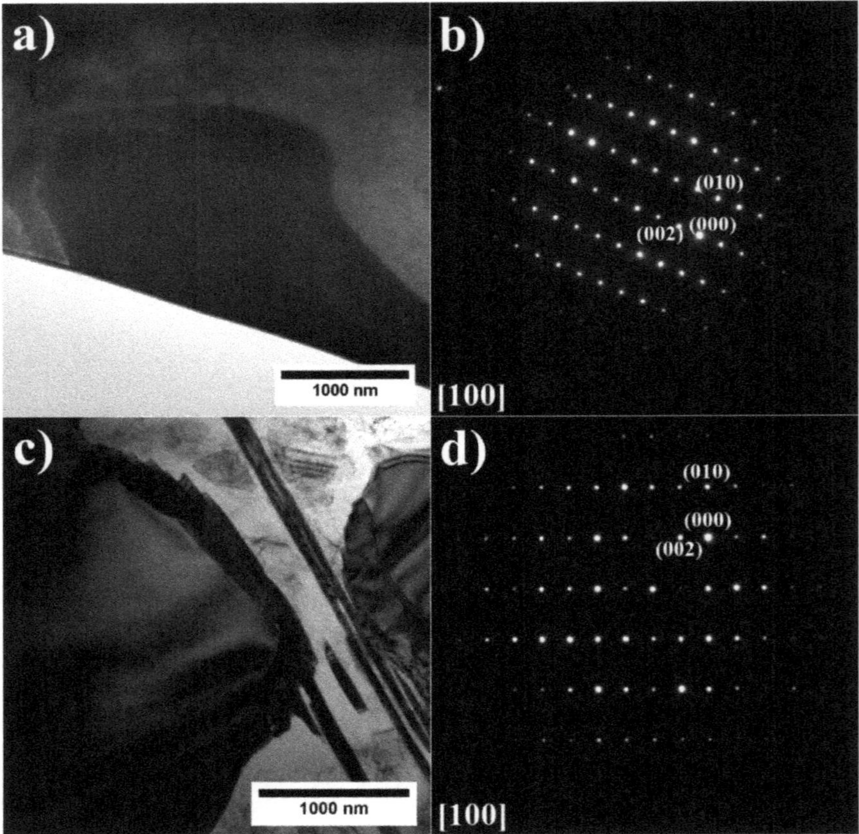


**Figure 10.** Laves phase precipitates with corresponding SAED patterns: (**a**,**b**) A3; (**c**,**d**) A6, TEM-BF.

Subsequently, the precipitates with plate-like morphologies, visible in variants A1–A3 and A6, were investigated using TEM. The TEM-BF images and the corresponding SAED patterns of superalloys A3 and A6 are presented in Figure 11.

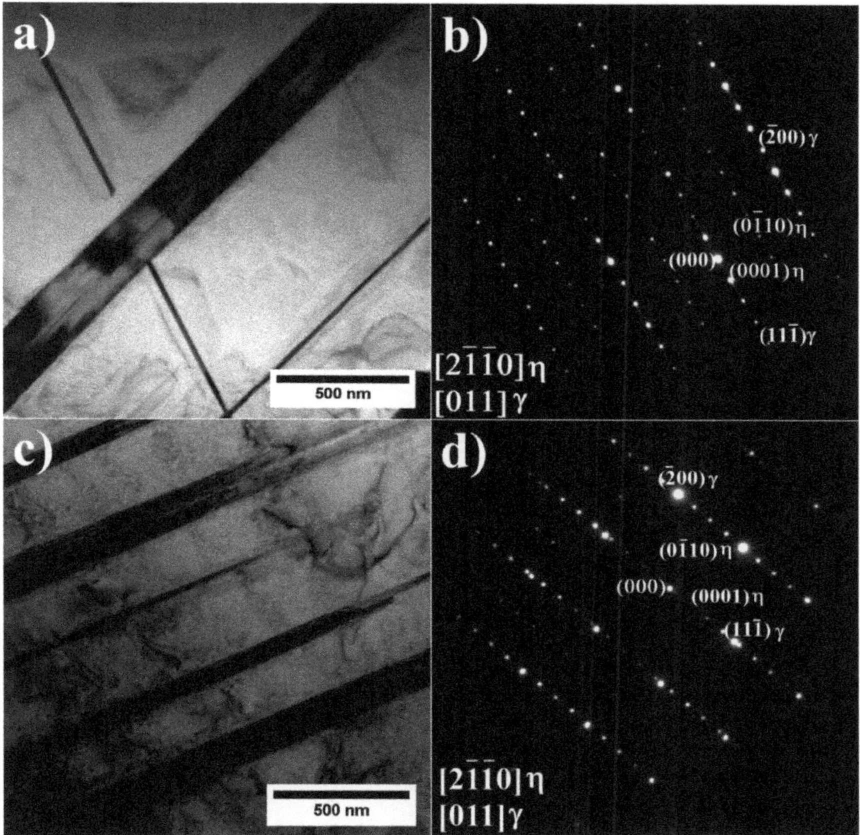


**Figure 11.** η phase precipitates with corresponding SAED patterns: (**a**,**b**) A3; (**c**,**d**) A6, TEM-BF.

It has been stated that those precipitates are the η ($Ni_3Ti$) phase with an ordered $DO_{24}$ hexagonal closed packed structure and nominal lattice parameter a = 5.096 Å and c = 8.3040 Å [32]. The SAED patterns confirmed that the η phase precipitates are incoherent with the γ matrix. The results suggest that large amounts of Ti and Ta favor the appearance of the η phase. It is especially visible in superalloys with a low Al/Ti ratio, as the addition of Ta causes the amount of η phase to increase compared to remelted IN740 [24]. The lack of η precipitates in superalloys A4 and A5 indicates that the increased Al concentration upon the addition of Ta at 2.0–3.0 wt% could prevent η phase formation. The results are in line with the conclusions formed by Bouse [33], who stated that the appearance of the η phase is favored by large amounts of Ti and Ta in Ni-based alloys. Seo et al. [34], based on microstructure observations of directionally solidified IN792 + Hf, indicated that the η phase was the final solidification product formed from the residual liquid after the γ-γ′ eutectic reaction. The authors stated that the concentration relationship (Ti + Ta + Hf)/Al in the residual liquid has a significant role in the nucleation of the η phase. During the solidification of eutectic γ-γ′ phase, the continual increase in the (Ti + Ta + Hf)/Al ratio in the residual liquid eventually led to the completion of the γ/γ′ eutectic reaction. Consequently, it caused η phase nucleation. According to theoretical calculations, the η phase has significant solubility for other alloying elements such as Co, Al, Cr, and W [35]. Kruk et al. [36] revealed by STEM-EDX analysis of the precipitates in the 718Plus alloy that Nb can also be present in η phase plates. Tan et al. [37] suggested that when the Ti/Al ratio in the liquid phase is higher than around 2.0, the η phase is favorable to form; however, when the ratio decreases below the key point, the eutectic $\gamma + \gamma'$ starts to precipitate. In the 151 superalloy, the η precipitates were strongly enriched in Nb [37]. The authors found that Nb has an atom size and similar chemical properties as Ti, enabling Nb to partially substitute Ti in the crystal structure, so a ratio of (Nb, Ti)/Al equal to 2.35 was used. Considering the alloying elements' segregation behavior and the phase transformation mechanisms, it can be stated that the additional segregation of Ti and Nb in the residual liquid phase is beneficial for η phase precipitate formation. Furthermore, the above-presented information and chemical similarities between Ta and Nb may explain the presence of the $\gamma$-$\gamma'$ eutectic and the lack of the η phase in the A4 and A5 alloys, i.e., in the alloys with a lower Ta concentration than the A6 alloy, in which the η phase is already present and the eutectic $\gamma$-$\gamma'$ is missing. Finally, near the η and Laves phase precipitates, the additional fine precipitates of the G phase are observed (Figure 12). They are complex silicides with a face-centered crystal structure and nominal lattice parameter of a = 11.2 Å [38]. The usual chemical composition is presented as $(Nb, Ti)_6(Ni, Co)_{16}Si_7$.

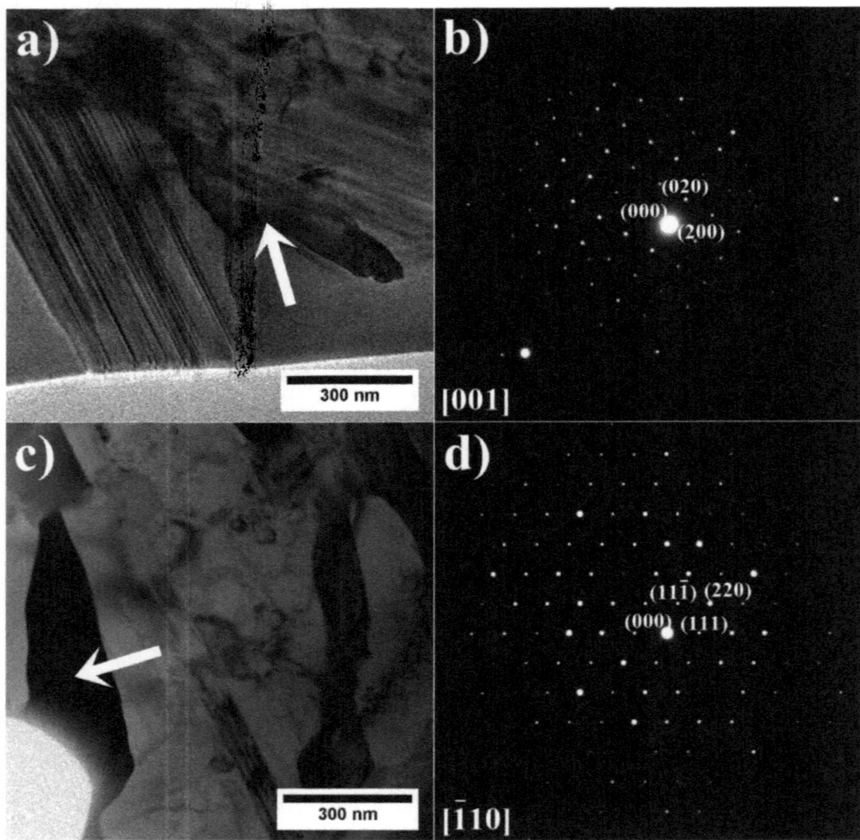


**Figure 12.** G phase precipitate with corresponding SAED patterns: (**a**,**b**) A3; (**c**,**d**) A6, TEM-BF.

It was believed that the G phase was only released in austenitic steels after irradiation for a long time. Powell et al. [39] showed that the G phase could precipitate in Fe-Ni alloys, with an increased concentration of Ti or Nb, after ageing in the range of 500–850 °C. The authors suggest that this phase was misidentified as the $M_6C$ carbide in earlier investigations due to its similar chemical composition and structure. It was found that in Fe-Ni alloys, the G phase first precipitates on NbC carbides at the grain boundaries and, after long-term ageing, inside the grains [39]. Xie et al. [40] stated that the weight fraction of the G phase in the IN740 after standard heat treatment is equal to 0.054%, whereas after 2000 h and ageing at 760 °C, it is almost 10 times higher (0.471%). The solidification simulation predicted the existence of the G phase in superalloys A1–A6 directly in the as-cast state, which is in line with TEM observations. In the remelted IN740, precipitates with high Si concentrations, suggesting the presence of the G phase, have also been observed [24]. Ecob et al. [41] stated that the oxygen concentration influences the stability of NbC carbides compared to the G phase. The authors state that oxygen and silicon strongly segregate. Near the carbides, there is increased oxygen concentration, resulting in an uneven distribution of oxygen in the material. The simultaneous Si enrichment in some regions creates potential G phase precipitation sites. This information is important in the context of superalloy production with elevated Nb and Si concentrations, cast not under vacuum conditions, which will naturally result in higher oxygen concentrations. Additionally, precipitates with dark phase contrast were found. The SAED patterns showed that the blocky-shaped precipitates, present in all castings, were TiN with a regular crystal structure and a nominal lattice parameter of a = 4.244 Å [42] (Figure 13). The primary carbides are found to nucleate on preexisting TiN precipitates, as presented in Figure 7.

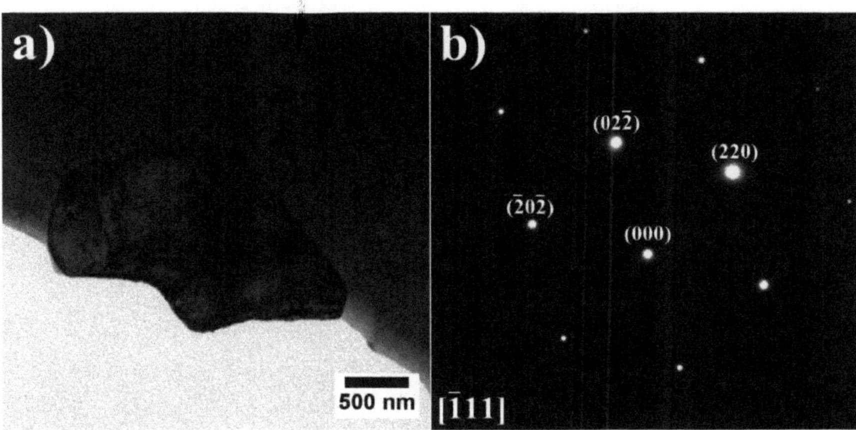


**Figure 13.** (**a**) morphology of the TiN; (**b**) corresponding SAED pattern, TEM-BF.

Sobczak et al. [43], during a systematic study on the HAYNES®282 superalloy's castability, carried out casting trials under non-vacuum conditions, similar to our investigation. The superalloy, which melted without deoxidizing, contained 0.38 wt% of oxygen and 0.02 wt% of nitrogen. To decrease the oxygen content, the molten superalloy was deoxidized using a NiMg alloy. It led to a three-fold decrease in oxygen (0.12 wt%). Cockcroft et al. [44] stated that the solubility of nitrogen in IN718 at 1427 °C, as determined from the solubility product of TiN, is approximately 38 ppm, which suggests that in the wrought commercial material at 60 ppm N, about 1/3 of the total N content will be present as solid TiN particles in the liquid. None of these phases were intended, with the precipitation induced by gaseous impurities (O and N), which cannot be completely removed when casting under non-vacuum conditions. Considering the thermodynamic stability of TiN, it should be underlined that it cannot be dissolved during solution heat treatment and it is insoluble up to its melting temperature.

In superalloys A7–A9, the interdendritic regions' microstructure is more complex than in the case of variants A1–A6. Sample A9 was chosen as a representative variant for more detailed analysis using STEM. The STEM-EDX maps of selected alloying elements were carried out in the interdendritic region (Figure 14).

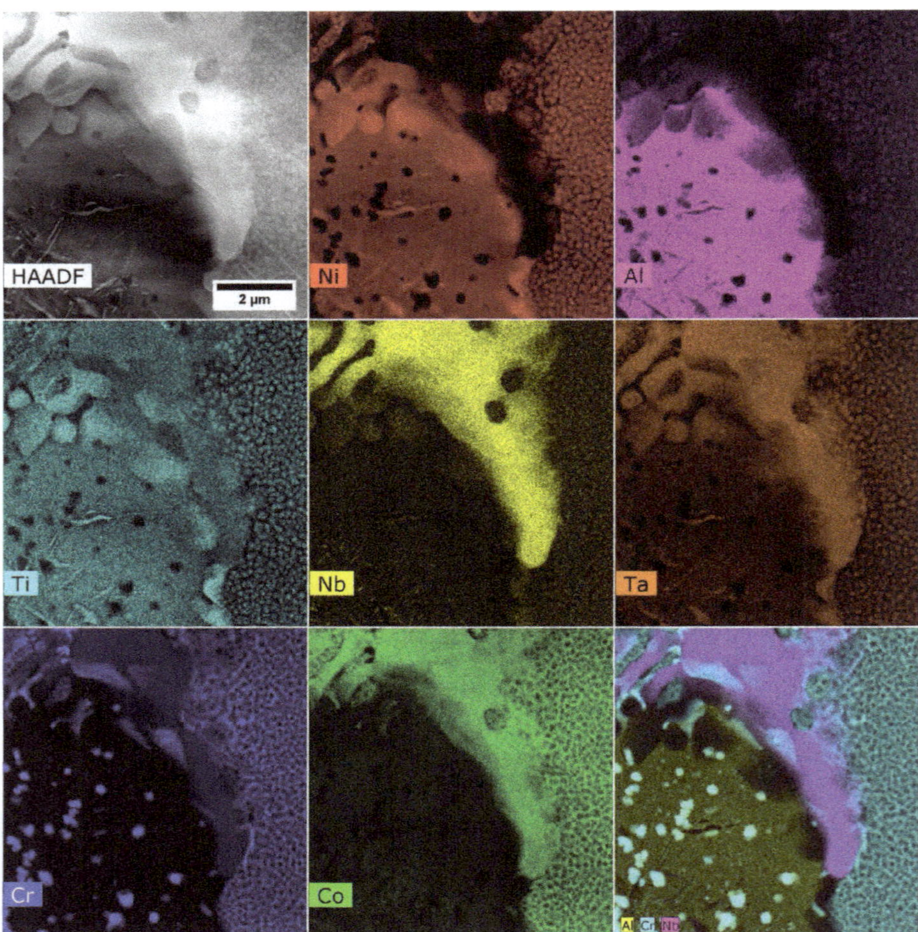


**Figure 14.** The distribution of the alloying elements in the interdendritic spaces of superalloy A9, STEM-HAADF.

The STEM-HAADF image shows the two-phase $\gamma + \gamma'$ region on the right side. The areas corresponding to the fine $\gamma'$ phase precipitates are enriched in Ni, Al, Ti, and Ta, and the matrix is enriched in Co and Cr. It was confirmed by SAED that the $\gamma$ matrix in this area is strengthened only by the coherent $\gamma'$ precipitates (Figure 15). In the central area, there is an island-like precipitate strongly enriched in Al. Inside, there are numerous fine precipitates with a plate-like morphology and characterized by increased Ni, Ti, Nb, and Ta concentration, as well as fine precipitates with a regular morphology, consisting of Cr. On the periphery of this island-like precipitate, there is a large Laves phase precipitate with an increased concentration of Co, Nb, Ti, and Ta.

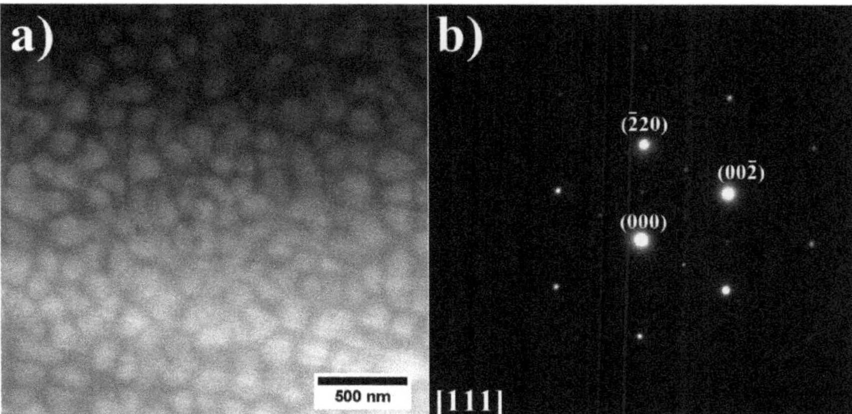


**Figure 15.** (**a**) Morphology of the $\gamma'$ precipitates in dendritic region; (**b**) corresponding SAED pattern. STEM-HAADF.

Based on the SAED of the island-like precipitate presented in Figure 16, the precipitate was identified as the β-NiAl phase, with a nominal lattice parameter of a = 2.88 Å [45].

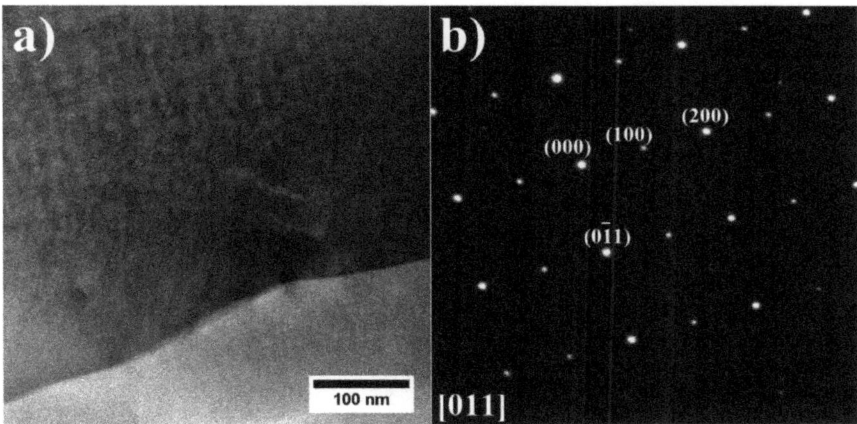


**Figure 16.** (**a**) Morphology of the β-NiAl phase island (dark part) in the interdendritic space of the superalloy A9; (**b**) corresponding SAED pattern, STEM-HAADF.

It should be noted that despite the high Al concentration in superalloys A7–A9, the $\gamma$-$\gamma'$ eutectic did not form. Based on the Scheil calculation, the β-NiAl phase should precipitate from the liquid at relatively high temperatures. The same conclusion was formed by Song et al. [46], who observed that β-NiAl forms earlier than the $\gamma$-$\gamma'$ eutectic, which can clarify why the eutectic islands are not present in superalloys A7–A9, despite the high Al contents. Based on solidification analysis of the Ru-containing SX superalloy, Cao et al. [47] stated that the concentration of Al, Ta, and Ru in the liquid phase significantly increased during the growth of $\gamma$ dendrites. EDX analysis confirmed that the main forming alloying elements of the β-NiAl phase were Ni, Al, Ta, and Ru. Therefore, the accumulation of Al,

Ta, and Ru in the remaining liquid was favorable for β-NiAl formation. Tan et al. [48] based on microstructural analysis of Ru-containing SX superalloy samples obtained after DSC investigations, indicated that the cooling rate plays an important role in the β-NiAl precipitates formation. No β-NiAl precipitates were found in the microstructure created using a cooling rate of 5 °C/min. Higher cooling rates (15 °C/min and 30 °C/min) led to quick superalloy solidification, followed by less diffusion in the unsolidified interdendritic liquid phase, which resulted in the more apparent segregation of Al and Ta to interdendritic regions. The solidification induces element segregation, which results in element consumption or aggregation in the residual liquid phase. The solidification path is controlled by the change in composition. The authors also concluded that with increasing concentration of refractory alloying elements, i.e., high supersaturation, some undesirable phases would occur in Ni-based superalloys. They suggest that Ru could strongly influence the segregation behavior of Ta. It is therefore speculated that β-NiAl/Ta-rich Laves eutectics may precipitate in high-Ta-containing superalloys that are highly supersaturated during solidification [48].

The morphology of Ti, Ta, and Nb-rich lamellar precipitates formed inside β-NiAl is shown in Figure 17. The SAED pattern confirms the presence of the η phase with a hexagonal ($DO_{24}$) crystal structure.

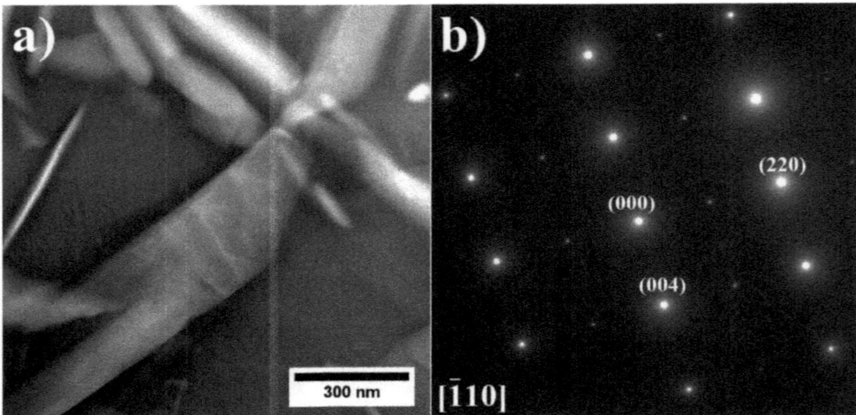


**Figure 17.** (**a**) Morphology of the η phase precipitate within β-NiAl; (**b**) SAED pattern of the η phase precipitate, STEM-HAADF.

η phase precipitates were confirmed in seven out of the nine produced superalloys. The difference, however, is that in the A7–A9 superalloys, the η phase precipitates do not occur near the dual-phase $\gamma/\gamma'$ region or Laves phase precipitates, but are within β-NiAl. However, thermodynamic simulations predicted the formation of this phase in superalloys with a high Al/Ti concentration ratio, when the residual liquid phase is below 10%. Similar precipitates with lamellar morphologies were found inside the β-NiAl islands in Co-based superalloys [49] and Ni-based superalloys [50]. Yuan et al. [51] even observed them in the IC10 $Ni_3Al$-based superalloy jointed by transient liquid phase (TLP) bonding. With the absence of Ti in both the superalloy and filler material, the monoclinic $Ni_3Ta$ phase was found to have precipitated via HRTEM. Pathare et al. [52], in the β-NiAl + 2 at% Ta alloy, identified plate-like precipitates as the intermetallic compound NiAlTa with a hexagonal C14 structure, using XRD. They also stated that at least three possible ternary compounds between Ni, Al, and Ta could form: (1) NiAlTa with a hexagonal $MgZn_2$ (C14)-type structure (space group $P6_3/mmc$), (2) $Ni_2AlTa$ with a cubic Heusler alloy structure ($Cu_2AlMn$-type) and nominal lattice parameter a = 5.949 Å, and (3) $(Al_{0.5}Ta_{0.5})Ni_3$ with a hexagonal $Ni_3Ti$-type ($DO_{24}$) structure, which has been confirmed in this work.

In addition to the plate-like precipitates, numerous fine Cr-rich (as confirmed on the EDX mapping) precipitates with a size of about 300 nm were also found inside β-NiAl (Figure 18). Chromium has limited solubility of approx. 1 at.%. in NiAl at lower temperatures. Maximum solubility additionally depends on the Ni/Al concentration ratio and is slightly greater in alloys containing less than 50% Al, and is related to the site preference of Cr for Al in NiAl [53]. Based on the SAED pattern, it is shown that the precipitates have a regular body-centered structure (BCC) with a nominal lattice parameter of a = 2.884 Å [54], which is nearly equal to the lattice parameter of the β-NiAl; therefore, only minor misfit occurs. Higher alloying levels of Cr lead to the formation of α-Cr precipitates upon slow cooling of castings. The orientation relationship between the α-Cr precipitates and the β-NiA1 matrix was determined to be cube-on-cube, i.e., <100> α {001} α//<100> β {001} β, as confirmed earlier by Cotton et al. [53]. The results of thermodynamic simulations did not show the presence of α-Cr in the as-cast A7–A9 superalloys. However, it should be noted that the simulation predicted Cr-rich phase precipitation (as shown in Figure 2 of the σ phase) at the end of solidification. The σ phase was not detected during STEM-HAADF observations.

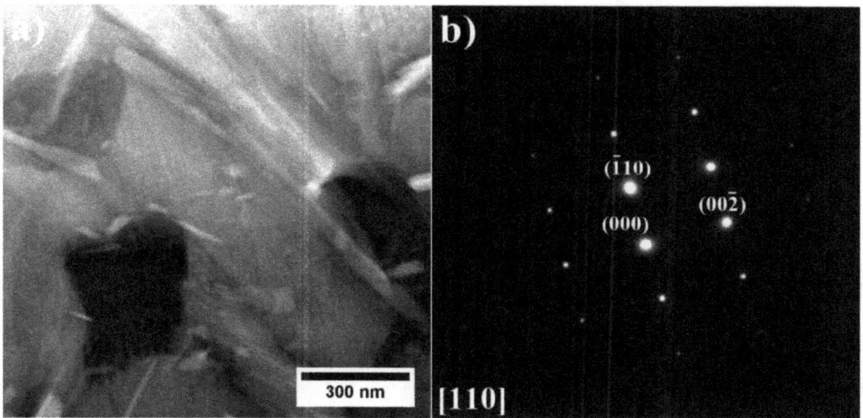


**Figure 18.** (**a**) Morphology of the α-Cr precipitates; (**b**) SAED pattern of the α-Cr precipitate, STEM-HAADF.

α-Cr has a higher melting temperature and a lower coefficient of thermal expansion compared to β-NiAl, reducing thermal stresses [53]. Based on heat-treating experiments of Ni-Al-Cr model alloys, Tian et al. [55] stated that the fine α-Cr precipitates strongly increase the hardness of NiAl thanks to the perfect lattice coherency at the α-Cr/NiAl interface. The softening that occurs after long-term annealing is caused by the loss of coherency induced by the attraction of matrix dislocations to the precipitate/matrix interface, followed by climbing around the precipitates.

### *3.3. Analysis of the A1–A9 Superalloys Phase Transformation Temperatures during Heating*

The dilatometric (DIL) study was carried out to analyze phase transformations occurring in the A1–A9 superalloys during heating. The obtained curves are shown in Figure 19. The dilatation effect observed above 700 °C in all superalloys is accompanied by an increase in the coefficient of thermal expansion (CTE). The $\gamma$ and $\gamma'$ phases dominate the microstructure of the A1–A9 superalloys, and it seems that changes originating from their behavior will be the most important. It could indicate that the gradual dissolution of the $\gamma'$ precipitates in the matrix causes its volume fraction to decrease. With increasing temperatures, the lattice parameters of the phases increase; however, the lattice parameter of the $\gamma$ matrix increases more rapidly than that of $\gamma'$ precipitates [56,57]. Consequently, this effect would be visible on the curves as an increase in the thermal expansion coefficient, as was indeed observed. The local highest point in this dilatation effect could correspond to the temperature at which the most intense dissolution process occurs, while the end may be associated with crossing the $\gamma'$ solvus temperature. The process of $\gamma'$ dissolution takes place in a wide temperature range, which has been confirmed in many superalloys [58–60]. During dilatometry experiments, Trexler et al. [61] observed that the GTD111 superalloy is characterized by a rapid change in CTE above 800 °C, at which the $\gamma'$ precipitates start to dissolve. The second dilatation effect was found in the range of 1170–1190 °C for superalloys A1–A6.

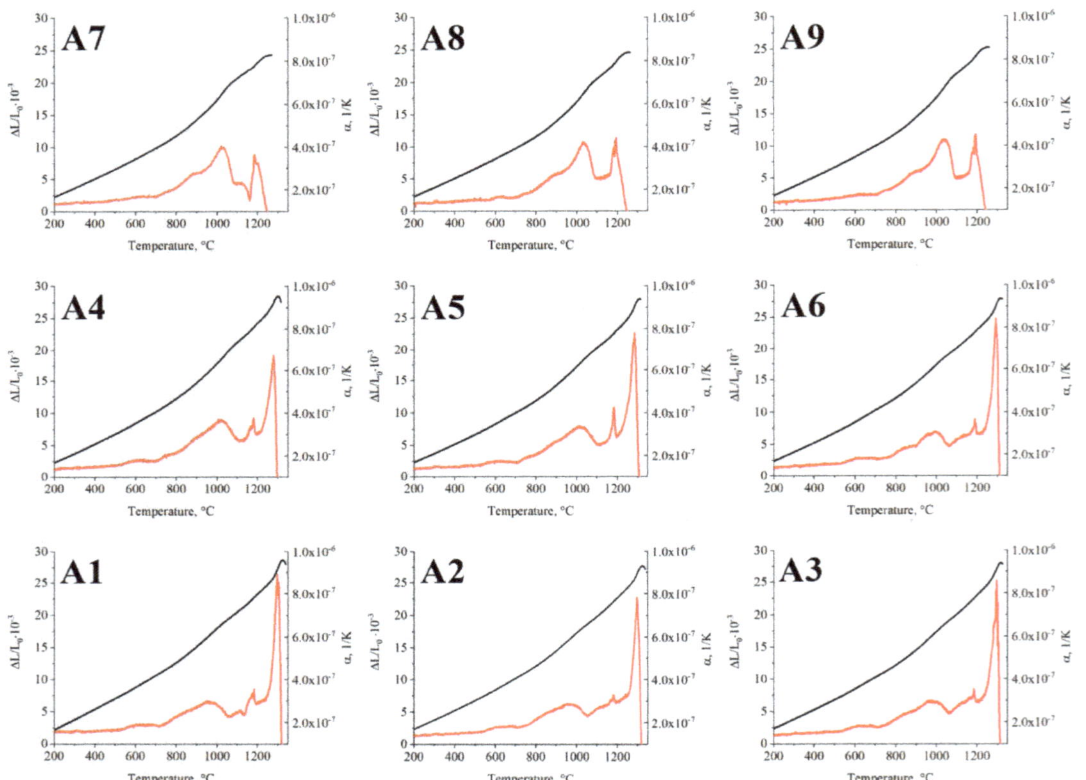


**Figure 19.** Dilatometry curves of the (**A1–A9**) superalloys.

Further heating significantly increased the length of the sample, accompanied by local melting. Since the effect occurs at a similar temperature for these six alloys, it must originate from the phase present in all of them. The Laves phase was found in all these alloys, characterized by a similar chemical composition and relatively high volume fraction. This effect will be discussed in greater detail in the calorimetry section. In superalloys A7–A9, Laves phase precipitation also occurs; however, the effect of their dissolution on the dilatometric curve most likely coincides with the temperature at which the sample starts to locally melt; therefore, it is not noticeable. Further heating of the samples led to a significant increase in the coefficient of thermal expansion, which is most likely related to the initiation of bulk sample melting.

The phase transformation temperatures in the as-cast superalloys were also analyzed via DSC (Figure 20) and the temperatures are summarized in Table 5. Knowledge of $\gamma'$

solvus temperatures is often important in controlling solution heat treatment temperatures before precipitation hardening. Moreover, the solvus temperature of the $\gamma'$ phase is closely related to the temperature-bearing capacity of Ni-based superalloys. Therefore it is crucial to investigate the effect of alloying elements on the solvus temperature when designing novel materials.

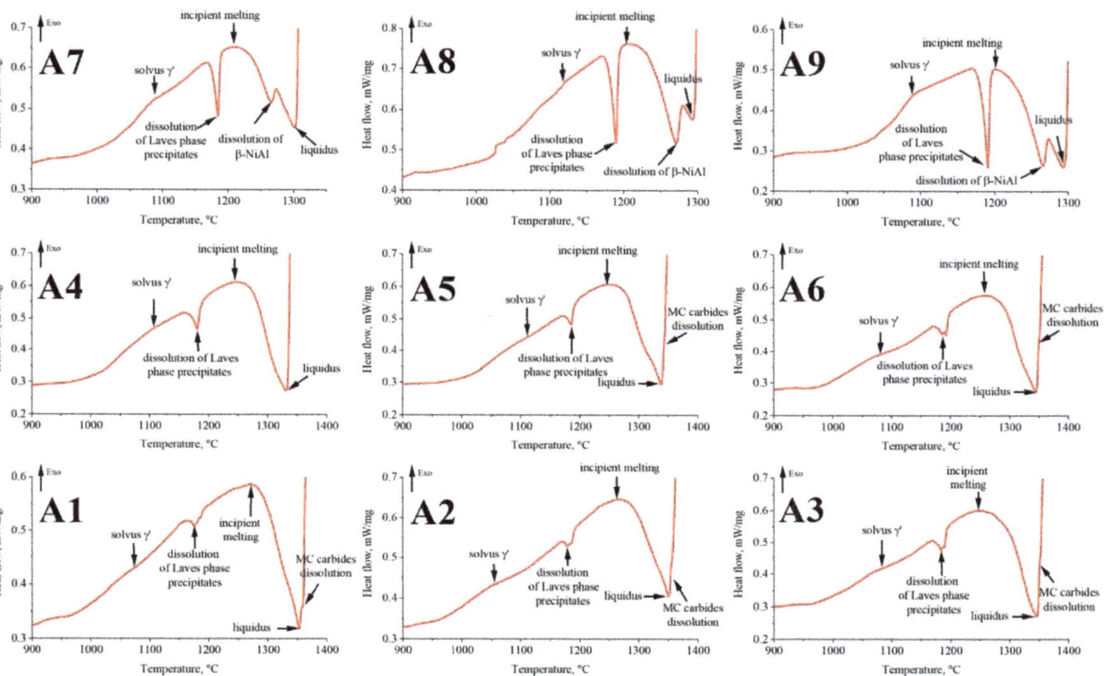


**Figure 20.** Representative DSC heating curves with a heating rate of 10 K min$^{-1}$ for the as-cast superalloys.

**Table 5.** Phase transformation temperatures of the A1–A9 superalloys registered during DSC heating.

| Alloy | $\gamma'$ solvus | Dissolution of Laves Phase | Incipient Melting (Solidus) | β-NiAl Dissolution | Liquidus | MC Carbides Dissolution |
|---|---|---|---|---|---|---|
| A1 | 1072 | 1176 | 1224 | not registered | 1353 | 1357 |
| A2 | 1055 | 1179 | 1263 | | 1351 | 1354 |
| A3 | 1084 | 1184 | 1247 | | 1346 | 1350 |
| A4 | 1106 | 1182 | 1245 | | 1330 | not registered |
| A5 | 1111 | 1185 | 1250 | | 1339 | 1343 |
| A6 | 1080 | 1193 | 1258 | | 1345 | 1350 |
| A7 | 1089 | 1185 | 1211 | 1267 | 1302 | not registered |
| A8 | 1117 | 1189 | 1208 | 1272 | 1293 | |
| A9 | 1087 | 1190 | 1201 | 1266 | 1292 | |

In low Al/Ti ratio superalloys, the solvus temperature is in the range of approx. 1055–1084 °C. In the second group, these values are slightly higher, i.e., 1080–1111 °C. This indicates that the Al concentration has a greater influence on the solvus temperature than Ta. This is most likely due to Ta being a significant element in other phases, like MC carbides and the η phase. In superalloys with a high Al/Ti concentration ratio, the $\gamma'$ solvus temperature is 1087–1117 °C. However, it should be noted that a large amount of Al forms β-NiAl precipitates, which are still stable at this temperature. The values of $\gamma'$

solvus determined by means of the DSC curves are very close to the end temperature of the dilatation effects. The $\gamma'$ solvus temperature of commercial superalloys IN740 (Ti = 1.8 wt%, Al = 0.9 wt%) and IN740H (Ti = 1.35 wt%, Al = 1.35 wt%) is lower, 860 °C and 975 °C, respectively, as determined by Xie [40] via thermodynamic simulations. It results from the lower amount of $\gamma'$-formers as compared to the A1–A9 superalloys. A very clear endothermic effect is observed in all superalloys above 1170 °C. Despite the differences in chemical composition, the temperature range in which this effect is observed is relatively narrow, i.e., 1176–1193 °C. In each of the individual groups (low, medium, and high Al/Ti ratio), the peak temperature increases with Ta contents. The values recorded on the DSC curves are very similar to those that correspond to the dilatation effect on the DIL curves. As previously, this effect most likely comes from the Laves phase. A similar endothermic peak within the temperature range of 1162–1190 °C was observed by Cao et al. [62] in the differential thermal analysis (DTA) thermograms of Alloy 718 in as-cast condition. The effect was assigned to the Laves phase eutectic reaction. Directly after initial melting, the liquation of the existing Laves eutectic and matrix was reported to rapidly increase with temperature. This phenomenon was responsible for the endothermic process after the Laves peak on the thermograms. The authors also investigated how earlier superalloy annealing influences the latter melting process of Alloy 718. They proved that the incipient melting temperature of a fully homogenized (T = 1190 °C, t = 48 h) superalloy was higher, approx. 1230 °C, which is the solidus temperature of Alloy 718. These results suggest that the solution heat treatment of the A1–A9 superalloys should include a step before the final temperature to prevent the local melting of Laves phase.

Further heating led to crossing the solidus temperature of the superalloys. In the A1–A3 superalloys, it is in the range of 1224–1263 °C. In the A4–A6 variants, it increases with the Ta concentration from 1245 °C to 1258 °C. In superalloys A7–A9, the solidus temperature decreased with increasing Ta contents from 1211 °C to 1201 °C. The β-NiAl phase, also present in the A7–A9 superalloys, dissolves at much higher temperatures. An inflection is registered in the range of 1266–1272 °C. During DSC experiments of as-cast superalloy SX, Song et al. [46] registered incipient melting at 1321 °C and β-NiAl dissolution at 1346 °C, when a significant amount of liquid phase was present. The high melting point of the β-NiAl phase can induce impediments in homogenizing heat treatment. Based on long-term annealing experiments, Song et al. [45] indicated that during high-temperature exposure, Ni and Al diffuse to the surrounding $\gamma$ and $\gamma'$ phases, and suggested that β-NiAl can be eliminated through exposure at 1100 °C for 1000 h.

Superalloys with a low Al/Ti ratio have the highest liquidus temperatures, whereas it is lower for each of the following two groups due to higher Al concentrations. As the Ta concentration increases, it begins to decrease in superalloys A1–A3 and A7–A9. The lower liquidus temperature in superalloys A4 and A5 in relation to A6 may result from the earlier dissolution of $\gamma/\gamma'$ eutectic islands and, in turn, the enrichment of the liquid in Al. The last recorded effect originates from the dissolution of MC carbides. In the A4 and A7–A9 superalloys, no thermal effects were recorded. However, this is not due to the lack of MC carbides, the presence of which was confirmed during microstructure observations. Liu et al. [63], who investigated the influence of temperature on the homogenization of melted Inconel 738LC, showed that MC carbides could survive at a temperature much higher than liquidus. At 1550 °C, the MC carbides did not completely dissolve in the liquid after about 5–10 min. For comparison, the liquidus temperature of Inconel 738LC superalloy is lower, 1336–1342 °C [64].

## 4. Conclusions

Based on the results and observations of as-cast model A1–A9 superalloys, characterized by various Al/Ti ratios and Ta concentrations, the following conclusions are proposed:

- The Scheil solidification simulation showed the formation of the $\gamma$, $\gamma'$, MC carbides and Laves phase in all castings. Additionally, β-NiAl was predicted in superalloys

with high Al/Ti ratios. G phase precipitates could form in the final stage of solidification in superalloys with low and medium Al/Ti ratios.
- For all castings, the XRD spectra revealed peaks corresponding to $\gamma$ matrix, $\gamma'$ precipitates, and MC. The $\beta$-NiAl phase is present in castings characterized by high Al/Ti ratios.
- As verified by SEM and TEM, the microstructure of alloys A1–A9 consists of a $\gamma$ matrix, $\gamma'$ phase, (Nb, Ti, Ta)C carbides, and Laves phase precipitates. The alloys with low and medium Al/Ti ratios, except variants A4 and A5, were additionally strengthened by plate-like $\eta$ phase precipitates.
- It was shown via SEM-EDX analysis that the initial Ta concentration in the superalloy has a strong influence on the MC carbides' composition, as it can partially replace Nb and Ti. The Ta/Nb and Ta/Ti concentration relationship in carbides increases with Ta content in the superalloy.
- The high initial Al/Ti concentration ratio in the A7–A9 superalloys led to the formation of the $\beta$-NiAl phase, strengthened locally by $\eta$ phase and $\alpha$-Cr precipitates.
- On the DIL curves of all alloys, dilatation effects originating from $\gamma'$ precipitates are observed. In the A7–A9 superalloys, the effect from Laves phase dissolution was registered right before sample melting.
- Based on the DSC curves, the $\gamma'$ solvus temperature in alloys with low and medium Al/Ti ratios was in the range of 1055–1084 °C and 1080–1111 °C, respectively. The liquidus temperature decreased with the increase in Al/Ti ratio.

**Author Contributions:** Conceptualization, M.G.-R. and Ł.R.; methodology, M.G.-R. and Ł.R.; software, Ł.R.; investigation, M.G.-R., Ł.R., K.C., R.C., O.M. and Z.P.; writing—original draft preparation, M.G.-R. and Ł.R.; writing—review and editing, R.C. and Z.P.; supervision, Z.P.; project administration, M.G.-R.; funding acquisition, M.G.-R. All authors have read and agreed to the published version of the manuscript.

**Funding:** This research was funded by the Polish National Science Centre (Preludium 14) under the grant for young scientists 2017/27/N/ST8/01801.

**Institutional Review Board Statement:** Not applicable.

**Informed Consent Statement:** Not applicable.

**Data Availability Statement:** Not applicable.

**Acknowledgments:** M.G.-R. thanks the European Virtual Institute on Knowledge-based Multifunctional Materials (KMM-VIN) for the fellowship to spend a research period at the Institute of Materials Research, Slovak Academy of Sciences. The authors wish to express appreciation to Grzegorz Cempura and Sebastian Lech (AGH University of Science and Technology in Kraków) for support with TEM/STEM experiments. O.M. was supported by the Scientific Grant Agency under contract VEGA project No. 2/0086/22.

**Conflicts of Interest:** The authors declare no conflict of interest.

## References

1. Reed, R.C. *The Superalloys Fundaments and Applications;* Cambridge University Press: Cambridge, UK, 2006.
2. Di Gianfrancesco, A. The fossil fuel power plants technology. In *Materials for Ultra-Supercritical and Advanced Ultra-Supercritical Power Plants;* Woodhead Publishing: Shaston, UK; Cambridge, UK, 2017. [CrossRef]
3. Rakoczy, Ł.; Grudzień, M.; Tuz, L.; Pańcikiewicz, K.; Zielińska-Lipiec, A. Microstructure and properties of a repair weld in a nickel based superalloy gas turbine component. *Adv. Mater. Sci.* **2017**, *17*, 55–63. [CrossRef]
4. deBarbadillo, J.J. Inconel 740H. In *Materials for Ultra-Supercritical and Advanced Ultra-Supercritical Power Plants;* Woodhead Publishing: Shaston, UK; Cambridge, UK, 2017. [CrossRef]
5. Baker, B.A. A new alloy designed for superaheater tubing in coal-fired ultra supercritical boilers. In Proceedings of the International Symposium on Superalloys 718, 625, 706 and Derivatives, TMS (TheMinerals, Metals & Materials Society), Pittsburgh, PA, USA, 2–5 October 2005; pp. 601–611.

6. de Barbadillo, J.J.; Baker, B.; Xie, X. Superalloys for advanced energy applications: Inconel alloy 740H—A case study on international government-industry-university collaboration. In Proceedings of the 13th International Symposium on Superalloys, TMS (The Minerals, Metals & Materials Society), Pittsburgh, PA, USA, 11–15 September 2016; pp. 217–226.
7. Xie, X.; Chi, C.; Yu, H.; Yu, Q.; Dong, J.; Zhao, S. Structure stability study on fossil power plant advanced heat-resistant steels and alloys in China. In Proceedings of the Advances in Materials Technology for Fossil Power Plants, Santa Fe, NM, USA, 31 August–3 September 2010; pp. 30–52.
8. Xie, X.; Zhao, S.; Dong, J.; Smith, G.; Patel, S. An investigation of structure stability and its improvement on new developed Ni-Cr-Co-Mo-Nb-Ti-Al superalloy. *Mater. Sci. Forum* **2005**, *475*, 613–618. [CrossRef]
9. Zhao, S.; Xie, X.; Smith, G.; Patel, S. Microstructural stability and mechanical properties of a new nickel-based superalloy. *Mater. Sci. Eng. A* **2003**, *355*, 96–105. [CrossRef]
10. Shin, G.S.; Yun, J.Y.; Park, M.C.; Kim, S.J. Effect of Mo on the thermal stability of $\gamma'$ precipitate in Inconel 740 alloy. *Mater. Charact.* **2014**, *95*, 180–186. [CrossRef]
11. Abbasi, M.; Kim, D.-I.; Shim, J.H.; Jung, W.-S. Effects of alloyed aluminium and titanium on the oxidation behavior of Inconel 740 superalloy. *J. All. Comp.* **2016**, *658*, 210–221. [CrossRef]
12. Ennis, P. Nickel-Base Alloys for Advanced Power Plant Components. In *Coal Power Plant Materials and Life Assessment*; Woodhead Publishing: Shaston, UK; Cambridge, UK, 2014; pp. 147–167. [CrossRef]
13. Rozman, K.; Detrois, M.; Jablonski, P. Mechanical performance of various Inconel 740/740H alloy compositions for use in A-USC castings. In Proceedings of the 9th International Symposium on Superalloy 718 & Derivatives: Energy, Aerospace, and Industrial Applications, Pittsburgh, PA, USA, 3–6 June 2018; pp. 611–627. [CrossRef]
14. Uhl, W.; Pirowski, Z.; Jaśkowiec, K. Trials to manufacture and test castings from Inconel alloy. *Arch. Foundry Eng.* **2010**, *10*, 73–78.
15. Detrois, M.; Rozman, K.; Jablonski, P.; Hawk, J. An alternative casting technique to improve the creep resistance of cast Inconel alloy 740H. *Metall. Mater. Trans.* **2020**, *51*, 3819–3831. [CrossRef]
16. Detrois, M.; Rozman, K.A.; Jablonski, P.D.; Hawk, J.A. Thermal processing design of cast Inconel Alloy 740H for improved mechanical performance. In Proceedings of the 9th International Symposium on Superalloy 718 & Derivatives: Energy, Aerospace, and Industrial Applications, Pittsburgh, PA, USA, 3–6 June 2018; pp. 829–846. [CrossRef]
17. Janowski, G.M.; Heckel, R.W.; Pletka, B.J. The effects of tantalum on the microstructure of two polycrystalline nickel-base superalloys: B1900+Hf and MAR-M247. *Metall. Trans. A* **1986**, *17*, 1891–1905. [CrossRef]
18. Rakoczy, Ł.; Rutkowski, B.; Grudzień-Rakoczy, M.; Cygan, R.; Ratuszek, W.; Zielińska-Lipiec, A. Analysis of $\gamma'$ precipitates, carbides and nano-borides in heat-treated Ni-based superalloy using SEM, STEM-EDX, and HRSTEM. *Materials* **2020**, *13*, 4452. [CrossRef]
19. Patel, S.J.; Smith, G.D. The role of niobium in wrought superalloys. In Proceedings of the International Symposium Niobium 2001, Orlando, FL, USA, 2–5 December 2001; pp. 1081–1108.
20. Szczotok, A.; Chmiela, B. Effect of heat treatment on chemical segregation in CMSX-4 nickel-base superalloy. *J. Mater. Eng. Perform.* **2014**, *23*, 2739–2747. [CrossRef]
21. Ojo, O.A.; Richards, N.L. On incipient melting during high temperature heat treatment of cast Inconel 738 superalloy. *J. Mater. Sci.* **2004**, *39*, 7401–7404. [CrossRef]
22. Xinxu, L.; Chonglin, J.; Yong, Z.; Shaomin, L.; Zhouhua, J. Segregation and homogenization for a new nickel-based superalloy. *Vacuum* **2020**, *177*, 109379. [CrossRef]
23. Du, Y.; Yang, Y.; Diao, A.; Li, Y.; Wang, X.; Zhou, Y.; Li, J.; Sun, X. Effect of solution heat treatment on creep properties of a nickel-based single crystal superalloy. *J. Mater. Res. Tech.* **2021**, *15*, 4702–4713. [CrossRef]
24. Grudzień-Rakoczy, M.; Rakoczy, Ł.; Cygan, R.; Kromka, F.; Pirowski, Z.; Milkovič, O. Fabrication and characterization of the newly developed superalloys based on Inconel 740. *Materials* **2020**, *13*, 2362. [CrossRef]
25. Walter, C.; Hallstedt, B.; Warnken, N. Simulation of the solidification of CMSX-4. *Mater. Sci. Eng. A* **2005**, *397*, 385–390. [CrossRef]
26. Sims, C.T.; Stoloff, N.S.; Hagel, W.C. *Superalloys II. High-Temperature Materials for Aerospace and Industrial Power*; John Wiley & Sons: New York, NY, USA, 1987.
27. Baeslack, W.A., III; West, S.L.; Kelly, T.J. Weld cracking in Ta-modified cast Inconel 718. *Scri. Metall.* **1988**, *22*, 729–734. [CrossRef]
28. Komura, Y.; Tokunaga, K. Structural studies of stacking variants in Mg-base Friauf-Laves phases. *Acta Cryst. Sect. B Struct. Cryst. Cryst. Chem.* **1980**, *36*, 1548–1554. [CrossRef]
29. Cieslak, M.J.; Knorovsky, A.; Headley, T.J.; Romig Jr, A.D. The solidification metallurgy of Alloy 718 and other Nb-containing superalloys. In *Superalloy 718—Metallurgy and Applications*; The Minerals, Metals and Materials Society: Pittsburgh, PA, USA, 1989; pp. 59–68.
30. Radhakrishna, C.; Prasad Rao, K. The formation and control of Laves phase in superalloy 718 welds. *J. Mater. Sci.* **1997**, *32*, 1977–1984. [CrossRef]
31. Kruk, A.; Cempura, G.; Lech, S.; Wusatowska-Sarnek, A.; Czyrska-Filemonowicz, A. Application of analytical electron microscopy and tomographic techniques for metrology and 3d imaging of microstructural elements in Allvac®718Plus™. In Proceedings of the 9th international symposium on Superalloy 718 and derivatives: Energy, aerospace and industrial applications, Pittsburgh, PA, USA, 3–6 June 2018; pp. 1035–1050. [CrossRef]
32. Laves, F.; Wallbaum, H.J. Die kristallstruktur von $Ni_3Ti$ und $Si_2Ti$. Zeitsch. für Kristall. *Cryst. Mater.* **1939**, *101*, 78–93. [CrossRef]

3. Bouse, G.K. Eta (η) and Platelet Phases in Investment Cast Superalloys. In Proceedings of the 8th International Symposium Superalloys 1996, Pittsburgh, PA, USA, 22–26 September 1996; pp. 163–172.

4. Seo, S.M.; Kim, I.S.; Lee, J.H.; Jo, C.Y.; Miyahara, H.; Ogi, K. Eta phase and boride formation in directionally solidified Ni-base superalloy IN792+Hf. *Metall. Mater. Trans. A* **2007**, *38*, 883–893. [CrossRef]

5. Willemin, P.; Durrand-Charre, M. Phase equilibria in multicomponent alloy systems. In Proceedings of the 6th International Symposium Superalloys 1988, Pittsburgh, PA, USA, 18–22 September 1988; pp. 723–732.

6. Kruk, A.; Cempura, G.; Lech, S.; Czyrska-Filemonowic, A. STEM-EDX and FIB-SEM tomography of Allvac 718Plus superalloy. *Arch. Metall. Mater.* **2016**, *61*, 535–542. [CrossRef]

7. Tan, Y.G.; Liu, F.; Zhang, A.-W.; Han, D.-W.; Yao, X.-Y.; Zhang, W.-W.; Sun, W.-R. Element Segregation and Solidification Behavior of a Nb, Ti, Al Co-Strengthened Superalloy 151. *Act. Metall. Sin.* **2019**, *32*, 1298–1308. [CrossRef]

8. Beattie, H.J.; Versnyder, F.L. A new complex phase in a high-temperature alloy. *Nature* **1956**, *178*, 208–209. [CrossRef]

9. Powell, D.J.; Pilkington, R.; Miller, D.A. The precipitation characteristics of 20%Cr/25% Ni-Nb stabilised stainless steel. *Acta Metall.* **1988**, *36*, 713–724. [CrossRef]

0. Xie, X.; Wu, Y.; Chi, C.; Zhang, M. Superalloys for Advanced Ultra-Super-Critical Fossil Power Plant Application. In *Superalloys*; Aliofkhazraei, M., Ed.; IntechOpen: London, UK, 2015; pp. 51–76. [CrossRef]

1. Ecob, R.C.; Lobb, R.C.; Kohler, V.L. The formation of G-phase in 20/25 Nb stainless steel AGR fuel cladding alloy and its effect on creep properties. *J. Mater. Sci.* **1987**, *22*, 2867–2880. [CrossRef]

2. Brager, A. An X-ray examination of titanium nitride III. Investigation by the powder method. *Acta Physicoch. URSS* **1939**, *11*, 617–632.

3. Sobczak, N.; Pirowski, Z.; Purgert, R.; Uhl, W.; Jaśkowiec, K.; Boroń, Ł.; Pysz, S.; Sobczak, J. Castability of Haynes 282 alloy. In Proceedings of the Advanced Ultrasupercritical Coal-fired Power Plants, Vienna, Austria, 12–20 September 2012; pp. 1–24.

4. Cockcroft, S.L.; Degawa, T.; Mitchell, A.; Tripp, D.W.; Schmalz, A. Inclusion precipitation in superalloys. In Proceedings of the 7th International Symposium Superalloys 1988, Pittsburgh, PA, USA, 20–24 September 1992; pp. 577–586.

5. Miracle, D.B. Overview No. 104 The physical and mechanical properties of NiAl. *Acta Metall. Mater.* **1993**, *41*, 649–684. [CrossRef]

6. Song, W.; Wang, X.G.; Li, J.G.; Meng, J.; Duan, T.F.; Yang, Y.H.; Liu, J.L.; Liu, J.D.; Pei, W.L.; Zhou, Y.Z.; et al. The formation and evolution of NiAl phase in a fourth generation nickel-based single crystal superalloy. *J. All. Comp.* **2020**, *848*, 156584. [CrossRef]

7. Cao, K.; Yang, W.; Zhang, J.; Liu, C.; Qu, P.; Su, H.; Zhang, J.; Liu, L. Solidification characteristics and as-cast microstructure of a Ru-containing nickel-based single crystal superalloy. *J. Mater. Res. Tech.* **2021**, *11*, 474–486. [CrossRef]

8. Tan, X.; Tan, X.; Liu, J.; Jin, T. Precipitation of β-NiAl/Laves eutectics in a Ru-containing single crystal Ni-based superalloy. *Met. Mater. Inter.* **2015**, *21*, 222–226. [CrossRef]

9. Lopez-Galilea, I.; Zenk, C.; Neumeier, S.; Huth, S.; Theisen, W.; Göken, M. The thermal stability of intermetallic compounds in an as-cast SX Co-base superalloy. *Adv. Eng. Mater.* **2015**, *17*, 741–747. [CrossRef]

0. Yu, L.; Zhao, Y.; Yang, S.; Sun, W.; Guo, S.; Sun, X.; Hu, Z. As-cast microstructure and solidification behavior of a high Al- and Nb containing superalloy. *J. Mater. Sci.* **2010**, *45*, 3448–3456. [CrossRef]

1. Yuan, L.; Xiong, J.; Zhao, W.; Shi, J.; Li, J. Transient liquid phase bonding of Ni3Al based superalloy using Mn-Ni-Cr filler. *J. Mater. Res. Tech.* **2021**, *11*, 1583–1593. [CrossRef]

2. Pathare, V.; Michal, G.M.; Vedula, K. Analysis of NiAlTa precipitates in β-NiAl + 2 at.% Ta alloy. *Scri. Metall.* **1987**, *21*, 283–288. [CrossRef]

3. Cotton, J.D.; Noebe, R.D.; Kaufman, M.J. The effects of chromium on NiAl intermetallic alloys: Part II. Slip systems. *Intermetallics* **1993**, *1*, 117–126. [CrossRef]

4. Wyckoff, R.W.G. *Crystal Structures 1*; Interscience Publishers: New York, NY, USA, 1963.

5. Tian, W.H.; Han, C.S.; Nemoto, M. Precipitation of α-Cr in B2-ordered NiAl. *Intermetallics* **1999**, *7*, 59–67. [CrossRef]

6. Pyczak, F.; Devrient, B.; Mughrabi, H. The effects of different alloying elements on the thermal expansion coefficients, lattice constants and misfit of nickel-based superalloys investigated by X-ray diffraction. In Proceedings of the 10th International Symposium Superalloys, Pittsburgh, PA, USA, 19–23 September 2004; pp. 827–836.

7. Rakoczy, Ł.; Milkovič, O.; Rutkowski, B.; Cygan, R.; Grudzień-Rakoczy, M.; Kromka, F.; Zielińska-Lipiec, A. Characterization of $\gamma'$ precipitates in cast Ni-based superalloy and their behaviour at high-homologous temperatures studied by TEM and in Situ XRD. *Materials* **2020**, *13*, 2397. [CrossRef]

8. Rakoczy, Ł.; Grudzień-Rakoczy, M.; Hanning, F.; Cempura, G.; Cygan, R.; Andersson, J.; Zielińska-Lipiec, A. Investigation of the $\gamma'$ precipitates dissolution in a Ni-based superalloy during stress-free short-term annealing at high homologous temperatures. *Metall. Mater. Trans. A* **2021**, *52*, 4767–4784. [CrossRef]

9. Rakoczy, Ł.; Cempura, G.; Kruk, A.; Czyrska-Filemonowicz, A.; Zielińska-Lipiec, A. Evolution of $\gamma'$ morphology and $\gamma/\gamma'$ lattice parameter misfit in a nickel-based superalloy during non-equilibrium cooling. *Int. J. Mater. Res.* **2019**, *110*, 66–69. [CrossRef]

0. Grosdidier, T.; Hazotte, A.; Simon, A. Precipitation and dissolution processes in $\gamma/\gamma'$ single crystal nickel-based superalloys. *Mater. Sci. Eng. A* **1998**, *256*, 183–196. [CrossRef]

1. Trexler, M.D.; Church, B.C.; Sanders, T.H., Jr. Determination of the Ni3(Ti, Al) dissolution boundary in a directionally solidified superalloy. *Scri. Mater.* **2006**, *55*, 561–564. [CrossRef]

2. Sponseller, D.L. Differential thermal analysis of nickel-base superalloys. In Proceedings of the 8th International Symposium Superalloys 1996, Pittsburgh, PA, USA, 22–26 September 1996; pp. 259–270.

63. Liu, L.; Zhen, B.L.; Banerji, A.; Reif, W.; Sommer, F. Effect of melt homogenization temperature on the cast structures of IN 738 LC superalloy. *Scri. Metall. Mater.* **1994**, *30*, 593–598. [CrossRef]
64. Zlá, S.; Dobrovská, J.; Smetana, B.; Žaludová, M.; Vodárek, V.; Konečná, K. Differential thermal analysis and phase analysis of nickel based superalloy IN 738LC. In Proceedings of the 19th International Conference on Metallurgy and Materials, Rožnov pod Radhoštěm, Roznov pod Radhostem, Czech Republic, 18–20 May 2010; pp. 790–795.

*Article*

# Analysis of $\gamma'$ Precipitates, Carbides and Nano-Borides in Heat-Treated Ni-Based Superalloy Using SEM, STEM-EDX, and HRSTEM


**Łukasz Rakoczy [1,*], Bogdan Rutkowski [1], Małgorzata Grudzień-Rakoczy [2], Rafał Cygan [3], Wiktoria Ratuszek [1] and Anna Zielińska-Lipiec [1]**

1 Faculty of Metals Engineering and Industrial Computer Science, AGH University of Science and Technology, Mickiewicza 30, 30-059 Kraków, Poland; rutkowsk@agh.edu.pl (B.R.); ratuszek@agh.edu.pl (W.R.); alipiec@agh.edu.pl (A.Z.-L.)

2 Łukasiewicz Research Network-Kraków Institute of Technology, Zakopiańska 73, 30-418 Kraków, Poland; malgorzata.grudzien@kit.lukasiewicz.gov.pl

3 Investment Casting Division, Consolidated Precision Products Corporation, Hetmańska 120, 35-078 Rzeszów, Poland; Rafal.Cygan@cppcorp.com

* Correspondence: lrakoczy@agh.edu.pl



Received: 20 September 2020; Accepted: 5 October 2020; Published: 8 October 2020



**Abstract:** The microstructure of a René 108 Ni-based superalloy was systematically investigated by X-ray diffraction, light microscopy, energy-dispersive X-ray spectroscopy, and electron microscopy techniques. The material was investment cast in a vacuum and then solution treated (1200 °C-2h) and aged (900 °C-8h). The $\gamma$ matrix is mainly strengthened by the ordered $L1_2$ $\gamma'$ phase, with the mean $\gamma/\gamma'$ misfit, $\delta$, +0.6%. The typical dendritic microstructure with considerable microsegregation of the alloying elements is revealed. Dendritic regions consist of secondary and tertiary $\gamma'$ precipitates. At the interface of the matrix with secondary $\gamma'$ precipitates, nano $M_5B_3$ borides are present. In the interdendritic spaces additionally primary $\gamma'$ precipitates, MC and nano $M_{23}C_6$ carbides were detected. The $\gamma'$ precipitates are enriched in Al, Ta, Ti, and Hf, while channels of the matrix in Cr and Co. The highest summary concentration of $\gamma'$-formers occurs in coarse $\gamma'$ surrounding MC carbides. Borides $M_5B_3$ contain mostly W, Cr and Mo. All of MC carbides are enriched strongly in Hf and Ta, with the concentration relationship between these and other strong carbide formers depending on the precipitate's morphology. The nano $M_{23}C_6$ carbides enriched in Cr have been formed as a consequence of phase transformation MC + $\gamma \rightarrow M_{23}C_6 + \gamma'$ during the ageing treatment.



**Keywords:** superalloy; HRSTEM; STEM-EDX; $M_{23}C_6$; nano-borides


## 1. Introduction

Ni-based alloys used for working at high homologous temperatures are the most advanced metallic construction materials. The unique combination of high strength and excellent oxidation resistance makes it an irreplaceable group of materials in aerospace, power, and nuclear industries. The usefulness of Ni-based superalloys during long-term service at harsh conditions depends strongly on the alloying elements, their concentrations, and the morphology of the main strengthening phases [1–4]. The development of the first Ni-based superalloys began over 100 years ago. It happened around 1918 with the patenting of the Ni-20Cr alloy (Nichrome) [5,6]. It became a progenitor for the next superalloys of the Nimonic and Inconel series [7]. In 1929, Bedford, Pilling, and Merica presented the results of the study on the influence of Al and Ti on the mechanical properties of the Nichrome superalloy [1,5]. They noticed that the addition of these elements significantly increases creep resistance. Despite difficulties in observing fine precipitates responsible for the strengthening, several subsequent Ni-based superalloys were developed [8]. In the following years, some of the nickel content was replaced with cobalt, thus increasing the high-temperature oxidation resistance. One of the main representatives of Ni-Cr-Co alloys was Nimonic 90 with 20% of cobalt [9]. In the 1930's the gas turbine design by Elling had an operating temperature which did not exceed 550 °C [10]. There was no huge demand for Ni-based superalloys because the austenitic stainless steels could still work safely at this temperature. Several years later engines constructed by Heinkel and Whittle led to breakthrough in the Ni-superalloys' development because the service temperature reached 780 °C [11,12]. Conventional stainless steels were not able to meet the growing requirements [13]. During the Second World War superalloys were therefore modified by introducing even higher amounts of aluminium and titanium. Meanwhile, at the end of the 1940s, it was noticed that the additions of refractory elements, like Mo, contribute to a significant solution and precipitation strengthening with carbides. In the 1940s, engine components were manufactured by forging and air casting and so the strength and metallurgical purity were not the highest. The introduction of vacuum casting technology in 1952 was considered to be the most essential step towards improving the mechanical properties and purity of both cast and wrought superalloys [4,14]. Before the advent of vacuum melting and casting techniques, cast superalloys did not gain wide acceptance for a turbine blade application. One of the first alloys to be vacuum cast was Inconel 713C and this became widely used in the aerospace industry. The high carbon content contributed to excessive primary carbides, poor ductility at low temperatures and problems with integrally bladed disc castings. To overcome this issue, a low carbon version of the alloy was made and is also widely used for turbine blades [14–16]. In the late 1950s, H.L. Eiselstein developed Inconel 718 precipitation strengthened by intermetallic phase $\gamma''$ ($Ni_3Nb$), which, along with the Waspaloy, was quickly put into continuous production [4,17]. In the 1960s, it began to be seen that chromium could be partially replaced to achieve even higher creep resistance. However, limiting its concentration sensitized superalloys to high-temperature corrosion and so more attention was paid to its influence on microstructure and properties [4,18]. Numerous chemical composition modifications broadened the knowledge of metallurgists and allowed to develop new grades of superalloys, but also this was associated with design faults. Excessive amounts of alloying elements led to the formation of detrimental topologically closed phases (TCP) [19]. These undesired results indicated that in such complex materials design, it is necessary to use additional tools, computer software [20]. In the late 1960s, the strengthening of superalloys by only increasing Al and Ti became impractical. Taking into account only weight fraction, they were not the dominant alloying elements. Their too high concentration prevents the production of complex castings due to brittleness and so Al + Ti concentration usually does not exceed 8% [21,22]. Thanks to lowering of the Ti concentration, the castability was considerably improved, which consequently led to design of the new superalloys with low Ti/Al ratio [23]. Additionally, the Mo and Ta concentrations were increased in order to strengthen the $\gamma$-matrix. The proposal of molybdenum replacement by tungsten was adopted during the development of the MAR M200 by Martin Metals Company. Too high creep rate led to premature failure, and in response to this result, the superalloy was modified by hafnium, and so the MAR M002 was introduced [14,24]. Hf leads to fine and stable MC carbides' formation. This contributed to the modification of many previously developed superalloys like Inconel 713C (to MM-004) and B1900 (to MM-007) [25,26]. Another, approach to overcoming the intermediate temperature ductility problem in cast superalloys has been the development of the so-called BC superalloys [14,27]. Maxwell [28] found that the high boron content (0.1–0.16%) with a low carbon concentration guarantees the improvement of the creep resistance of MAR M200 superalloy at 760 °C, without reducing other properties. Quichou [29] confirmed that BC superalloys were characterized by better castability properties than IN792, IN792+Hf, and IN100. This impact was the result of borides' formation in the grain boundaries. A similar effect was obtained in many other superalloys, thanks to which boron is still an essential alloying element [30–32]. In the following years, the design of polycrystalline alloys assumed the use of all previously known mechanisms. After several decades, the archetype of the superalloys, two-component Nichrome, was modified to such an extent that the alloying elements exceeded twelve. The best examples of this are the microstructurally complex superalloys as MAR-M247 and Inconel 792+Hf [33–35].

To understand the behaviour of the especially complex Ni-based superalloys at high-homologous temperatures it is necessary to characterize their microstructures in fully heat-treatment condition. Detection and characterization of precipitates, even nano-precipitates, is essential in the study of creep and fatigue degradation. Without the knowledge about the phase composition of material, it is problematic to predict the microstructure evolution during long-term service. The main aim of this work was to perform the comprehensive characterization of equiaxed Ni-based superalloy René 108 from micro- to atomic-scale resolution by using analytical microscopy techniques.

## 2. Experimental Procedure

The experimental material was a René 108 Ni-based superalloy casting. It was heat treated in a vacuum with the following parameters: (a) solution: 1200 °C-2h; (b) ageing: 900 °C-8h. Chemical composition designated by optical emission spectroscopy (OES) is presented in Table 1.

**Table 1.** Chemical composition of René 108 determined by optical emission spectroscopy (OES), wt%.

| Element | Cr | W | Co | Al | Ta | Hf | Ti | Mo | C | B | Zr | Ni |
|---|---|---|---|---|---|---|---|---|---|---|---|---|
| Concentration | 11.4 | 8.9 | 8.2 | 6.4 | 3.6 | 1.5 | 0.8 | 0.5 | 0.08 | 0.02 | 0.0125 | Bal. |

Microstructural observations and analyses were performed using: thermodynamic simulations, X-ray diffraction (XRD), light microscopy (LM), scanning electron microscopy (SEM), scanning-transmission electron microscopy (STEM), high-resolution scanning-transmission electron microscopy (HRSTEM), dispersive X-ray spectroscopy (EDX). Thermo-Calc software (ThermoCalc Software AB, version 2020a, Solna, Sweden) (database TCNI6:Ni-Alloys) was applied to calculate the maximum solubility of selected alloying elements (Cr, Co, W, Al, Ti, Ta, Hf, Mo, B, C) in the Ni-matrix at 1200 °C and 900 °C. XRD study was on a Siemens/Bruker D5005 diffractometer using Co $K_\alpha$ ($\lambda$ = 1.79 Å) radiation source in Bragg-Brentano geometry. The samples were wire cut from the part into a plate with dimensions 20×20×10 mm and then were ground and polished on a diamond suspension. The XRD patterns were acquired at a scan rate of 0.04 °/s with 2θ (Bragg angle) and a scan range from 20° to 130°. The angles were read off from the positions of the peaks on a diffractogram and the interplanar spacings $d_{khl}$ were calculated using the Bragg-Wulff equation (Equation (1)). Deconvolution for the double-peaks was performed by the least-square method. Based on the computed lattice $a_\gamma$ and $a_{\gamma'}$ parameters (Equation (2)), the $\gamma/\gamma'$ misfit coefficient (Equation (3)) was determined.

$$d_{hkl} = \frac{\lambda}{2\sin\theta} \tag{1}$$

$$a = \frac{\lambda\sqrt{h^2+k^2+l^2}}{2\sin\theta} \tag{2}$$

$$\delta = \frac{2(a_{\gamma'} - a_\gamma)}{a_{\gamma'} + a_\gamma} \tag{3}$$

The samples (30×30×30 mm) for LM and SEM observations, cut from the investment casting, were ground, polished (diamond suspensions: 3 µm, 1 µm; colloidal silica: 0.25 µm), and electrochemically etched in 10% oxalic acid ($C_2H_2O_4$). Initial characterization of morphology and distribution of precipitates was carried out on a Leica light microscope (Leica Microsystems, Wetzlar, Germany). For SEM-EDX analyses, 20 kV accelerating voltage was set on a Phenom XL (Phenom-World, Eindhoven, Netherlands) apparatus equipped with Energy Dispersive X-ray Spectroscopy detector. Segregation coefficient (concentration of alloying element "i" in the center of dendrite arm divided by concentration in the interdendritic space, determined by SEM-EDX, was calculated in accordance with

Equation (4), based on the 20 measurements in various locations (each area 10 × 10 μm). Quantitative chemical analysis was carried out using the ZAF correction.

$$k^i = \frac{C^i_D}{C^i_{ID}} \tag{4}$$

where:

$C^i_D$—concentration of "*i*" alloying element in dendritic region, at%
$C^i_{ID}$ —concentration of "*i*" alloying element in interdendritic space, at%

Image analysis of the precipitates was performed by ImageJ commercial software (National Institutes of Health and the Laboratory for Optical and Computational Instrumentation, University of Wisconsin, 1.51j8, Madison, WI, USA). The relative volume of carbides and $\gamma'$ precipitates was determined by the planimetric method. According to the Cavalieri-Hacquert principle, the estimator of the relative volume occupied by a given component of the alloy structure is the fraction of the area occupied by this component on the unit plane of the sample:

$$V_v = A_A \tag{5}$$

$$A_A = \left(\frac{\Sigma A_i}{A}\right) * 100\% \tag{6}$$

where:

$V_v$—total volume of the phase object per unit volume of the alloy, $\mu m^3/\mu m^3$
$A_A$—total field flat sections on the individual phase of the image per unit area, $\mu m^2/\mu m^2$
$A_i$—total field of flat sections on the individual i-phase, $\mu m^2$
$A$—total image area, $\mu m^2$

For the calculation of carbides' volume fraction, the 5 images captured by LM under magnification x100 were used (area 1.18 $mm^2$). In the case of $\gamma'$ 10 images were used at ×20000 magnification (area 179.8 $\mu m^2$). All images were put to binarisation and a despeckle filter, which removed noise without blurring edges. The area (A) and perimeter (P) of each $\gamma'$ precipitate (from the area range 0.06–5.0 $\mu m^2$) were measured to calculate the distribution of $\gamma'$ shape factor $\xi = (4\Pi^*A)/(P^2)$. The mean size of the $\gamma'$ was represented by the equivalent radius for the area. Stereological analysis for $\gamma'$ precipitates was carried for more than 1650 precipitates in a dendritic region. Specimens for STEM and HRSTEM investigation were prepared by ion polishing. Slices were first ground mechanically to a thickness of about 0.05 mm, and then 3 mm discs were punched from these ground samples and dimpled using a Gatan Dimple Grinder (Gatan, Inc., Pleasanton, CA, USA) on each side. The last step was thinning by $Ar^+$ ion beam (PIPS of Gatan). Before loading into the microscope, the thin foils were cleaned by a plasma cleaner for removing surface contamination. Investigations were performed using a probe Cs-corrected FEI Titan Cubed G-2 60-300 (Thermo Fisher Scientific, Eindhoven, Netherlands) with ChemiSTEM system operating at 300 kV. Images in atomic scale resolution were analyzed by Fast Fourier Transformation (FFT) to obtain position of diffraction peaks. The patterns were solved with JEMS software (JEMS-SWISS, Jongny, Switzerland).

## 3. Results and Discussion

### 3.1. Calculation of the Alloying Elements Solubility in Nickel by Thermo-Calc

Table 2 summarizes atomic radius [36] and solubility data for various elements in Ni at 1200 °C and 900 °C calculated by thermodynamic simulations with Thermo-Calc. It is helpful information in the phase analysis of Ni-based superalloy by advanced electron microscopy. Co and Ni exhibit complete solid solution in the face-centered cubic structure. Besides these, the elements characterized by the highest solubility in the Ni matrix include Cr and Mo. At 1200 °C it is 49.53 at% and 25.97 at%, respectively. As the temperature drops to 900 °C, the solubility of Cr and Mo in $\gamma$ matrix decreases, however, these values are still high. The main strengthening phase in cast nickel superalloys is $\gamma'$ with the stoichiometric formula $Ni_3Al$. The solubility of Al in Ni for cooling 1200 °C → 900 °C changes from 18.38 at% to 14.38 at%, while Al position in the unit cell can be replaced by other elements as Ti and Ta. Hf is characterized by remarkably low solubility in nickel, which drops from 1.24 at% at 1200 °C to just 0.26 at% at 900 °C. In contrast, carbon causes Hf, as also Ta, Ti, and W to form carbides. From all alloying elements, boron has the lowest solubility in Ni, both at 1200 °C (0.36 at%) and 900 °C (0.019 at%). If its content in the material exceeds the maximum solubility in Ni, it creates favorable conditions for borides' formation. To summarise, the limited solubility in Ni of several alloying elements results in strengthening phases in the René 108 microstructure, i.e., intermetallic phases, carbides and borides.

**Table 2.** Solubility of selected alloying elements in nickel at 1200 °C and 900 °C. Calculated by Thermo-Calc.

| Element | Radius, Å | Solubility in Nickel, wt%/at% | |
|---|---|---|---|
| | | 1200 °C | 900 °C |
| Ni | 1.49 | | |
| Cr | 1.66 | 46.50/49.53 | 38.49/41.40 |
| Co | 1.52 | 100/100 | 100/100 |
| W | 1.93 | 39.01/16.96 | 35.31/14.84 |
| Al | 1.18 | 9.38/18.38 | 7.17/14.38 |
| Ta | 2.00 | 20.19/7.58 | 12.87/4.57 |
| Hf | 2.08 | 3.67/1.24 | 0.78/0.26 |
| Ti | 1.76 | 11.98/14.30 | 8.58/10.32 |
| Mo | 1.90 | 36.45/25.97 | 27.77/19.04 |
| B | 0.87 | 0.067/0.36 | 0.019/0.10 |
| C | 0.67 | 0.426/2.05 | 0.162/0.79 |
| Zr | 2.06 | 1.047/0.68 | 0.591/0.38 |

### 3.2. *Characterization of Phase Compositions and $\gamma/\gamma'$ Lattice Misfit by XRD*

The XRD pattern for heat-treated René 108 is presented in Figure 1. It reveals peaks of three phases: $\gamma$ matrix, intermetallic phase $\gamma'$ and MC carbides. In Table 3, the interplanar spacing $d_{hkl}$ and lattice parameters of detected phases are shown as measured on the XRD pattern. The mean lattice parameter of the matrix and precipitates determined using analyzed reflections were 3.59 Å (±0.01 Å) and 3.60 Å (±0.02 Å), respectively. The mean misfit coefficient calculated by Equation. (3) based on all double peaks is positive, $\delta = +0.6\%$ (±0.5%). Note that the misfit value is an average value because of the relatively high level of alloying elements segregating in the dendritic structure of equiaxed superalloys. Lattice misfit is related to the morphology of precipitates. Superalloys with spherical precipitates are characterized by misfit ±0–0.2%, and with cubic ±0.5–1% and plate-like precipitates > ±1.25% [37]. Lattice misfit has long been associated with alloy strengthening, and so low misfit is preferred to withstand coarsening. Typical heat treatment (solution + ageing) is generally carried out to favour cubic morphology of $\gamma'$ precipitates.

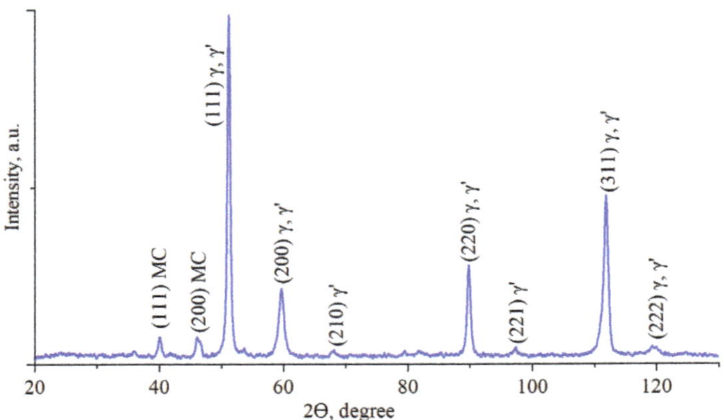


**Figure 1.** XRD pattern of René 108 (Co $K_\alpha$ radiation), λ = 1.79 Å.

**Table 3.** The measured values of $d_{hkl}$ spacing and lattice parameters for γ, γ′, MC carbides in the René 108.

| (hkl) | $d_{hkl}$, Å | a, Å |
|---|---|---|
| | γ matrix FCC Fm$\bar{3}$m (225): PDF No. 47-1417 | |
| (111) | 2.07 | 3.59 |
| (200) | 1.80 | 3.60 |
| (220) | 1.27 | 3.58 |
| (311) | 1.08 | 3.58 |
| (222) | 1.04 | 3.59 |
| | γ′ FCC ordered $L1_2$: PDF No. 65-3245 | |
| (111) | 2.09 | 3.62 |
| (200) | 1.81 | 3.63 |
| (210) | 1.60 | 3.58 |
| (220) | 1.27 | 3.59 |
| (221) | 1.20 | 3.60 |
| (311) | 1.08 | 3.60 |
| (222) | 1.04 | 3.60 |
| | MC FCC Fm$\bar{3}$m (225): PDF No. 74-1223 | |
| (111) | 2.61 | 4.52 |
| (200) | 2.26 | 4.52 |

### 3.3. Microstructure of René 108 Superalloy by LM and SEM

The unetched microstructure of the superalloy reveals the carbides' distribution and their morphology (Figure 2a). Their volume fraction is 0.88% (±0.16%). The high standard deviation resulted from their uneven distribution in the casting volume. Carbides are characterized by a complex morphology, namely small regular blocks, elongated parallelograms with smooth edges, and "Chinese script". The last one usually consisted of three separate parts: the central core terminated with an angular head and perpendicularly arranged elongated arms. The core and arms were made of fine, irregular carbides, unlike thick plates and rods. Etching revealed the dendritic structure of the casting and significant local microsegregation of microstructural components (Figure 2b). Despite the heat treatment (solution + ageing), the microstructure is still very heterogeneous. The high content of alloying elements, with their limited solubility in the $\gamma$ matrix, has led to the formation of numerous constituents. The dendritic areas are characterized by a relatively homogeneous microstructure with $\gamma'$ precipitates of various sizes. In the interdendritic spaces, carbides and eutectic islands $\gamma$-$\gamma'$ are additionally present.

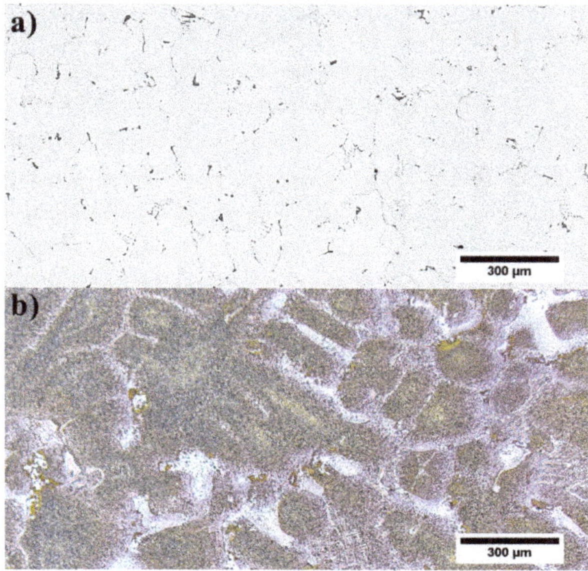


**Figure 2.** Microstructure of René 108: **a**) unetched; **b**) etched, light microscopy (LM).

Significant differences in the microstructure of dendritic areas in relation to interdendritic spaces indicate that segregation of the alloying elements is present. To show this quantitatively, SEM-EDX measurements in these both regions were carried out. The data were used to calculate the segregation coefficient $k^i = \frac{C^i_D}{C^i_{ID}}$ of each alloying element. The ratio of the concentration of an alloying element in the centerline of dendrite arms to the concentration in the interdendritic spaces was calculated and is presented in Figure 3. Results confirm the irregular distribution of elements. The $k^i$ values, when lower than one, indicate that the element segregates to interdendritic spaces. When $k^i > 1$, the element is more concentrated in the dendritic regions. Elements that strongly segregate to interdendritic spaces are Hf, Ta, Ti, and Al. When $k^i$ values are below one, corresponding to W, Co, and Cr, these elements are more concentrated in dendrites. The most uniform distribution is observed for Ni and Mo.

Using the backscattered detector, SEM images with a strong dependence on the average atomic number Z were obtained. The microstructure of the $\gamma'$ in dendritic regions is presented in Figure 4. Large secondary $\gamma'$ precipitates are characterised by a complex morphology. Locally at their edges there are single precipitates. During homogenization, secondary $\gamma'$ precipitates were dissolving in the matrix. Some of those that were not completely dissolved, coagulated, and coalesced. During cooling, they became nucleation sites for re-precipitating from the supersaturated $\gamma$ solution. The misfit stresses between the $\gamma$ and $\gamma'$ phases promoted nucleation in the stress fields of subsequent precipitates in these sites. Finer secondary $\gamma'$ precipitates had a cubic-like morphology. The differences in their morphology are due to the various temperatures and time ranges in which they formed, as well as the low coagulation rate during ageing. Inside the $\gamma$ channels, between secondary $\gamma'$ precipitates, spherical tertiary $\gamma'$ precipitates were revealed, formed independently from the supersaturated matrix during cooling after the ageing treatment. The volume fraction of the $\gamma'$ precipitates within the dendrites calculated using SEM-BSE images is 54.99% (±3.65%). The $\gamma'$ mean radius distribution presented in Figure 5a clearly reveals three classes which belong to large and fine secondary $\gamma'$ and also tertiary $\gamma'$. Based on the area and perimeter of each precipitate, its shape coefficient was calculated. All obtained values are summarised in Figure 5b. Two curves are fitted to the distribution. The green one corresponds to the large secondary $\gamma'$ precipitates and the second pink curve to cubic secondary and spherical tertiary $\gamma'$. According to the shape factor equation $\xi = (4\Pi^*\text{Area})/(\text{Perimeter}^2)$, the cubic precipitates have a coefficient of $\zeta$ 0.785, while the spherical $\zeta$ of 1.0. Variation in nature (size, shape, and volume fraction) of $\gamma'$ precipitates is observed among others due to the effects of microsegregation and local cooling rates during casting solidification.

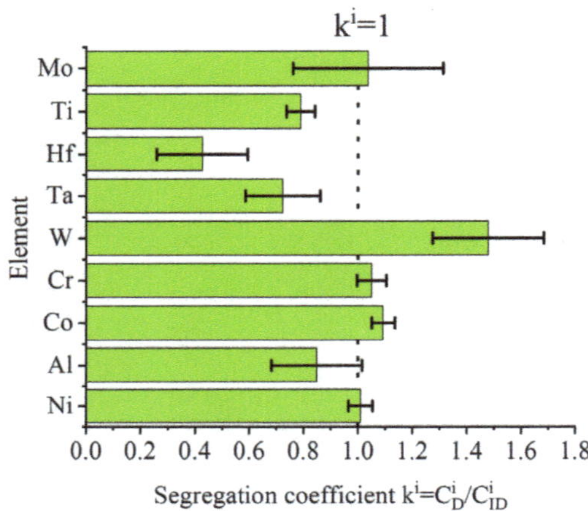


**Figure 3.** Segregation coefficient $k^i = \frac{C^i_D}{C^i_{ID}}$ of selected alloying elements.

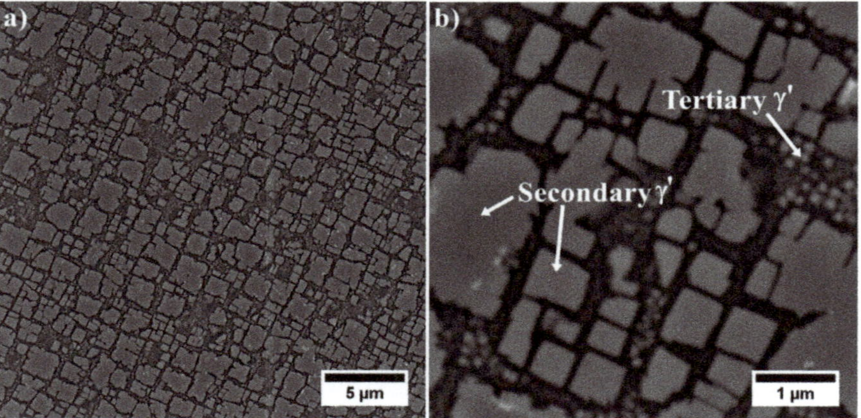


**Figure 4.** Microstructure of precipitates in a dendrite arm: (**a**) secondary $\gamma'$; (**b**) secondary and fine tertiary $\gamma'$.

In the interdendritic spaces, the primary, secondary, and tertiary $\gamma'$ precipitates are characterized by much greater variability in distribution and morphology (Figure 6). These factors do not allow us to compare effectively stereological parameters with precipitates inside the dendrites. The primary $\gamma'$ precipitates created at the end of casting's solidification, through the $L \rightarrow \gamma + \gamma'$ phase transformation have an especially complex morphology. It is suggested that this phase transformation was monovariant, namely progressing in the temperature range. The dissolution of large eutectics is challenging during industrial solid solution treatment, while increasing the solution temperature may lead to undesirable incipient melting. The eutectic $\gamma$-$\gamma'$ islands are an unwanted component in superalloys, but their complete dissolution in the matrix during solution treatment is not possible. However, the high solution temperature changed the thermodynamic conditions and lead to partial dissolution, which consequently influenced the differentiation of the size of the secondary precipitates of the $\gamma'$ phase around the eutectic $\gamma$-$\gamma'$ islands.

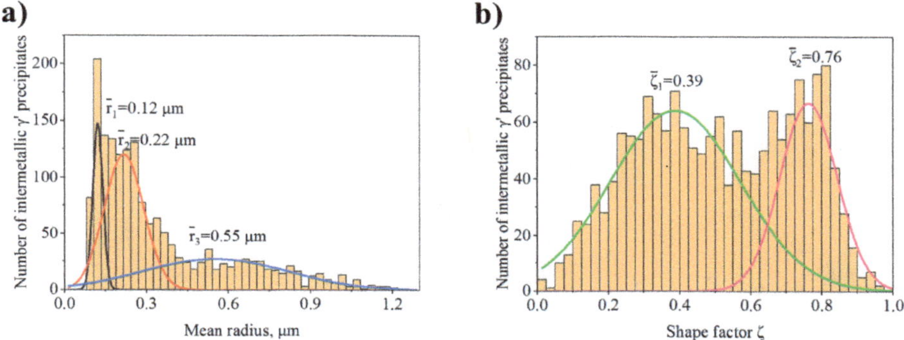


**Figure 5.** (**a**) the $\gamma'$ mean radius distribution; (**b**) shape coefficient $\zeta$ distribution.

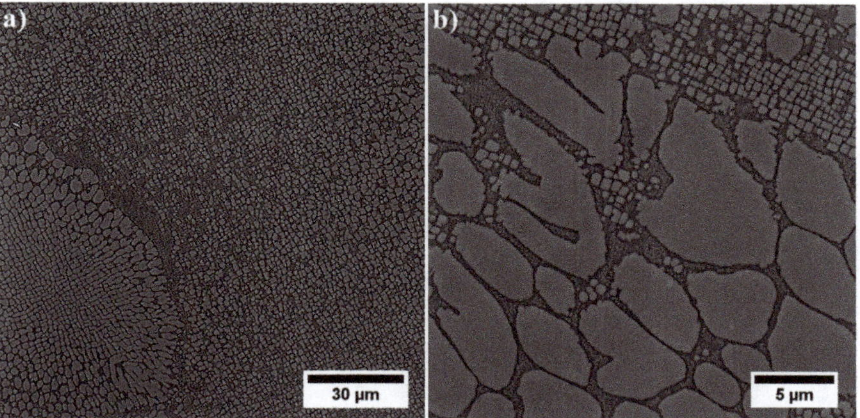


**Figure 6.** Microstructure of $\gamma'$ precipitates in interdendritic space: (**a**) difference in size and morphology of precipitates; (**b**) large primary $\gamma'$ precipitates.

MC precipitates appear as "white" precipitates and are concentrated in the interdendritic areas. For all of them, the Z-contrast is almost the same. Semi-quantitative SEM-EDX analyses of MC carbides were carried out to show the relationships between carbide formers, because it is widely known that carbides of metals belonging to the same class show considerable inter-solubility. Results from selected locations are presented in Figure 7 and Table 4. In chemical compositions of all carbides, strong carbide formers dominate, especially hafnium and tantalum. The mutual relation between these elements change, depending on the shape and size of a precipitate. The following relationships were calculated between the elements with the highest affinity for carbon: Ta/Hf, (Ta + Hf)/(W + Ti) and (Ta/Hf)/(W + Ti). Blocky shaped carbides with sharp edges located in the close vicinity of eutectic $\gamma$-$\gamma'$ islands are the only ones characterized by a low Ta/Hf ratio and a very high (Ta + Hf)/(W + Ti) (more than a dozen). The carbides surrounded by the coarse $\gamma'$ phase had the Ta/Hf and (Ta + Hf)/(W + Ti) coefficients in the range 2.0–3.0 and the (Ta/Hf)/(W + Ti) $\leq$ 0.1. The morphology of the last group of carbides is similar to the previously described, with one general difference, lack of surrounding coarse $\gamma'$. Relation Ta/Hf is 0.29–0.75, whereas (Ta + Hf)/(W + Ti) is in the range 4.75–8.06 and the (Ta/Hf)/(W + Ti) $\leq$ 0.05. An additional 40 measurements were performed to present graphically the concentration relationships between strong carbide formers in the carbides' composition (Figure 8). Relationships in the corners of the graph correspond with these in Table 4.

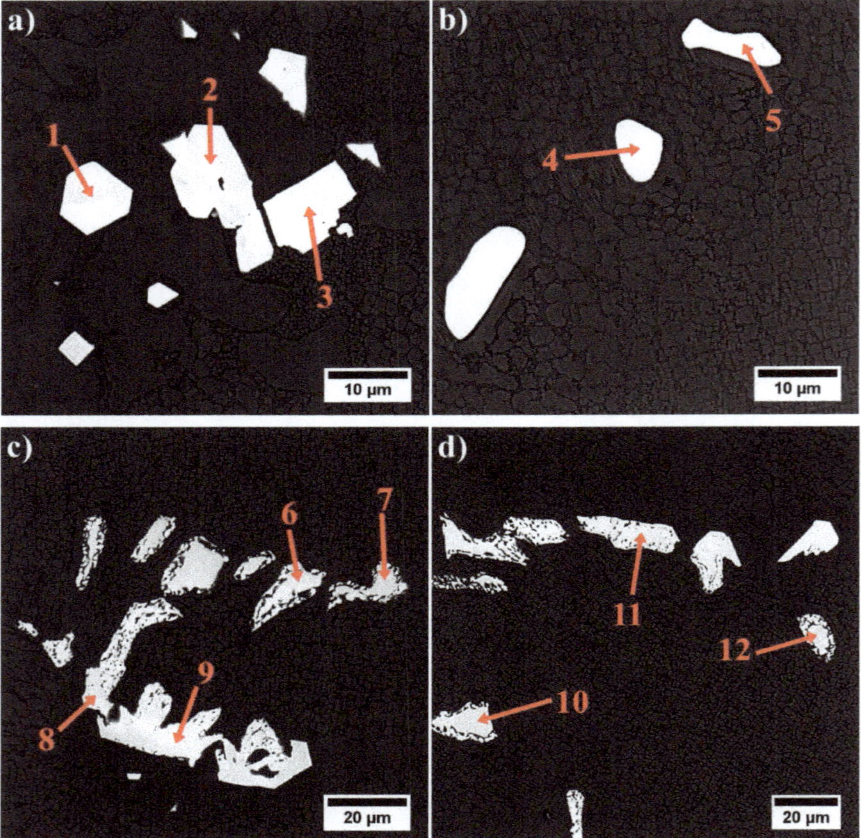


**Figure 7.** Microstructure of MC carbides and location of SEM-EDX analysis points: (**a**) blocky carbides and points no. 1–3; (**b**,**c**) carbides surrounded by the coarse $\gamma'$ and points no. 4–9; (**d**) morphologically complex carbides and points no. 10–12.

At a grain boundary, continuous or semi-continuous layers of fine precipitates with irregular morphology are observed (Figure 9a,b). Crosswise to the grain boundary, a linear analysis of the distribution of selected alloying elements in one of these precipitates was performed (Figure 9c,d). Enrichment in chromium and tungsten is revealed, and also slightly increased Mo content. Alloying elements like Cr, W, and Mo are generally known to have a high affinity for boron/carbon, and the solid solubility of boron/carbon in the $\gamma$ phase is very low (as presented in Table 2). It is likely that borides and/or secondary carbides can easily form at the grain boundaries in René 108.

**Table 4.** Results of SEM-EDX analysis in MC carbides, at%.

| No. | Ta | Hf | Ti | Ni | W | Co | Cr | Mo | Ta/Hf | (Ta + Hf)/(W + Ti) | (Ta/Hf)/(W + Ti) |
|---|---|---|---|---|---|---|---|---|---|---|---|
| 1 | 12.2 | 75.0 | 1.1 | 6.9 | 1.0 | 1.4 | 1.0 | 1.4 | 0.16 | 42.95 | 0.08 |
| 2 | 16.3 | 73.3 | 1.2 | 5.4 | 1.2 | 1.0 | 0.9 | 0.7 | 0.22 | 37.99 | 0.09 |
| 3 | 16.6 | 72.3 | 2.0 | 6.0 | 0.8 | 0.7 | 1.1 | 0.5 | 0.23 | 32.32 | 0.08 |
| 4 | 42.5 | 21.1 | 17.4 | 7.7 | 5.9 | 2.4 | 1.8 | 1.3 | 2.01 | 2.73 | 0.09 |
| 5 | 42.7 | 16.8 | 19.1 | 8.2 | 7.1 | 2.4 | 2.3 | 1.4 | 2.54 | 2.27 | 0.10 |
| 6 | 43.9 | 19.9 | 19.5 | 6.9 | 5.5 | 0.9 | 1.6 | 1.8 | 2.20 | 2.56 | 0.09 |
| 7 | 42.8 | 18.3 | 18.4 | 7.9 | 7.0 | 2.2 | 1.8 | 1.7 | 2.35 | 2.40 | 0.09 |
| 8 | 23.4 | 54.5 | 6.5 | 5.7 | 5.9 | 1.4 | 1.0 | 1.6 | 0.43 | 6.26 | 0.03 |
| 9 | 17.5 | 61.1 | 4.2 | 5.9 | 5.5 | 2.9 | 2.1 | 1.0 | 0.29 | 8.06 | 0.03 |
| 10 | 44.9 | 21.5 | 18.9 | 5.9 | 5.2 | 1.1 | 1.0 | 1.5 | 2.09 | 2.76 | 0.09 |
| 11 | 30.9 | 41.3 | 10.0 | 6.9 | 5.2 | 2.6 | 2.0 | 1.1 | 0.75 | 4.75 | 0.05 |
| 12 | 20.2 | 19.9 | 6.4 | 5.6 | 1.1 | 1.9 | 1.6 | 20.2 | 2.15 | 2.49 | 0.08 |

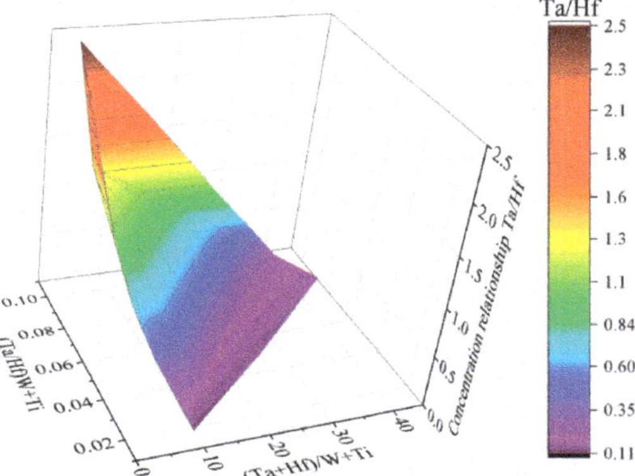


**Figure 8.** Graphical representation of selected concentration relationships between alloying elements in carbides. Based on the SEM-EDX point analysis.

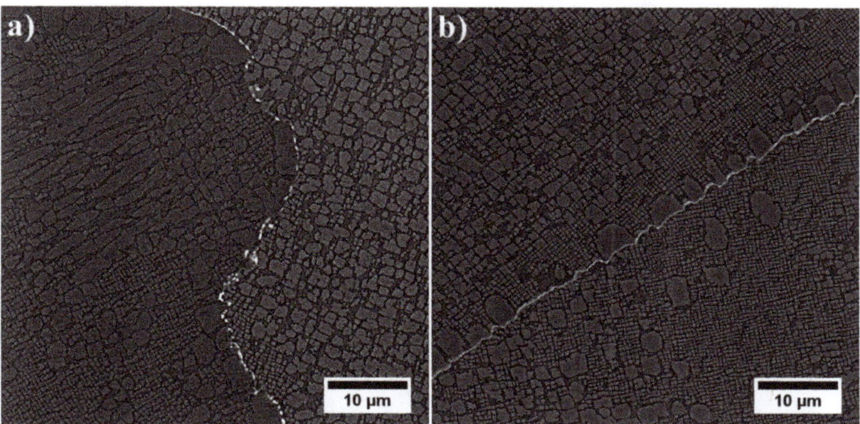


**Figure 9.** *Cont.*

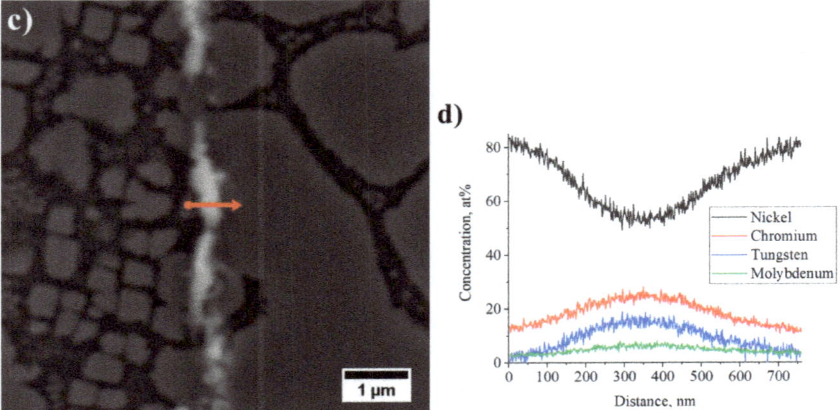


**Figure 9.** (**a**,**b**) microstructure of precipitates at the grain boundaries; (**c**) location of linear SEM-EDX analysis; (**d**) distribution of selected alloying elements at a grain boundary.

### *3.4. Characterization of Strengthening Phases by STEM and HRSTEM*

#### 3.4.1. Dendritic Regions

The $\gamma$ phase constituting the matrix of René 108 superalloy is strengthened mainly with the $\gamma'$ phase. In the narrow channels of the $\gamma$ matrix, between the large secondary $\gamma'$ precipitates, nanometric tertiary $\gamma'$ precipitates are present (Figure 10). The interfaces $\gamma/\gamma'$ are additionally occupied by bright nano-precipitates. The degree of strengthening and mechanical properties depends on the chemical composition of the precipitates and the matrix, as well as the mutual crystallographic relationship, which was also analysed.

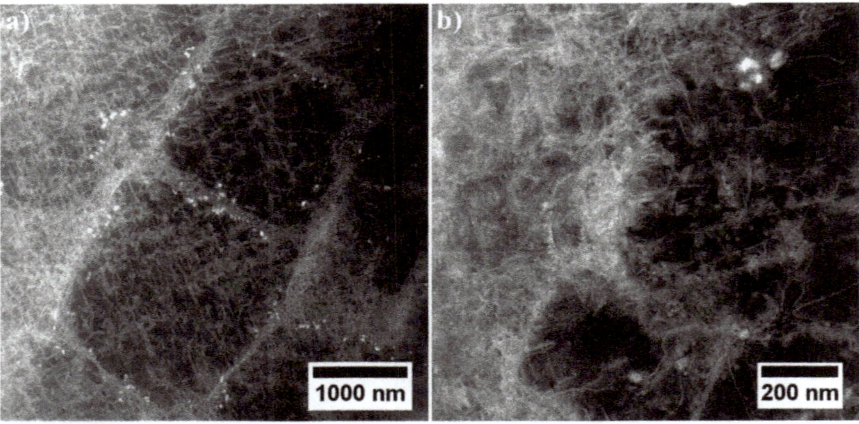


**Figure 10.** Microstructure of secondary $\gamma'$ with nano-precipitates at the $\gamma/\gamma'$ interface: (**a**) cubic-shaped $\gamma'$ precipitates; (**b**) complex-shaped $\gamma'$ precipitate, STEM-HAADF.

The STEM-HAADF microstructure of dendritic region contains nano-precipitates at the interfaces $\gamma/\gamma'$. Bright contrast indicates that they consist mainly of elements with higher atomic numbers Z than those in phases $\gamma$ and $\gamma'$ (Figure 10a). During observations of thin foils at higher magnification, it was confirmed that these precipitates differ in morphology (Figure 11), with the one precipitate type being polygon-shaped with a 50 nm diameter, and the other being diamond-shaped (50 × 25 nm).

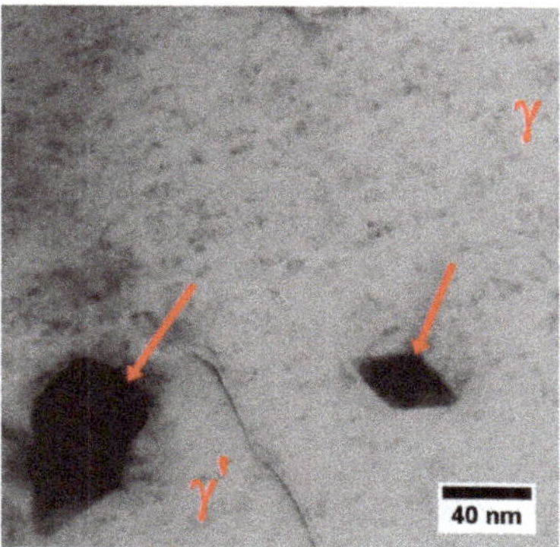


**Figure 11.** Nano-precipitates in dendritic regions, STEM-BF.

Nanostructure of the precipitates and their interfaces with the γ′ is shown in Figure 12. Based on the FFT image analysis, both precipitates are confirmed to be $M_5B_3$ borides. This phase is characterized by a body-centered tetragonal crystal structure with the space group I4/mcm and lattice parameters of a = 5.46 Å and c = 10.64 Å [38]. It has a $D8_1$-type structure (Strukturbericht notation) based on the information of atomic occupations in $M_5B_3$ phase. The atomic structure of boride obtained along the four-fold [001]$M_5B_3$ direction of the $M_5B_3$ phase is presented in Figure 12b. Although B atoms are invisible in the HAADF images due to their weak scattering ability, the structural characteristics of the $M_5B_3$ phase can be displayed. Imaging $M_5B_3$ along [001] zone axis reveals only one type of Wyckoff site. The relationship the γ/γ′ and $M_5B_3$ boride phases can be written as: [001]γ, γ′//[001]$M_5B_3$; (200)γ, γ′//(130)$M_5B_3$ and [001]γ, γ′//[$3\bar{1}0$]$M_5B_3$; (020)γ, γ′//(006)$M_5B_3$. The orientation relationship between $M_5B_3$ and γ/γ′ has also been investigated by other authors. Han [39] and Sheng [40] observed the relationship [001]γ//[001]$M_5B_3$ in model superalloys, while Zhang [41], based on René 80 superalloy study, indicated that [010]γ//[130]$M_5B_3$ and [001]γ//[001]$M_5B_3$.

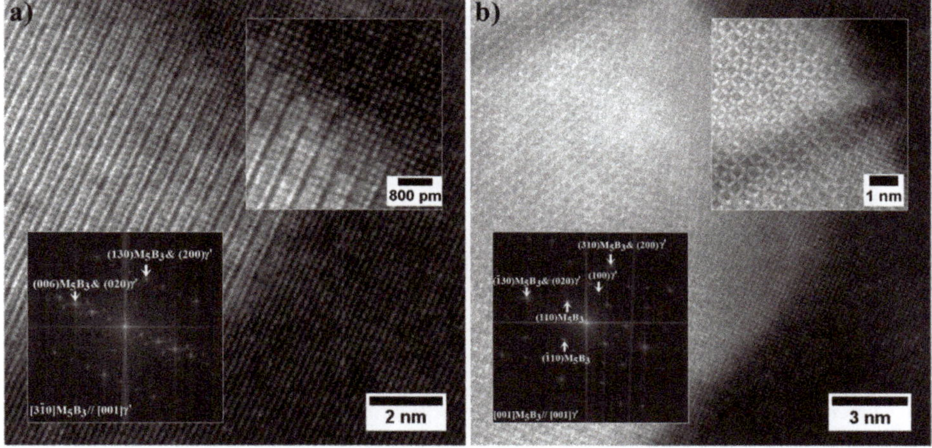


**Figure 12.** Nanostructure of the $M_5B_3$ boride with corresponding Fast Fourier Transformation (FFT) image: (**a**) Zone axis [$3\bar{1}0$]$M_5B_3$; (**b**) Zone axis [001]$M_5B_3$, HRSTEM-HAADF.

The STEM-EDX mapping of alloying elements covering the $\gamma$ matrix channel and $\gamma'$ precipitates with nano-precipitates at the $\gamma/\gamma'$ interface is presented in Figure 13. Additionally, the quantitative analysis was carried out in 5 regions (Table 5). It was found that the main elements strengthening the $\gamma$ matrix (precipitates-free zone, PFZ) are Cr and Co, with concentrations of over 35 at%. Among the elements which usually form (with Ni) the $\gamma'$ precipitates are Al, Ti, Ta, and Hf. Their total concentration in the secondary $\gamma'$ phase is 17 at%, while in the tertiary is slightly lower, 15.4 at%.

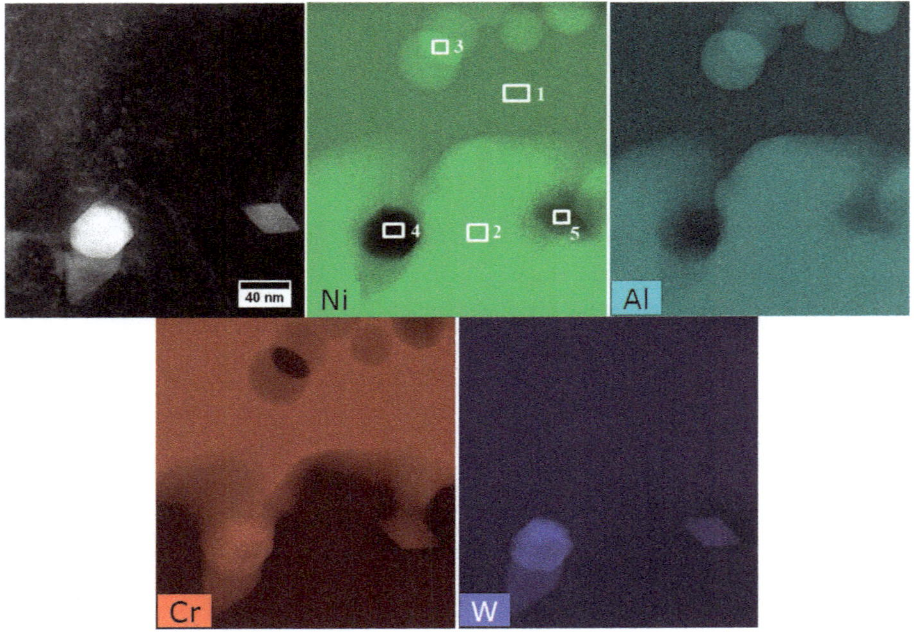


**Figure 13.** Distribution of selected alloying elements in a dendritic region, STEM-EDX.

**Table 5.** Chemical composition of the precipitates in dendritic regions, STEM-EDX, at%.

| Area | Phase | Ni | Cr | Co | W | Al | Ta | Hf | Ti | Mo |
|---|---|---|---|---|---|---|---|---|---|---|
| 1 | Matrix $\gamma$ | 52.1 | 21.5 | 16.9 | 4.7 | 3.4 | 0.3 | 0.2 | 0.2 | 0.7 |
| 2 | Secondary $\gamma'$ | 69.8 | 2.9 | 6.8 | 3.2 | 13.7 | 1.6 | 0.6 | 1.1 | 0.3 |
| 3 | Tertiary $\gamma'$ | 69.4 | 2.8 | 7.3 | 4.8 | 13.4 | 0.9 | 0.3 | 0.8 | 0.5 |
| 4 | Boride $M_5B_3$ | 10.0 | 35.5 | 3.2 | 42.0 | 2.3 | 1.2 | 0.3 | 0.4 | 5.2 |
| 5 | | 50.9 | 12.4 | 5.9 | 17.1 | 9.4 | 1.3 | 0.4 | 0.7 | 1.9 |

The mapping and analysis in selected areas confirmed that the partitioning of the alloying elements occurs between $\gamma$ and $\gamma'$ phases. The additional chemical composition measurements by STEM-EDX of $\gamma$ matrix and secondary $\gamma'$ phases were performed in 10 other locations to obtain better statistics; the results are presented in Figure 14. The concentration of selected alloying element ($i$ = Ni, Cr, Co, etc.) in $\gamma$ and secondary $\gamma'$ is presented as $C_i^{\gamma}$ and $C_i^{\gamma'}$, respectively. $C_i^{n}$ is the nominal concentration of an element in the bulk superalloy. The elements most strongly segregating to the intermetallic $\gamma'$ precipitates, i.e., those placed in the first quarter, are Ni, Al, Ta, Ti, and Hf. Alloying elements in the $\gamma'$ phase alter the formation energies of anti-phase boundaries (APBs), super lattice intrinsic stacking faults (SISFs), and complex stacking faults (CSFs), and may also directly influence the strength and plasticity of the $\gamma'$ phase. The strengthening effect of the intermetallic $\gamma'$ precipitates could be improved by the presence of elements like Ta and Ti because Ta-Ta and Ti-Ti bonds have higher energies than Al-Al bond [42].

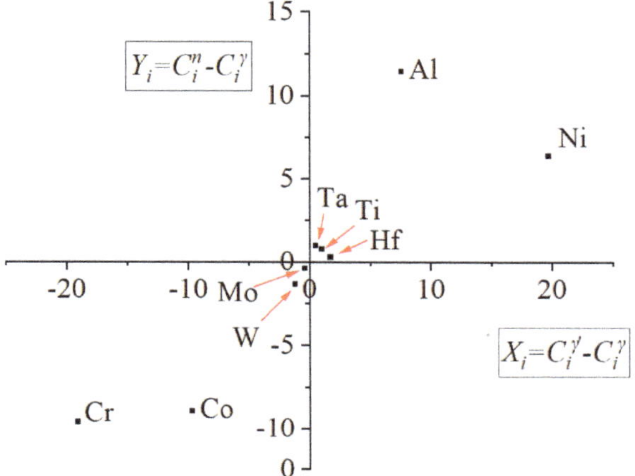


**Figure 14.** Plot $C_i^n - C_i^\gamma$ vs. $C_i^{\gamma'} - C_i^\gamma$ calculated based on the composition of the bulk superalloy, secondary $\gamma'$ and the matrix.

In accord with results of STEM-EDX analyses, in areas no. 4 and 5 (Table 5) borides are enriched in W, Cr, and Mo. Therefore they can be described by the formula (W, Cr, Mo)$_5$B$_3$. The concentration relationship of the two dominant W/Cr elements in the examined areas is 1.2, and 1.4, respectively, and the W/Mo relationship is 8.1 and 9.0. These relations are therefore very similar, and the differences in the relative concentration results from the small thickness of the precipitate with a diamond shape. The distribution of selected elements presented in Figure 15 shows the sharp increase of tungsten concentration at the interface $\gamma$/M$_5$B$_3$. In the case of Cr, the concentration growth is more smoother and occurs over a longer distance.

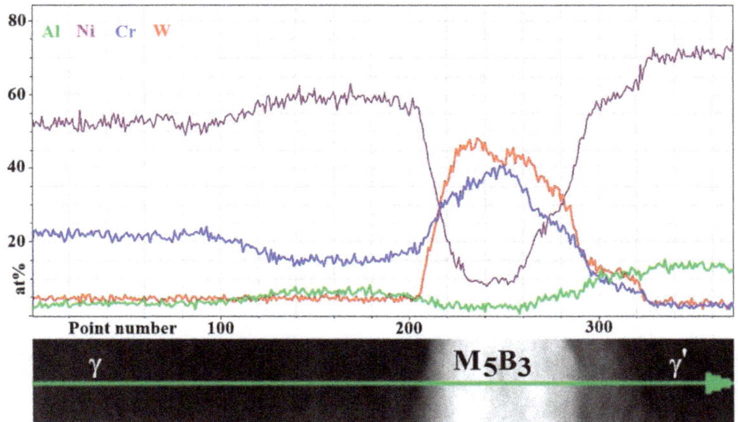


**Figure 15.** Distribution of selected alloying elements in $\gamma$, M$_5$B$_3$ and $\gamma'$ phases, STEM-EDX.

### 3.4.2. Interdendritic Spaces

Numerous Hf and Ta-rich carbides, as well as the γ-γ′ eutectic islands, were revealed during the SEM studies of interdendritic spaces. Further observations with transmission electron microscopy were done to show the remaining phases. The microstructure of the interdendritic spaces is much more complex than that of the dendritic regions. The difference is due both to the more complex morphology of the precipitates and their phase contrast. One similarity to the dendritic regions is numerous $M_5B_3$ borides at the γ/γ′ interfaces (Figure 16). It should be noted that boron may reduce the solubility of carbon in the γ matrix, which promotes carbides' precipitation.

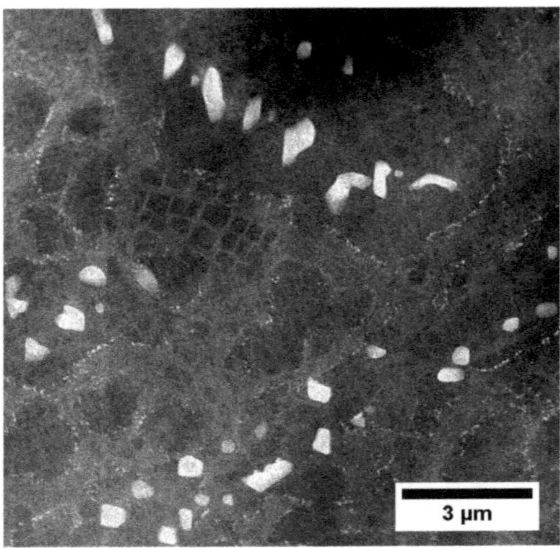


**Figure 16.** Microstructure of precipitates in the interdenritic spaces, STEM-HAADF.

The chemical composition of primary and secondary (cubic) γ′ was investigated. The locations of mapping, distribution of selected alloying elements, and the areas subjected to quantitative analysis are shown in Figure 17 and Table 6. In areas no. 1 and 2 involving primary γ′ precipitates, the total content of γ′-formers (Al, Ti, Ta, Hf) is 15.5 at% and 16.1 at%, respectively. In the areas no. 3–4, which correspond to the secondary γ′ precipitates, the values are lower and ranged from 13.3–14.9 at%.

**Table 6.** Chemical composition of the primary and secondary γ′ precipitates in interdendritic space, STEM-EDX, at%.

| Area | Phase | Ni | Cr | Co | W | Al | Ta | Hf | Ti | Mo |
|---|---|---|---|---|---|---|---|---|---|---|
| 1 | Primary γ′ | 71.0 | 3.5 | 7.1 | 2.6 | 11.6 | 1.4 | 1.1 | 1.4 | 0.3 |
| 2 | | 70.2 | 3.6 | 7.0 | 2.8 | 12.3 | 1.5 | 1.0 | 1.3 | 0.3 |
| 3 | Secondary γ′ | 71.8 | 2.7 | 6.5 | 2.7 | 11.2 | 1.3 | 1.0 | 1.4 | 0.6 |
| 4 | | 74.1 | 2.7 | 6.5 | 3.1 | 10.0 | 1.3 | 0.9 | 1.1 | 0.5 |

STEM-EDX mapping was performed in the region which includes a fragment of the "Chinese script" carbide, surrounded by a coarse γ′ precipitate (Figures 18 and 19). Additionally, three quantitative analyses of the coarse γ′ chemical composition were carried out. Based on the results in Table 7, the total content of Al, Ta, Ti, and Hf in the γ′ phase precipitates in an area no. 1–3 is 17.3 at%, 16.3 at%, and 16.9 at%, respectively. The Cr distribution revealed the nano-precipitates not observed by SEM. They are present locally on the edges of the MC carbides and are composed of mainly of Cr, whose concentration is clearly higher than of W and Mo. Thus, the Cr/W relation in region no. 4 and 5 is 21.9

and 13.3, respectively. It should be noted that similar layers enriched in Cr were not found on the edge of blocky carbides near eutectic γ-γ′ islands, which indicates differences in thermodynamical stability at elevated temperatures.

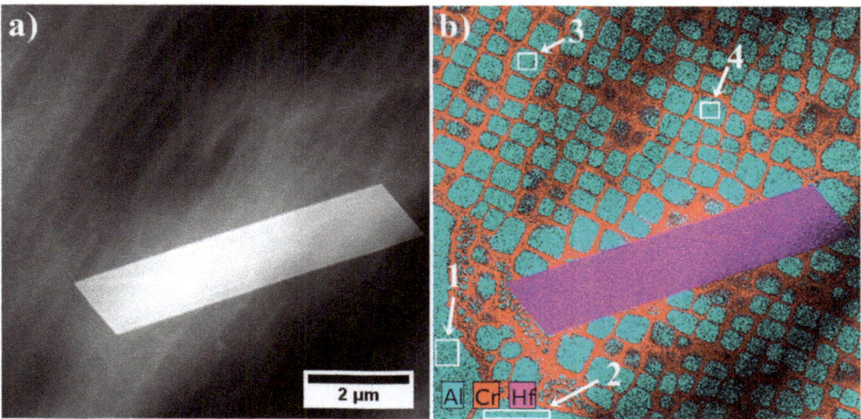


**Figure 17.** (**a**) location of mapping; (**b**) distribution of selected alloying elements in the area of primary γ′, secondary γ′, and MC carbide, STEM-EDX.

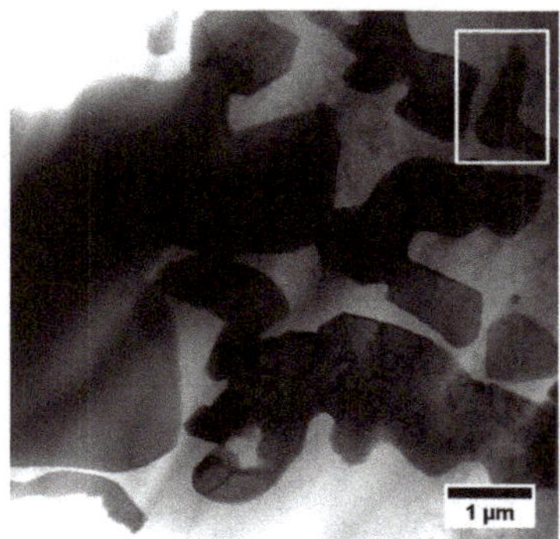


**Figure 18.** Microstructure of "Chinese script" carbide in interdendritic space, STEM-HAADF.

**Table 7.** Chemical composition of the coarse secondary γ′ and $M_{23}C_6$ (at the MC carbide edge), at%.

| Area | Phase | Cr | Co | W | Al | Hf | Ta | Ti | Mo | Ni |
|---|---|---|---|---|---|---|---|---|---|---|
| 1 | Coarse secondary γ′ | 3.0 | 7.5 | 3.3 | 11.8 | 1.1 | 2.7 | 1.7 | 0.5 | 68.4 |
| 2 | | 3.5 | 7.3 | 2.7 | 10.9 | 1.0 | 2.6 | 1.8 | 0.8 | 69.6 |
| 3 | | 3.4 | 7.3 | 2.9 | 11.5 | 0.9 | 2.7 | 1.8 | 0.6 | 68.8 |
| 4 | $M_{23}C_6$ | 43.7 | 4.5 | 2.0 | 6.2 | 3.3 | 6.5 | 2.7 | 1.9 | 29.4 |
| 5 | | 53.3 | 6.3 | 4.0 | 2.5 | 2.8 | 5.3 | 1.6 | 2.2 | 22.0 |

Locally on the edges of MC carbides, nano-layers with width 5–15 nm are present (Figure 20a,b). Reflections assigned to the FFT image correspond to the $M_{23}C_6$ carbide, which possesses a cubic crystal structure composed of 116 atoms ($D8_4$ type). In order to reveal the distribution of selected elements at the $M_{23}C_6$/γ′ and $M_{23}C_6$/MC interfaces, additional linear measurements were performed (Figure 20c). The Cr smooth distribution changes at the interfaces γ′/$M_{23}C_6$ and $M_{23}C_6$/MC, indicating that the presence of $M_{23}C_6$ carbides can be explained by the interaction between the MC carbide and the matrix during the heat treatment.

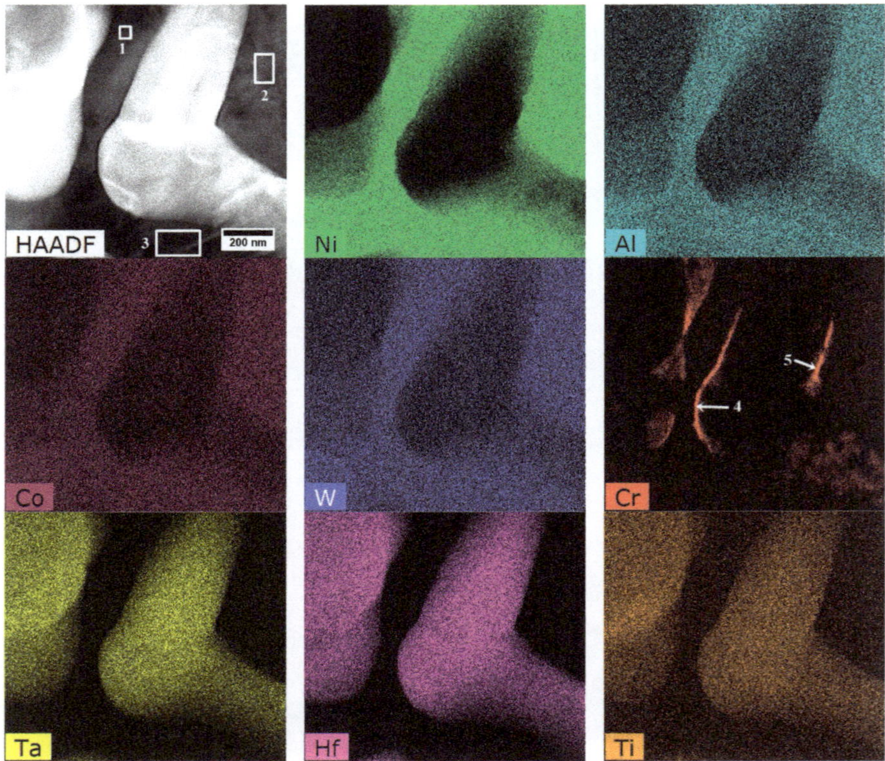


**Figure 19.** Distribution of selected alloying element in the fragment of "Chinese script" carbide, STEM-EDX.

$M_{23}C_6$ carbides are formed in medium- and high-chromium superalloys during casting's crystallization, heat treatment or annealing in the range of 760-980 °C. The alloying elements replacing chromium in the crystal lattice could increase the lattice parameter [8]. The position of M is mainly occupied by Cr. However, numerous alloying elements, including Mo and W, lead to a much more complex composition, e.g., $Cr_{21}(Mo, W)_2C_6$ and $(Ni, Co, Fe, Cr)_{21}(Mo, W)_2C_6$. In superalloys with increased iron content, it can be expected that they will partially replace chromium. In the solid-state, the $M_{23}C_6$ carbides are precipitating directly from the carbon-enriched $\gamma$ matrix or by the decomposition reaction of the MC carbide, usually along grain boundaries, at twin boundaries, and on stacking faults [37,43]. Long-term service or heat treatment of superalloys at elevated temperatures may lead to the decomposition of MC carbides into more stable $M_{23}C_6$ and $M_6C$ with lower carbon content, according to the following reactions [23]:

$$MC + \gamma \rightarrow M_{23}C_6 + \gamma', \quad \text{(I)}$$

$$MC + \gamma \rightarrow M_{23}C_6 + \eta, \quad \text{(II)}$$

$$MC + \gamma \rightarrow M_6C + \gamma', \quad \text{(III)}$$

The similarity in crystal structure sometimes makes the $M_6C$ and $M_{23}C_6$ carbides challenging to distinguish. The $M_6C$ usually replace $M_{23}C_6$ carbides in alloys containing approximately over 6 wt% Mo plus its atomic equivalent in W (weight per cent Mo plus one half the weight per cent W) [9,44,45]. The combined HRSTEM and STEM-EDX analyses show that I-type phase transformation occurred in René 108 during heat-treatment. The same transformation was also observed in crept Inconel 713C superalloy at 982 °C [46,47]. The Cr-rich $M_{23}C_6$ carbides then precipitated along the grain boundaries in the form of blocks surrounded by the $\gamma'$ precipitates. The degradation of the microstructure after annealing of the GTD111 superalloy at lower temperatures, 927 °C and 871 °C, mainly included the type II transformation [48]. The η phase (tetragonal structure with a = b = 5.09 Å, c = 8.29 Å) was particularly favoured by the enrichment of MC carbides in Ti and Ta because the Ti/Al concentration ratio in the superalloy exceeded 1.5. Relatively high Ti/Al ratio in superalloys promotes η phase formation, which, due to significant difference in lattice parameter compared to γ matrix, sensitizes the η/γ interface to crack nucleation and consequently decreases creep rupture life [42,49,50]. In this study, the Ti/Al relationship is 0.125, which is not favourable for the II-type phase transformation. The last possible phase transformation includes the formation of $M_6C$. The nano $M_6C$ carbides discontinuously precipitated in the γ matrix or along $\gamma/\gamma'$ interfaces of the crept K416B superalloy (1100 °C) were found by Xie [51]. Based on the thermodynamic analyses, solubility of C in the γ phase decreases during creep, which leads to its segregation in the stress concentration areas and then combining with carbide-formers such as W.

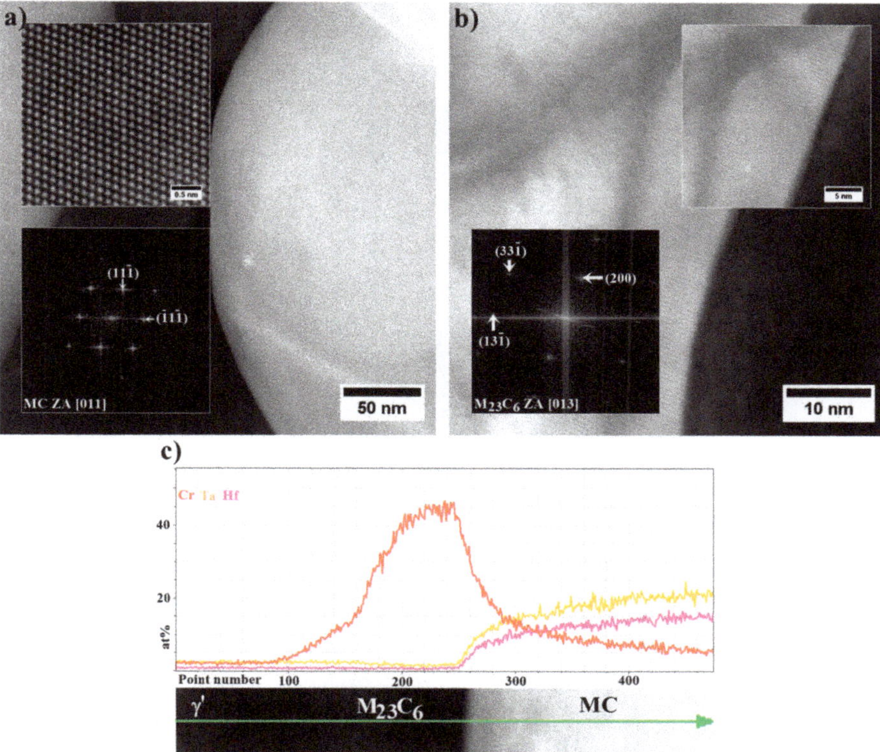


**Figure 20.** (**a**) atomic structure and FFT image of MC carbide (STEM-HAADF); (**b**) the nano-precipitate of $M_{23}C_6$ at the MC edge (STEM-BF); (**c**) linear distribution of Cr, Hf and Ta (STEM-EDX).

During the SEM observation, precipitates with a clear and bright contrast (high Z-number) are revealed along the grain boundaries. This area has also been subjected to detailed STEM and HRSTEM studies. Figure 21 shows the microstructure of the precipitates along the grain boundary and the distribution of Ni, W and Cr in this area. The morphology and size of these precipitates are the same as these at the $\gamma/\gamma'$ interfaces. The second similarity is enrichment in the same alloying elements. Their nanostructure with the corresponding FFT image is presented in Figure 22. The $M_5B_3$ borides are confirmed. Imaging this phase using tilting of grain boundary region revealed numerous precipitates. On the SEM-BSE, such precipitates' agglomeration is shown as a continuous or semi-continuous layer.

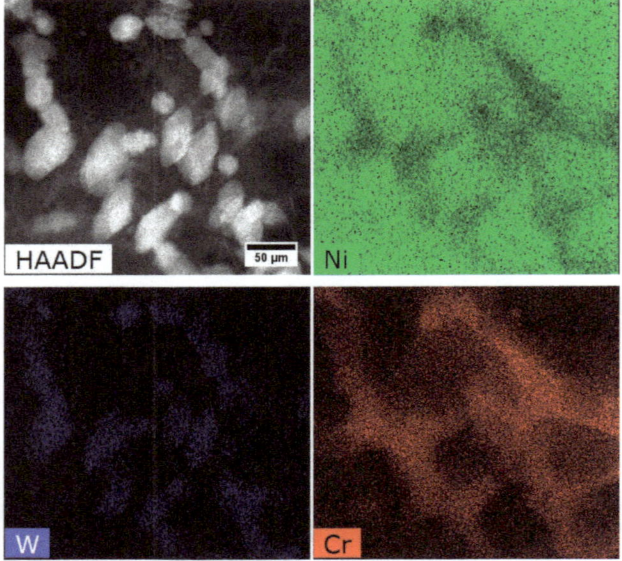


**Figure 21.** Distribution of selected alloying elements at a grain boundary, STEM-EDX.

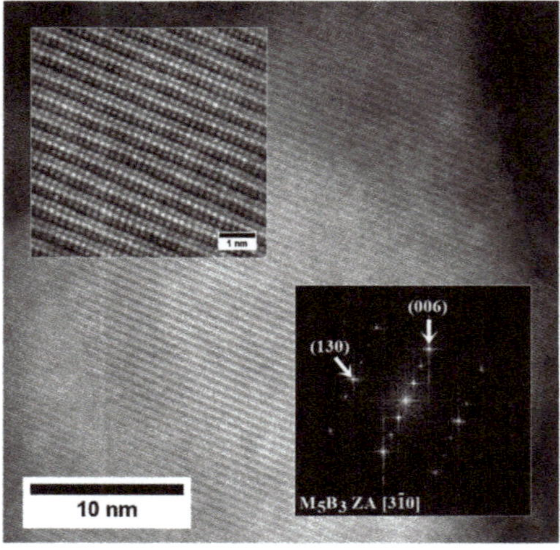


**Figure 22.** Nanostructure of the $M_5B_3$ boride located at grain boundary, HRSTEM-HAADF.

It has been shown that the addition of boron to Ni-based superalloys could significantly influence the grain boundary precipitation [31,52,53]. B atoms are larger than those of the other interstitial atoms and smaller than those of the substitutional atoms in Ni-based superalloys. So, the lattice distortion caused by boron is considerable. The small diameter of boron allows it to fill vacancies at the grain boundaries, reducing the diffusivity in these regions [54]. During the ageing-treatment, boron segregation to the grain boundaries occurs and a large number of solute atoms, such as tungsten, chromium, and molybdenum, distribute at the grain boundaries. Because W, Cr, and Mo are more inclined to dissolve in the $\gamma$ matrix when intermetallic $\gamma'$ phase is formed, these elements would be expelled and gather at the $\gamma/\gamma'$ interfaces. As a result, B would interact with W, Cr, and Mo and form $M_5B_3$ borides. Their formation at grain boundaries would pin these boundaries and impede grain-boundary sliding mechanism at high-temperatures. Xiao suggested that any strong interaction between B atoms and dislocation cores could hinder dislocation motion and thus increase resistance to fatigue failure [55].

## 4. Conclusions

- The FCC $\gamma$ matrix is strengthened mainly by coherent precipitates characterised by ordered $L1_2$ crystal structure. The mean misfit coefficient between matrix and precipitates is $\delta = +0.6\%$.
- Dendritic structure with significant segregation of alloying elements and microstructural constituents is observed.
- The mean volume fraction of MC carbides is around 0.8% (LM), while $\gamma'$ precipitates in dendritic regions around 54.99% (SEM-BSE).
- In dendritic regions the $\gamma'$ precipitates have a complex morphology with two classes of shape factor for which the mean $\xi$ values are 0.39 and 0.76. Too high diversity in size and morphology of precipitates in interdendritic spaces did not allow their effective comparison.
- The MC carbides are preferentially precipitated in the interdendritic areas. The "M" position is occupied mainly by Ta, Hf, Ti, and W, while the mutual concentration relationships depend on the morphology.
- The MC carbide degradation occurred during ageing according to the phase transformation reaction: $MC + \gamma \rightarrow M_{23}C_6 + \gamma'$. The $M_{23}C_6$ carbides are revealed at the MC edges as nano-layers with width 5-15 nm.
- The $M_5B_3$ borides characterised by body-centered tetragonal I4/mcm crystal structure have polygon and rhombus forms in thin foils. They are preferentially formed at the interfaces of secondary $\gamma'$ with matrix as nano-precipitates, both in the dendritic and interdendritic regions, and also on the grain boundaries. The STEM-EDX analysis revealed that they are enriched mainly in W, Cr, and Mo.

**Author Contributions:** Conceptualization, Ł.R.; methodology, Ł.R., B.R.; investigation, Ł.R., B.R., M.G.-R., W.R., A.Z.-L.; writing—original draft preparation, Ł.R., M.G.-R.; writing—review and editing, R.C., A.Z.-L.; supervision, A.Z.-L., R.C.; project administration, Ł.R.; funding acquisition, Ł.R. All authors have read and agreed to the published version of the manuscript.

**Funding:** This work was supported by the Polish National Science Centre (Preludium 13) under grant 2017/25/N/ST8/02368. Ł.R. thanks Krzysztof Chruściel, M.Sc. (AGH) for support in the XRD analysis.

**Conflicts of Interest:** The authors declare no conflict of interest.

## References

1. Reed, R.C. *The Superalloys: Fundamentals and Applications*; Cambridge University Press: Cambridge, UK, 2006.
2. Grudzień-Rakoczy, M.; Rakoczy, Ł.; Cygan, R.; Kromka, F.; Pirowski, Z.; Milkovič, O. Fabrication and characterization of the newly developed superalloys based on Inconel 740. *Materials* **2020**, *13*, 2362. [CrossRef] [PubMed]

3. Rogalski, G.; Świerczyńska, A.; Landowski, M.; Fydrych, D. Mechanical and microstructural characterization of TIG welded dissimilar joints between 304L austenitic stainless steel and Incoloy 800HT nickel alloy. *Metals* **2020**, *10*, 559. [CrossRef]
4. Rakoczy, Ł.; Cempura, G.; Kruk, A.; Czyrska-Filemonowicz, A.; Zielińska-Lipiec, A. Evolution of $\gamma'$ morphology and $\gamma/\gamma'$ lattice parameter misfit in a nickel-based superalloy during non-equilibrium cooling. *Int. J. Mater. Res.* **2019**, *110*, 66–69. [CrossRef]
5. Sims, C.T. A History of superalloy metallurgy for superalloy metallurgists. In Proceedings of the Fifth International Symposium on Superalloys, Pittsburgh, PA, USA, 7–11 October 1984; pp. 399–419. [CrossRef]
6. Reed, R.C.; Rae, C.M.F. 22—Physical metallurgy of the nickel-based superalloys. In *Physical Metallurgy*, 5th ed.; Laughlin, D.E., Hono, K., Eds.; Elsevier: Oxford, UK, 2014; pp. 2215–2290. ISBN 978-0-444-53770-6.
7. Satyanarayana, D.; Prasad, N.E. Nickel-based superalloys technologies. In *Aerospace Materials and Material Technologies*; Eswara Prasad, N., Wanhill, R.J.H., Eds.; Springer: Singapore, 2017; pp. 199–228. [CrossRef]
8. Kishawy, H.A.; Hosseini, A. Superalloys. In *Machining Difficult-to-Cut Materials. Basic Principles and Challenges*; Springer: Cham, Switzerland, 2019. [CrossRef]
9. Sabol, G.P.; Stickler, R. Microstructure of nickel-based superalloys. *Phys. Stat. Sol.* **1969**, *35*, 11–52. [CrossRef]
10. Geddes, B.; Leon, H.; Huang, X. *Superalloys: Alloying and Performance*; ASM International: Materials Park, OH, USA, 2010; ISBN 978-1-61503-040-8.
11. Bradley, A.J.; Taylor, A. An X-ray analysis of the nickel–aluminium system. *Proc. R. Soc. Lond. Ser. A Math. Phys. Sci.* **1937**, *159*, 56–72.
12. Calmon, J. From Sir Frank Whittle to the year 2000—what is new in propulsion? *Aeronaut. Astronaut.* **1988**, *6*, 32–46. [CrossRef]
13. Ennis, P. 6- Nickel-base alloys for advanced power plant components. In *Coal Power Plant Materials and Life Assessment*; Shimbli, A., Ed.; Woodhead Publishing: Shaston/Cambridge, UK, 2014; pp. 147–167. [CrossRef]
14. White, C.H. Nickel base alloys. In *The Development of Gas Turbine Materials*, 1st ed.; Meetham, G.W., Ed.; Springer: Dordrecht, The Netherlands, 1981; pp. 89–119. [CrossRef]
15. Rakoczy, Ł.; Cygan, R. Analysis of temperature distribution in shell mould during thin-wall superalloy casting and its effect on the resultant microstructure. *Arch. Civ. Mech. Eng.* **2018**, *18*, 144–1450. [CrossRef]
16. Rakoczy, Ł.; Grudzień, M.; Cygan, R.; Zielińska-Lipiec, A. Effect of cobalt aluminate content and pouring temperature on macrostructure, tensile strength and creep rupture of Inconel 713 castings. *Arch. Metall. Mater.* **2019**, *63*, 1537–1545. [CrossRef]
17. Eiselstein, H.L. Metallurgy of a columbium-hardened nickel-chromium-iron alloy. In *Advances in the Technology of Stainless Steels and Related Alloys*; ASTM STP 369: West Conshohocken, PA, USA, 1965; pp. 62–77. [CrossRef]
18. Smith, S.R.; Carter, W.J.; Mateescu, G.D.; Kohl, F.J.; Fryburg, G.C.; Stearns, C.A. ESCA study of oxidation and hot corrosion of nickel-base superalloys. *Oxid Met.* **1980**, *14*, 415–435. [CrossRef]
19. Wilson, A.S. Formation and effect of topologically close-packed phases in nickel-base superalloys. *Mat. Sci. Tech.* **2017**, *33*, 1108–1118. [CrossRef]
20. Morinaga, M.; Yukawa, N.; Adachi, H.; Ezaki, H. New phacomp and its applications to alloy design. In Proceedings of the Fifth International Symposium on Superalloys, Pittsburgh, PA, USA, 7–11 October 1984; pp. 523–532.
21. Muktinutalapati, R.N. Materials for gas turbines—An overview. In *Advances in Gas Turbine Technology*; Benini, E., Ed.; IntechOpen: Rijeka, Croatia, 2011; pp. 293–314. [CrossRef]
22. Rakoczy, Ł.; Milkovič, O.; Rutkowski, B.; Cygan, R.; Grudzień-Rakoczy, M.; Kromka, F.; Zielińska-Lipiec, A. Characterization of $\gamma'$ precipitates in cast Ni-based superalloy and their behaviour at high-homologous temperatures studied by TEM and in Situ XRD. *Materials* **2020**, *13*, 2397. [CrossRef] [PubMed]
23. Donachie, M.J.; Donachie, S.J. *Superalloys: A Technical Guide*, 2nd ed.; ASM International: Materials Park, OH, USA, 2002.
24. Burt, H.; Dennison, J.P.; Elliot, I.C.; Wilshire, B. The effect of hot isostatic pressing on the creep and fracture behaviour of the cast superalloy Mar 002. *Mat. Sci. Eng.* **1982**, *53*, 245–250. [CrossRef]
25. Maslenkov, S.B.; Burova, N.N.; Khangulov, V.V. Effect of hafnium on the structure and properties of nickel alloys. *Met. Sci. Heat Treat.* **1980**, *22*, 283–285. [CrossRef]
26. Dahl, J.M.; Danesi, W.F.; Dunn, R.G. The partitioning of refractory metal elements in hafnium-modified cast nickel-base superalloys. *Metall. Mater. Trans. B* **1973**, *4*, 1641. [CrossRef]

27. da Silva Costa, A.M.; Nunes, C.A.; Baldan, R.; Coelho, G.C. Thermodynamic evaluation of the phase stability and microstructural characterization of a cast B1914 superalloy. *J. Mat. Eng. Perform.* **2014**, *23*, 819–825. [CrossRef]
28. Maxwell, D.H.; Baldwin, J.F.; Radavich, J.F. New concept in superalloy ductility. *Metall. Metal. Form.* **1975**, *42*, 332.
29. Ouichou, L.; Lavaud, F.; Lesoult, G. Influence of the chemical composition of nickel-base superalloys on their solidification behavior and foundry performance. In Proceedings of the Superalloys 1980 (Fourth International Symposium), The Seven Springs International Symposium Committee, Champion, PA, USA, 21–25 September 1980; pp. 235–244.
30. Rakoczy, Ł.; Grudzień-Rakoczy, M.; Cygan, R. The influence of shell mold composition on the as-cast macro- and micro-structure of thin-walled IN713C superalloy casting. *J. Mat. Eng. Perform.* **2019**, *28*, 3974–3985. [CrossRef]
31. Kontis, P.; Mohd Yusof, H.A.; Pedrazzini, S.; Danaie, M.; Moore, K.L.; Bagot, P.A.J.; Moody, M.P.; Grovenor, C.R.M.; Reed, R.C. On the effect of boron on grain boundary character in a new polycrystalline superalloy. *Act. Mat.* **2016**, *103*, 688–699. [CrossRef]
32. Hu, X.; Zhu, Y.; Sheng, N.; Ma, X.L. The Wyckoff positional order and polyhedral intergrowth in the $M_3B_2$- and $M_5B_3$-type boride precipitated in the Ni-based superalloys. *Sci. Rep.* **2014**, *4*, 7367. [CrossRef]
33. Strunz, P.; Petrenec, M.; Polák, J.; Gasser, U.; Farkas, G. Formation and dissolution of $\gamma'$ precipitates in IN792 superalloy at elevated temperatures. *Metals* **2016**, *6*, 37. [CrossRef]
34. Jena, A.K.; Chaturvedi, M.C. The role of alloying elements in the design of nickel-base superalloys. *J. Mat. Sci.* **1984**, *19*, 3121–3139. [CrossRef]
35. Rakoczy, Ł.; Grudzień, M.; Zielińska-Lipiec, A. Contribution of microstructural constituents on hot cracking of MAR-M247 nickel based supealloy. *Arch. Met. Mat.* **2018**, *63*, 181–189. [CrossRef]
36. Clementi, E.; Raimondi, D.L.; Reinhardt, W.P. Atomic screening constants from SCF functions. II. Atoms with 37 to 86 electrons. *J. Chem. Phys.* **1967**, *47*, 1300–1307. [CrossRef]
37. Davis, J.R. *Heat-Resistant Materials*; ASM International: Materials Park, OH, USA, 1997.
38. Bertaut, F.; Blum, P. Etude des borures de chrome. *Compt. Rend. Hebd. Seances Acad. Sci.* **1953**, *236*, 1055–1056.
39. Han, F.F.; Chang, J.X.; Li, H.; Lou, L.H.; Zhang, J. Influence of Ta content on hot corrosion behaviour of a directionally solidified nickel base superalloy. *J. All. Comp.* **2015**, *619*, 102–108. [CrossRef]
40. Sheng, N.; Li, B.; Liu, J.; Jin, T.; Sun, X.; Hu, Z. Influence of the substrate orientation on the isothermal solidification during TLP bonding single crystal superalloys. *J. Mater. Sci. Technol.* **2014**, *30*, 213–216. [CrossRef]
41. Zhang, H.R.; Ojo, O.A. TEM analysis of Cr-Mo-W-B phase in a DS nickel based superalloy. *J. Mater. Sci.* **2008**, *43*, 6024–6028. [CrossRef]
42. Long, H.; Mao, S.; Liu, Y.; Zhang, Z.; Han, X. Microstructural and compositional design of Ni-based single crystalline superalloys. *J. All. Comp.* **2018**, *743*, 203–220. [CrossRef]
43. Ducki, K. *Microstructural Aspects of Deformation, Precipitation and Strengthening Processes in Austenitic Fe-Ni Superalloy*; Silesian University of Technology: Gliwice, Poland, 2010. (In Polish)
44. Grudzień, M.; Tuz, L.; Pańcikiewicz, K.; Zielińska-Lipiec, A. microstructure and properties of a repair weld in a nickel based superalloy gas turbine component. *Adv. Mater. Sci.* **2017**, *17*, 55–63. [CrossRef]
45. Hanning, F.; Khan, A.K.; Steffenburg-Nordenström, J.; Ojo, O.; Andersson, J. Investigation of the effect of short exposure in the temperature range of 750–950 °C on the ductility of Haynes® 282® by advanced microstructural characterization. *Metals* **2019**, *9*, 1357. [CrossRef]
46. Rakoczy, Ł.; Grudzień, M.; Cygan, R. Influence of melt-pouring temperature and composition of primary coating of shell mold on tensile strength and creep resistance of Ni-based superalloy. *J. Mat. Eng. Perform.* **2019**, *28*, 3826–3834. [CrossRef]
47. Grudzień, M.; Cygan, R.; Pirowski, Z.; Rakoczy, Ł. Microstructural characterization of Inconel 713C superalloy after creep testing. *Trans. Found. Res. Inst.* **2018**, *58*, 39–45. [CrossRef]
48. Choi, B.G.; Kin, I.S.; Kim, D.H.; Seo, S.M.; Jo, C.Y. Eta phase formation during thermal exposure and its effect on mechanical properties in Ni-base superalloy GTD 111. In Proceedings of the 10th International Symposium Superalloys, Pittsburgh, PA, USA, 19–23 September 1996; pp. 163–172.

49. Nunomura, Y.; Kaneno, Y.; Tsuda, H.; Takasugi, T. Phase relation and microstructure in multi-phase intermetallic alloys based on $Ni_3Al$-$Ni_3Ti$-$Ni_3V$ pseudo-ternary alloy system. *Intermetallics* **2004**, *12*, 389–399. [CrossRef]
50. Xu, Y.; Yang, C.; Xiao, X.; Cao, X.; Jia, G.; Shen, Z. Evolution of microstructure and mechanical properties of Ti modified superalloy Nimonic 80A. *Mat. Sci. Eng. A* **2001**, *530*, 315–326. [CrossRef]
51. Xie, J.; Yu, J.; Sun, X.; Jin, T. Thermodynamics analysis and precipitation behavior of fine carbide in K416B Ni-based superalloy with high W content during creep. *Trans. Nonferrous Met. Soc. China* **2015**, *25*, 1478–1483. [CrossRef]
52. Li, Q.; Lin, X.; Wang, X.; Yang, H.; Song, M.; Huang, W. Research on the grain boundary liquation mechanism in heat affected zones of laser forming repaired K465 nickel-based superalloy. *Metals* **2016**, *6*, 64. [CrossRef]
53. Kang, B.; Han, C.; Shin, Y.; Youn, J.; Kim, Y. Effects of boron and zirconium on grain boundary morphology and creep resistance in Nickel-based superalloy. *J. Mat. Eng Perform.* **2019**, *28*, 7025–7035. [CrossRef]
54. Tytko, D.; Choi, P.P.; Klöwer, J.; Kostka, A.; Inden, G.; Raabe, D. Microstructural evolution of a Ni-based superalloy (617B) at 700 °C studied by electron microscopy and atom probe tomography. *Acta Mater.* **2012**, *60*, 1731–1740. [CrossRef]
55. Xiao, L.; Chen, D.L.; Chaturvedi, M.C. Effect of boron on fatigue crack growth behavior in superalloy IN718 at RT and 650 °C. *Mater. Sci. Eng. A* **2006**, *428*, 1–11. [CrossRef]

 materials

MDPI

*Article*

# Repair of a Cracked Historic Maryan Bell by Gas Welding

**Dariusz Bartocha and Czesław Baron ***

Department of Foundry, Silesian University of Technology, 7 Towarowa St., 44-100 Gliwice, Poland; dariusz.bartocha@polsl.pl
* Correspondence: czeslaw.baron@polsl.pl

**Abstract:** In this article, the range of works connected with the repair of a historical Maryan bell from 1639 are presented. The first attempts to repair damaged bells occurred in the 1930s in Poland. However, this process was stopped because of extensive technological difficulties. Welding and soldering-welding were the basic methods. There is one difference between these two methods—connecting surfaces are melted during the welding process but only heated until the melting temperature of the material added to the connection (that is the solder) during the soldering-welding process. It was important to heat the bell to the proper temperature during welding. Uneven heating causes the enlargement of existing cracks or the appearance of new ones, or even the complete destruction of the bell. Nowadays, a method of even heating using a special heating mat has been devised. Thanks to this method it is possible to control the heating and cooling process. The most important task during the whole operation of bell welding was obtaining the original sound. During this research, the chemical composition was examined to prepare a welding rod with a suitable chemical composition. After the repair process, an analysis of the sound of the bell was conducted. It was shown that the repair of bells is possible when correct thermal parameters are used. The most highly recommended technique for repairing bells is gas welding.

**Keywords:** high-tin bronzes; microstructure; welding of bell; bell's sound

**Citation:** Bartocha, D.; Baron, C. Repair of a Cracked Historic Maryan Bell by Gas Welding. *Materials* **2021**, *14*, 2504. https://doi.org/10.3390/ma14102504

Academic Editor: Sergei Yu Tarasov

Received: 2 April 2021
Accepted: 9 May 2021
Published: 12 May 2021

**Publisher's Note:** MDPI stays neutral with regard to jurisdictional claims in published maps and institutional affiliations.



## 1. Introduction

There are a lot of churches with very old, historical bells in Poland, as it has historically been a Catholic country. Unfortunately, their strength is decreasing and cracks and scratches are have appeared. These defects make them useless, because not only does the sound become worse, but reacting to the damage too late can cause the complete destruction of the bell as well.

The lifetime of the bell was determined to be 200–300 years [1] on the basis of data in the literature. After that time the probability of the bell cracking is increased. Of course, it depends on many factors, such as the frequency of bell work and the bell's rotation on its suspension. Unless the bell is rotated, the clapper hits the same place and it may cause cracks. It is important to control the thickness of that place. If the thickness falls more than 10%, it is necessary to rotate the bell to allow it to hit another place. Constant hits on the same place also causes changes to the inner structure of the material. It becomes harder and loses strength properties. Concurrently, inner stresses increase. When stresses exceed the material strength limit, the bell will be damaged, and cracks and scratches will appear. It is hard to notice such a crack on a bell placed high in a tower, but it is possible to hear the change in the sound, which is usually much worse than the original sound.

The repair of the bell is possible up to its complete destruction. Repairing the bell is an expensive and time-consuming process. However, attempts to repair these bells are not rare, because the bell is a precious item not only thanks to its material value but also its historical and artistic value as well. Despite the avoidance of the repair of cracked bells in Poland for a long time, this problem has been considered in many other countries [2,3].

It is necessary to examine the chemical composition of the alloy used to produce the bell to repair the crack. The oldest bells are made of gunmetal, while the younger are made of bronze. Copper at a concentration of about 80% and tin with a concentration between 19% and 21% are the main components in both alloys. Gunmetal also contains zinc, lead, carbon, and iron. Trace amounts of silver and gold are also possible to observe in both alloys because of the tradition to add these elements into the liquid metal to ennoble the material [4–6].

The melting temperature of bronze is about 850–950 °C. The alloys with high tin content are characterized by great strength, but concurrently low toughness—they are very brittle (Figure 1). This, in combination with high thermal expansion and high diversity of the microstructure component properties, has a negative influence on weldability. It is proper to heat the whole bell at the adequate speed until it reaches the temperature of 350–450 °C, before welding. This allows differences in temperature between the welded place and the rest of the bell which are too large to be avoided. What's more, it is also important to cool it slowly and evenly (the speed of the cooling process should be slower than heating). This is connected with the risk of inducing residual heat as a result of differences in the cooling rates of different parts of the structure (Figure 2a,b shows different phases according to different cooling rates). If the bell had not been heated, only the welded part would have shrunk and new cracks would have appeared. On the other hand, a cooling process which is too fast may cause new stresses, with new cracks as the result.

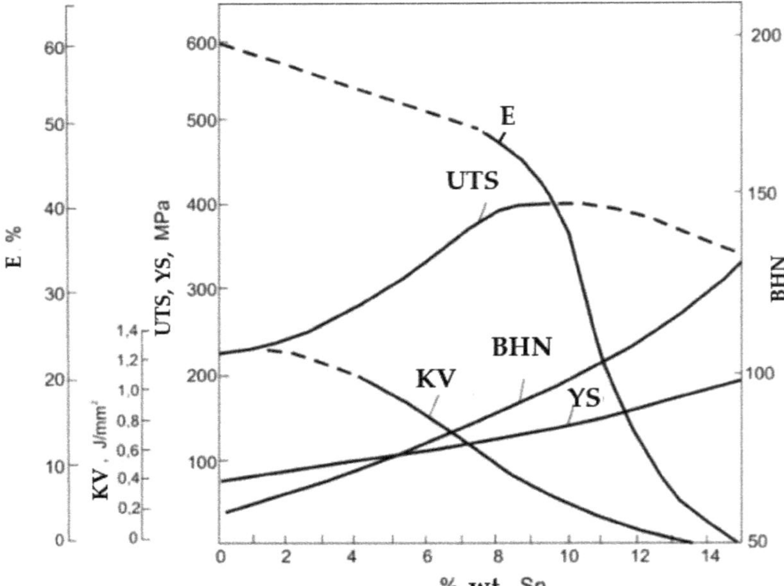


**Figure 1.** The influence of tin on chosen mechanical properties, the tin bronze hardness change dependent on tin concentration. (E—elongation, UTS—ultimate tensile strength, BHN—Brinell hardness number, KV—impact strength, YS—yield strength) [7].

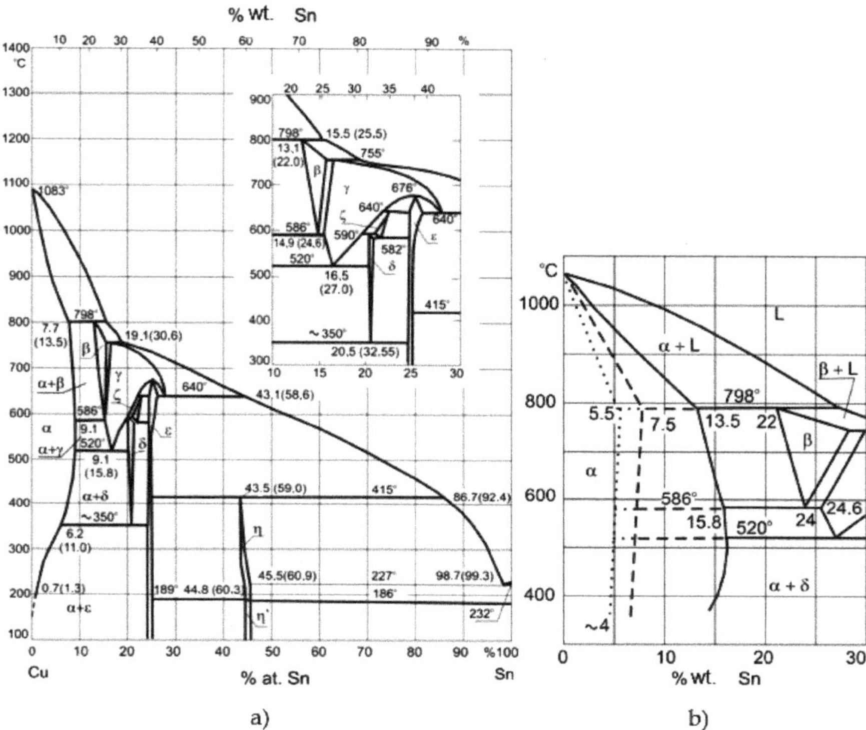


**Figure 2.** (**a**) Phase diagram Cu-Sn, (**b**) Metastable phases Cu-Sn; dashed line—casting solidified in sandy mold, dotted line—casting solidified in a metal mold [8,9].

## 2. Historical Background

In Poland, the first attempts to repair damaged bells were observed in the 1930s. Welding and soldering-welding were the base methods. There is one difference between these two methods: connecting surfaces are melted during welding process and only heated until the melting temperature of the material added to the connection (that is solder) during the soldering-welding process. Additionally, a welding rod with a chemical composition similar to indigenous metal was used during welding process, whilst sticks made of brass alloy (known as bronzite) were used for soldering-welding.

First, it was desirable to estimate the size of the crack. A simple penetration study was conducted with the use of chalk and kerosene. Chalk was rubbed into the inner side of the bell and the external side was lubricated with kerosene. The greasy spot was used to show the range of the crack. After estimating the size of the damage, the place of repair should be properly prepared. In both cases, the method of action was similar. A groove with a v–shape was cut along the crack and metal was poured in. In this procedure the hole at the end of the crack must be remembered. Its task was to limit the increase of the crack during bell heating. An acetylene torch was used as the welding tool in both methods. Uniformity of the weld obtained was the main difference between welding and the soldering-welding process. The weld had nearly the same chemical composition as welded material during the welding process. The weld had different chemical composition to the repaired bell during the soldering-welding process, which caused worse sound.

The position heated by charcoal (in the past it was often used as a fuel) was used to heat the bell before repair. This solution was connected with uneven bell heating and cooling. It may have caused new cracks or increased the old ones as a result.

## 3. Maryan Bell Crack Repairing

The company Rduch Bells & Clocks and Foundry Department collectively made a decision to repair a cracked Maryan bell from the 17th century at the request of urban restorer in Krosno–Marta Rymar. The bell hangs in a church tower. It is the smallest bell of three cast in 1639, weighing 580 kg. It is characterized by producing a "G#" sound. The bell was cast by two bell founders, Szepan Meutel and Jerzy Olivier, for a special order from the great philanthropist Robert Wojciech Portius. The Maryan bell is one of three treated bells. On the tower there are also the Urban bell and the Jan bell. The bells are tuned to a major scale. This means that they sound happy, merry, and concurrently noble. The lack of one bell or unclear sound will cause the whole set to lose its musical value. This is why it was so important not only to repair the crack but to do it in such a way as to avoid changes to the sound.

The most important task during welding process was to obtain the original sound. Fortunately, in 2013, during the change of the clapper, acoustic measurements were conducted. Thanks to this it was possible to obtain the sound before and after the damage to the bell (the damage occurred in 2017—exact date is unknown). This was the base for further activities. The work was divided into a few steps:

1. The first step was to take a material sample to determine the averaged chemical composition of the alloy, which was analyzed with a glow–discharge spectrometer LECO GDS500A (LECO Corporation, St. Joseph, MI, USA, 2011) (Table 1);
2. In the next step a series of welding rods with the same chemical composition were prepared on the basis of these results, which were used during bell welding process;
3. The next step was to determine the size and range of the crack, and penetration research was conducted;
4. The next step was preparing the bell for the welding process by properly bevelling the sides of the bell;
5. The next step was heating and keeping the bell at the proper temperature;
6. The next step was obtaining the required temperature to conduct welding process;
7. After welding slow cooling was conducted to avoid stresses;
8. The last step was analysis of the sound of the repaired bell.

**Table 1.** Chemical composition of the Maryan bell (wt. %).

| Sn | Pb | Sb | Zn | Fe | Ni | Ag | Cu |
|---|---|---|---|---|---|---|---|
| 15.2 | 2.84 | 2.69 | 0.35 | 0.03 | 0.41 | 0.15 | bal. |

Accuracy in all of these activities allowed us to obtain the ideal sound from the repaired bell. The samples of material obtained were examined with the use of a spectrometer to determine averaged chemical composition (presented in Table 1). The analysis of structure was also conducted with the use of a scanning microscope to determine the distribution and size of solid and gaseous inclusions. (Figure 3a,b). The chemical composition in particular places (with visible solid and gaseous inclusions) was presented in Table 2.

Metallographic microsections showed the original structure of the bell. Unfortunately many gaseous (Figure 3a, 6) and non–metallic inclusions were observed. A large amount of carbon (Figure 3b, 1) can indicate residue of charcoal, which was used as fuel during the melting process. Zinc inclusions were also observed (Figure 3a, 5).This negatively influenced the welding process.

A series of welding rods were made after chemical composition determination and consultations with the company conducting the welding process. Their composition was selected to be as compatible as possible with the examined material of the bell. A set of molds was worked out and prepared in the Foundry Department, and thanks to them welding rods of different lengths and diameters were produced (Figure 4).

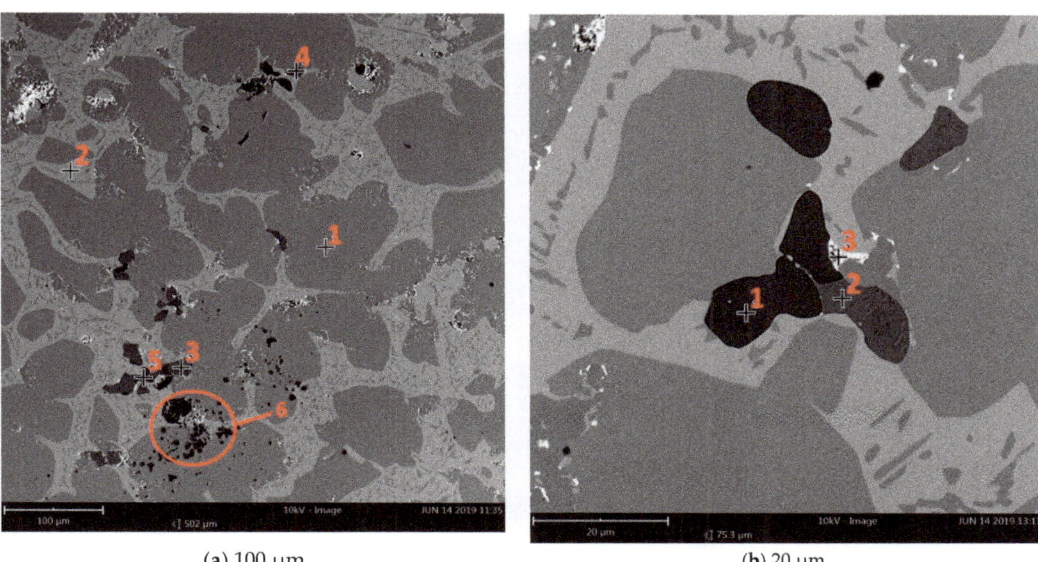


(**a**) 100 µm (**b**) 20 µm

**Figure 3.** Location of the measurement points for scanning microscope examination with visible impurities and inclusions. (**a**) 1–6 measurement points for magnification 530×, (**b**) 1–3 measurement points for magnification 800×.

**Table 2.** Chemical composition in particular examined places presented in Figure 3.

| Figure 3a | | | Figure 3b | | |
|---|---|---|---|---|---|
| Number | Element Symbol | Atomic Concentration | Number | Element Symbol | Atomic Concentration |
| 1 | Cu | 21.43 | 1 | C | 75.57 |
| | C | 70.37 | | Zn | 12.53 |
| | Sn | 1.77 | | Cu | 3.56 |
| | O | 6.44 | | S | 5.12 |
| | | | | O | 3.22 |
| 2 | Cu | 28.54 | 2 | Cu | 17.76 |
| | Sn | 6.34 | | C | 75.99 |
| | C | 57.87 | | S | 3.79 |
| | O | 7.25 | | O | 2.47 |
| 3 | Cu | 24.57 | 3 | Pb | 9.95 |
| | C | 64.84 | | C | 64.53 |
| | S | 6.57 | | Cu | 7.85 |
| | O | 3.84 | | O | 17 |
| | Sb | 0.19 | | Sn | 0.68 |
| 4 | Cu | 25.7 | | | |
| | C | 63.68 | | | |
| | S | 7.44 | | | |
| 5 | Zn | 24.46 | | | |
| | C | 57.29 | | | |
| | S | 13.12 | | | |
| | Cu | 2.88 | | | |
| | O | 2.25 | | | |

**Figure 4.** Welding rods of different lengths and diameters (the length of rods was 40 cm; the diameters were 8 mm and 6 mm).

The size and range of the crack were examined with the use of penetration testing (Figure 5). Penetrator was used for this examination by covering the crack and film, which helped to determine the range of the crack.

**Figure 5.** Penetration testing.

After crack range determination, mechanical treatment of the damaged place was conducted to remove the external oxidized surface (Figure 5). This phase was performed in such way to obtain the best access to whole crack by the welder during the welding process (Figure 6). It was found during mechanical treatment that the bell's structure is very porous, especially the external surface. This worsened the welding process. The welding process was conducted with the use of an oxyacetylene torch. During this process the bell edges were melted with the welding rods made earlier. Welding was conducted with the use of the "up method". Better efficiency of welding and very good penetration of the whole thickness of the connected parts were obtained thanks to this method. It was possible to perform the weld with a single torch cut due to this method.

**Figure 6.** Mechanical treatment of the damaged place.

It was important to heat and keep the bell at the proper temperature during the welding process. This temperature was obtained by using heating mats and aluminosilicate fiber isolation. The whole process of heating was under the control of and recorded by a computer program (Figure 7);the heating rate was ~10 °C/h. The time of cooling after performing the weld was longer than the heating time, and the cooling rate was ~7 °C/h. Thus slow cooling was conducted to avoid stresses, which could have caused the bell to crack again.

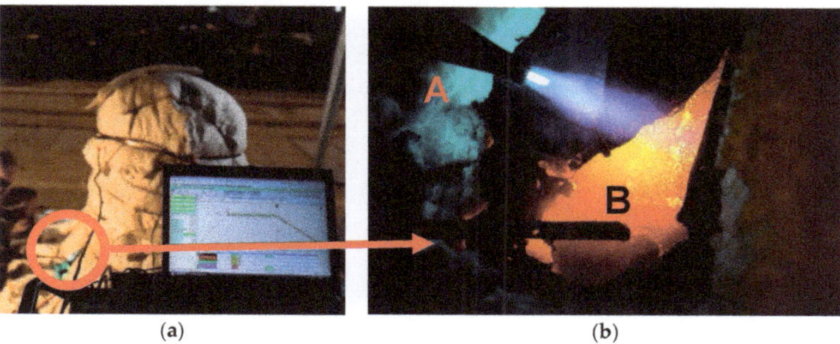


**Figure 7.** Welding process (**a**) the bell protected by fibro isolation; red circle–the place of welding (**b**) close–up of the place of welding. A—acetylene torch, B—bronze rod.

After the cooling process, the place of welding was ground (Figure 8) and the sound of bell was examined.

**Figure 8.** Welded bell attached to the tower.

## 4. Sound Analysis of the Bell

In 2013, before the crack, the bell's sound was analyzed. After welding and the thermal stabilization process, this analysis was repeated. The sound of the bell, a G#4 note, did not worsen or even improve as a result of the negligible reduction of the main aliquots frequency.

The target frequencies for the first partials are in the ratios 0.5:1.0:1.2:1.5:2.0 and these needed to be quite closely matched. The first aliquot, called the hum, is not prominent and the perceived pitch is usually that of the second aliquot, called the prime, perhaps because it is reinforced by the harmonically-related aliquots with relative frequencies 2, 3 and 4. The tone of the bell is complex, however, particularly because of the presence of the minor-third (From Old French tierce, from Latin tertia) interval of 1.2 [10].

The lower (tone lower than the prime about the octave) and upper (tone higher than the prime about the octave) octaves with prime, tierce and quint were found to be beautifully harmonious after the repair. There was no distortionary vibration and the bell sustained its note for a long time.

The frequency spectrums of the Maryan bell before and after the crack are presented in Figure 9. The units of the amplitude in the figure are arbitrary; they are measured as voltages from a microphone, i.e., sound pressure levels on a linear scale. The amplitude of the spectrum is described in decibel scale. The program Wavanal [11] was used to analyze the sound of bell examined to determine the spectrum of emitted sound waves. This program was developed by W. A. Hibbert [12] for the sound analysis of bells to determine the influence of side tones on the height of the perceived strike tone (pitch tone). The possibility of fast and precise determination of their frequency was crucial, and the Wavanal program enabled it. This program allows a Fourier transform of sound waves directly recorded by a microphone joined to a computer or saved in a sound file recorded with other devices to be performed (Figure 10a,b). This program has received recognition among many bell makers as a great device for the evaluation of a bell's sound and the process of tuning it up.

Determined by the Wavanal program, values of frequency of basic side tones (aliquots) the lower octave (hum), prime (fundamental), minor tierce, quint, and upper octave (nominal) of the bells examined are presented in Table 3 and as a diagram in Figure 11.

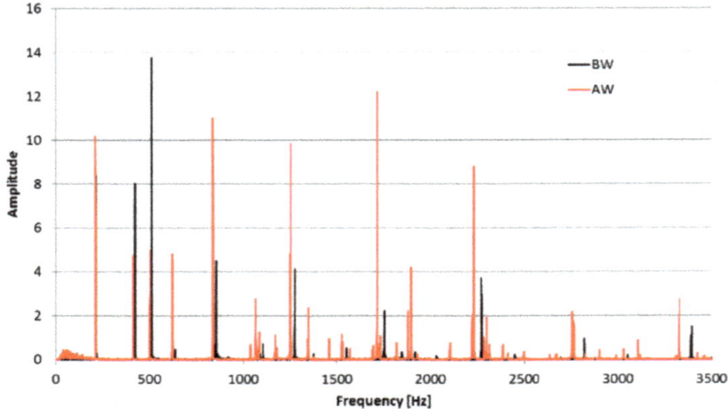


**Figure 9.** The spectrum of the St. Maryan bell's sound, before (BW) and after (AW) welding.

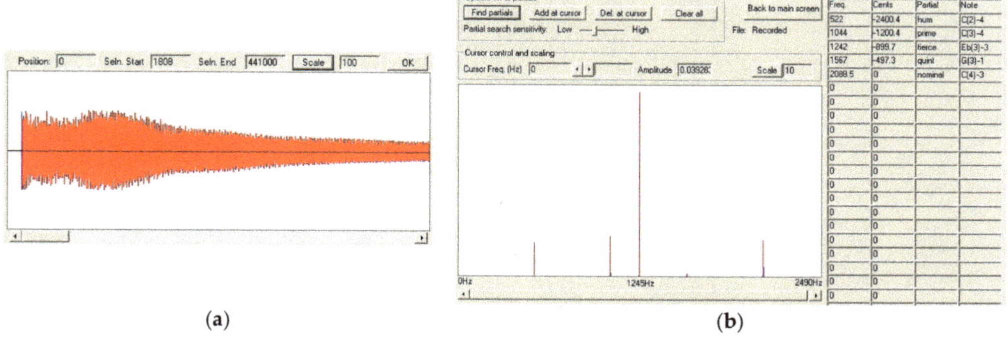

(a) (b)

**Figure 10.** The analysis of sound wave emitted by bell C3 in program Wavanal; (**a**) the shape of recorded wave, (**b**) its spectrum.

**Table 3.** Main partials of the St. Maryan bell's sound frequencies before and after welding in comparison to harmonic tones for the musical note G#4.

| Partials Tone | BW (Hz) | AW (Hz) | G#4 (Hz) (ET) |
|---|---|---|---|
| Hum | 212.5 | 208.5 | 207.6 |
| Prime | 422 | 413 | 415.3 |
| Tierce | 512 | 502.5 | 493.8 |
| Quint | 635 | 619 | 622.3 |
| Nominal | 853.5 | 836 | 830.6 |

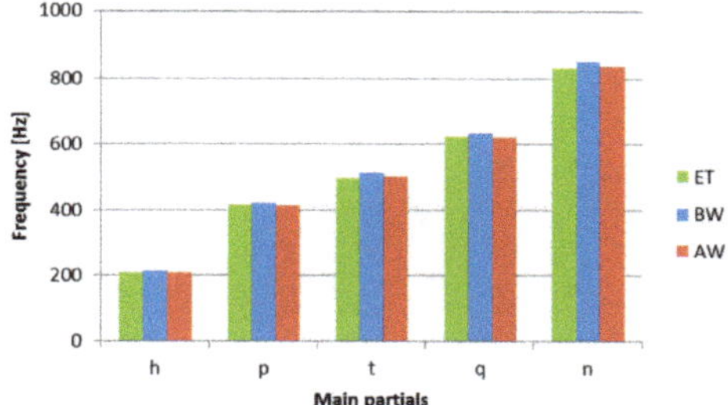


**Figure 11.** The main partials of the St. Maryan bell's sound frequencies, before (BW), after (AW) welding and for musical note G#4 according to the equal temperament scale (ET).

## 5. Conclusions

Based on the experience gained during the work and research carried out on the Maryan bell, the following conclusions can be drawn:

1. If the welding process is carried out with the correct parameters, especially with thermal ones, and with monitoring and control of the heating and cooling rates, the repair even of contaminated high tin bronze bells is possible;
2. After repair by welding, the bell has a "better" sound, and this is probably due to a kind of heat treatment performed, including a cycle of slow heating and cooling which improves the properties of the material; and
3. The most highly recommended technique for repairing bells is gas welding, due to the relatively low temperature in the bonding area and the efficiency of the process.

**Author Contributions:** Conceptualization, methodology, software, formal analysis, writing—review and editing, D.B.; conceptualization, methodology, investigation, writing—original draft preparation, C.B. Both authors have read and agreed to the published version of the manuscript.

**Funding:** This research received no external funding.

**Institutional Review Board Statement:** Not applicable.

**Informed Consent Statement:** Not applicable.

**Data Availability Statement:** The data presented in this study are available in article.

**Acknowledgments:** The authors would like to acknowledge to Rduch Bells and Clocks and Jan Felczynski's Bell Foundry for materials and technical support.

**Conflicts of Interest:** The authors declare no conflict of interest.

## References

1. Szupp, B. Repair of church bells by welding. *Weld. Cut. Metals* **1936**, *12*, 198–205.
2. Ernesto Ponce, L. Restoration of ancient bronze bells. Part II: Welding. *Ingeniare Rev. Chil. Ing.* **2015**, *23*, 30–37. [CrossRef]
3. Raimbault, L. Process for Welding Bells. Patent No FR2703615A1, 14 October 1994. France.
4. Strafford, K.N.; Newell, R.; Audy, K.; Audy, J. Analysis of Bell Material from the Middle Ages to the Recent Time. *Endeavour* **1996**, *20*, 22–27. [CrossRef]
5. Audy, J.; Audy, K. Analysis of bell materials: Tin bronzes. *China Foundry* **2008**, *5*, 199–204.
6. Bartocha, D.; Baron, C. The "Secret" of Traditional Technology of Casting Bells. *Arch. Foundry Eng.* **2015**, *15*, 5–10.
7. Kurski, K. *Cooper and Its Technical Alloys*; Wydawnictwo Śląsk: Katowice, Poland, 1967.
8. Górny, Z.; Sobczak, J.J. *Modern Casting Materials Based on Non-Ferrous Metals*; ZA-PIS: Kraków, Poland, 2005.
9. Górny, Z. *Foundry Non-Ferrous Metal Alloys*; WNT: Warszawa, Poland, 1992.
10. Fletcher, N.H. The nonlinear physics of musical instruments. *Rep. Prog. Phys.* **1999**, *62*, 723–764. [CrossRef]
11. Available online: http://www.hibberts.co.uk (accessed on 5 October 2020).
12. Hibbert, W.A. The Quantification of Strike Pitch and Pitch Shifts in Church Bells. Ph.D Thesis, The Open University Milton Keynes, Milton Keynes, UK, 2008, in press.

MDPI
St. Alban-Anlage 66
4052 Basel
Switzerland
Tel. +41 61 683 77 34
Fax +41 61 302 89 18
www.mdpi.com

*Materials* Editorial Office
E-mail: materials@mdpi.com
www.mdpi.com/journal/materials

www.ingramcontent.com/pod-product-compliance
Lightning Source LLC
LaVergne TN
LVHW070132170726
843515LV00007B/1779

* 9 7 8 3 0 3 6 5 4 3 9 4 9 *